Montego
Owners
Workshop
Manual

John S Mead

Models covered
Austin (Rover) Montego 1.3 & 1.6 litre models with
petrol engines, Saloon and Estate, including
special/limited editions
1275 cc & 1598 cc

Does not cover 2.0 litre petrol or Diesel engine models

(1066-9U8)

ABCDE
FGHIJ

2

Haynes Publishing Group
Sparkford Nr Yeovil
Somerset BA22 7JJ England

Haynes Publications, Inc
861 Lawrence Drive
Newbury Park
California 91320 USA

Acknowledgements

Thanks are due to Champion Spark Plug who supplied the illustrations showing spark plug conditions, to Holt Lloyd Limited who supplied the illustrations showing bodywork repair, and to Duckhams Oils, who provided lubrication data. Thanks are also due to BL Cars Limited for the supply of technical information and to Unipart for their assistance. Sykes-Pickavant Limited provided some of the workshop tools. Special thanks are due to all those people at Sparkford who helped in the production of this manual.

© **Haynes Publishing Group 1992**

A book in the **Haynes Owners Workshop Manual Series**

Printed by J. H. Haynes & Co. Ltd., Sparkford, Nr Yeovil, Somerset BA22 7JJ, England

ISBN 1 85010 871 4

British Library Cataloguing in Publication Data
A catalogue record for this book is available from the British Library

Restoring and Preserving our Motoring Heritage

Few people can have had the luck to realise their dreams to quite the same extent and in such a remarkable fashion as John Haynes, Founder and Chairman of the Haynes Publishing Group.

Since 1965 his unique approach to workshop manual publishing has proved so successful that millions of Haynes Manuals are now sold every year throughout the world, covering literally thousands of different makes and models of cars, vans and motorcycles.

A continuing passion for cars and motoring led to the founding in 1985 of a Charitable Trust dedicated to the restoration and preservation of our motoring heritage. To inaugurate the new Museum, John Haynes donated virtually his entire private collection of 52 cars.

Now with an unrivalled international collection of over 210 veteran, vintage and classic cars and motorcycles, the Haynes Motor Museum in Somerset is well on the way to becoming one of the most interesting Motor Museums in the world.

A 70 seat video cinema, a cafe and an extensive motoring bookshop, together with a specially constructed one kilometre motor circuit, make a visit to the Haynes Motor Museum a truly unforgettable experience.

Every vehicle in the museum is preserved in as near as possible mint condition and each car is run every six months on the motor circuit.

Enjoy the picnic area set amongst the rolling Somerset hills. Peer through the William Morris workshop windows at cars being restored, and browse through the extensive displays of fascinating motoring memorabilia.

From the 1903 Oldsmobile through such classics as an MG Midget to the mighty 'E' Type Jaguar, Lamborghini, Ferrari Berlinetta Boxer, and Graham Hill's Lola Cosworth, there is something for everyone, young and old alike, at this Somerset Museum.

Haynes Motor Museum

Situated mid-way between London and Penzance, the Haynes Motor Museum is located just off the A303 at Sparkford, Somerset (home of the Haynes Manual) and is open to the public 7 days a week all year round, except Christmas Day and Boxing Day.

Contents

Spark plug condition and bodywork repair colour pages between pages 32 and 33

1.6 litre Austin Montego

General dimensions, weights and capacities

For information applicable to later models, see Supplement at end of manual

Dimensions

Turning circle (between kerbs):	
Saloon	34 ft 4 in (10.46 m)
Estate	34 ft 6 in (10.52 m)
Wheelbase	101.5 in (2578 mm)
Overall length	176 in (4470 mm)
Overall width (excluding mirrors)	67.3 in (1710 mm)
Overall height:	
Saloon	55.2 in (1402 mm)
Estate	57.0 in (1445 mm)
Ground clearance (nominal)	6.5 in (165 mm)
Track:	
Front	57.7 in (1465 mm)
Rear	56.7 in (1439 mm)

Weights

	Saloon	Estate
Kerb weight:		
1.3	2140 lb (970 kg)	
1.6	2170 lb (985 kg)	2315 lb (1050 kg)
1.6 L	2250 lb (1020 kg)	2370 lb (1075 kg)
1.6 HL	2260 lb (1025 kg)	2395 lb (1085 kg)
1.6 L automatic	2290 lb (1040 kg)	2410 lb (1095 kg)
1.6 HL automatic	2300 lb (1045 kg)	2435 lb (1105 kg)
Maximum roof rack weight (distributed)	155 lb (70 kg)	as Saloon
Towing hitch downward load	75 to 100 lb (35 to 45 kg)	as Saloon

Capacities

Engine oil (refill with filter change):	
1.3 litre models	5 pt (2.8 litres)
1.6 litre models	6.75 pt (3.8 litres)
Manual gearbox refill:	
4-speed	2.75 pt (1.6 litres)
5-speed	3.5 pt (2.0 litres)
Automatic transmission:	
Drain and refill	4 pt (2.3 litres)
Total capacity	10.5 pt (6.0 litres)
Final drive unit	1.25 pt (0.71 litres)
Cooling system:	
1.3 litre models	11.6 pt (6.6 litres)
1.6 litre models	15.0 pt (8.5 litres)
Fuel tank	11.0 gal (50 litres)
Fuel octane rating (all models)	97 RON (4-star)

About this manual

Its aim

The aim of this manual is to help you get the best value from your vehicle. It can do so in several ways. It can help you decide what work must be done (even should you choose to get it done by a garage), provide information on routine maintenance and servicing, and give a logical course of action and diagnosis when random faults occur. However, it is hoped that you will use the manual by tackling the work yourself. On simpler jobs it may even be quicker than booking the car into a garage and going there twice, to leave and collect it. Perhaps most important, a lot of money can be saved by avoiding the costs a garage must charge to cover its labour and overheads.

The manual has drawings and descriptions to show the function of the various components so that their layout can be understood. Then the tasks are described and photographed in a step-by-step sequence so that even a novice can do the work.

Its arrangement

The manual is divided into thirteen Chapters, each covering a logical sub-division of the vehicle. The Chapters are each divided into Sections, numbered with single figures, eg 5; and the Sections into paragraphs (or sub-sections), with decimal numbers following on from the Section they are in, eg 5.1, 5.2, 5.3 etc.

It is freely illustrated, especially in those parts where there is a detailed sequence of operations to be carried out. There are two forms of illustration: figures and photographs. The figures are numbered in sequence with decimal numbers, according to their position in the Chapter – eg Fig. 6.4 is the fourth drawing/illustration in Chapter 6. Photographs carry the same number (either individually or in related groups) as the Section or sub-section to which they relate.

There is an alphabetical index at the back of the manual as well as a contents list at the front. Each Chapter is also preceded by its own individual contents list.

References to the 'left' or 'right' of the vehicle are in the sense of a person in the driver's seat facing forwards.

Unless otherwise stated, nuts and bolts are removed by turning anti-clockwise, and tightened by turning clockwise.

Vehicle manufacturers continually make changes to specifications and recommendations, and these, when notified, are incorporated into our manuals at the earliest opportunity.

We take great pride in the accuracy of information given in this manual, but vehicle manufacturers make alterations and design changes during the production run of a particular vehicle of which they do not inform us. No liability can be accepted by the authors or publishers for loss, damage or injury caused by any errors in, or omissions from, the information given.

Introduction to the Montego

Introduced in Spring 1984, the Montego has many design features of its sister car the Maestro, both being conceived and developed as members of a fully integrated range of medium-sector vehicles.

1.3 litre versions are powered by the proven A+ series overhead valve engine used in the Metro and Maestro range. The 1.6 litre S series engine is an all-new belt-driven overhead camshaft unit, used for the first time in Montego models. Also available is a 2.0 litre version which is the subject of a separate publication.

All models are available with 4- or 5-speed manual gearboxes with a fully automatic transmission being offered as an option on 1.6 litre versions.

The Montego range features the now almost universal medium saloon configuration of front-wheel-drive, independent front/semi-independent rear suspension and servo-assisted self-adjusting brakes to give precise roadholding and smooth ride characteristics.

A computer-controlled engine management system, together with 12 000 mile service intervals, make this a very economical and attractive package which should prove highly successful in a very competitive sector of the market.

BL and BL Cars Limited are now known as the Rover Group (previously Austin Rover Group). All instances in which we recommend owners seek the advice of a BL dealer should now, of course, be taken to mean seek the advice of a Rover Group dealer.

Buying spare parts and vehicle identification numbers

Buying spare parts

Spare parts are available from many sources, for example: BL garages, other garages and accessory shops, and motor factors. Our advice regarding spare part sources is as follows:

Officially appointed BL garages – This is the best source of parts which are peculiar to your car and are not generally available (eg complete cylinder heads, internal gearbox components, badges, interior trim etc). It is also the only place at which you should buy parts if your vehicle is still under warranty – non-BL components may invalidate the warranty. To be sure of obtaining the correct parts it will always be necessary to give the storeman your car's vehicle identification number, and if possible, to take the 'old' part along for positive identification. Many parts are available under a factory exchange scheme – any parts returned should always be clean. It obviously makes good sense to go straight to the specialists on your car for this type of part for they are best equipped to supply you.

Other dealers and accessory shops – These are often very good places to buy materials and components needed for the maintenance of your car (eg oil filters, spark plugs, bulbs, drivebelts, oils and grease, touch-up paint, filler paste etc). They also sell general accessories, usually have convenient opening hours, charge lower prices and can often be found not far from home.

Motor factors – Good factors will stock all of the more important components which wear out relatively quickly (eg clutch components, pistons, valves, exhaust systems, brake pipes/seals and pads etc).

Motor factors will often provide new or reconditioned components on a part exchange basis – this can save a considerable amount of money.

Vehicle identification numbers

Modifications are a continuing and unpublicised process in vehicle manufacture, quite apart from major model changes. Spare parts manuals and lists are compiled upon a numerical basis, the individual vehicle numbers being essential to correct identification of the component required.

When ordering spare parts, always give as much information as possible. Quote the car model, year of manufacture, body and engine numbers as appropriate.

The vehicle identification number is stamped on a plate attached to the front body panel, or to the left-hand front door pillar. The number is repeated on the right-hand front strut turret.

The engine number is stamped on a plate attached to the cylinder block below No 1 spark plug.

The gearbox number is stamped on the bottom face of the gearbox casing.

The automatic transmission number is stamped on the top face of the final drive housing.

The body number plate is located on the right-hand side of the spare wheel well.

Tools and working facilities

Introduction

A selection of good tools is a fundamental requirement for anyone contemplating the maintenance and repair of a motor vehicle. For the owner who does not possess any, their purchase will prove a considerable expense, offsetting some of the savings made by doing-it-yourself. However, provided that the tools purchased meet the relevant national safety standards and are of good quality, they will last for many years and prove an extremely worthwhile investment.

To help the average owner to decide which tools are needed to carry out the various tasks detailed in this manual, we have compiled three lists of tools under the following headings: *Maintenance and minor repair, Repair and overhaul,* and *Special.* The newcomer to practical mechanics should start off with the *Maintenance and minor repair* tool kit and confine himself to the simpler jobs around the vehicle. Then, as his confidence and experience grow, he can undertake more difficult tasks, buying extra tools as, and when, they are needed. In this way, a *Maintenance and minor repair* tool kit can be built-up into a *Repair and overhaul* tool kit over a considerable period of time without any major cash outlays. The experienced do-it-yourselfer will have a tool kit good enough for most repair and overhaul procedures and will add tools from the *Special* category when he feels the expense is justified by the amount of use to which these tools will be put.

It is obviously not possible to cover the subject of tools fully here. For those who wish to learn more about tools and their use there is a book entitled *How to Choose and Use Car Tools* available from the publishers of this manual.

Both UNF and metric threads to ISO standards are used on the Montego range.

Maintenance and minor repair tool kit

The tools given in this list should be considered as a minimum requirement if routine maintenance, servicing and minor repair operations are to be undertaken. We recommend the purchase of combination spanners (ring one end, open-ended the other); although more expensive than open-ended ones, they do give the advantages of both types of spanner.

Combination spanners - 10, 11, 12, 13, 14 & 17 mm
Combination spanners − $\frac{7}{16}$, $\frac{1}{2}$, $\frac{9}{16}$, $\frac{5}{8}$, $\frac{3}{4}$, $\frac{3}{16}$, $\frac{7}{8}$ and $\frac{15}{16}$ in AF
Adjustable spanner - 9 inch
Gearbox/drain plug key
Spark plug spanner (with rubber insert)
Spark plug gap adjustment tool
Set of feeler gauges
Brake adjuster spanner
Brake bleed nipple spanner
Screwdriver - 4 in long x $\frac{1}{4}$ in dia (flat blade)
Screwdriver - 4 in long x $\frac{1}{4}$ in dia (cross blade)
Combination pliers - 6 inch
Hacksaw (junior)
Tyre pump
Tyre pressure gauge
Oil can
Fine emery cloth (1 sheet)
Wire brush (small)
Funnel (medium size)
Oil filter removal tool (1.6 litre models)

Repair and overhaul tool kit

These tools are virtually essential for anyone undertaking any major repairs to a motor vehicle, and are additional to those given in the *Maintenance and minor repair* list. Included in this list is a comprehensive set of sockets. Although these are expensive they will be found invaluable as they are so versatile - particularly if various drives are included in the set. We recommend the $\frac{1}{2}$ in square-drive type, as this can be used with most proprietary torque wrenches. If you cannot afford a socket set, even bought piecemeal, then inexpensive tubular box spanners are a useful alternative.

The tools in this list will occasionally need to be supplemented by tools from the *Special* list.

Sockets (or box spanners) to cover range in previous list
Reversible ratchet drive (for use with sockets)
Extension piece, 10 inch (for use with sockets)
Universal joint (for use with sockets)
Torque wrench (for use with sockets)
'Mole' wrench - 8 inch
Ball pein hammer
Soft-faced hammer, plastic or rubber
Screwdriver - 6 in long x $\frac{5}{16}$ in dia (flat blade)
Screwdriver - 2 in long x $\frac{5}{16}$ in square (flat blade)
Screwdriver - 1$\frac{1}{2}$ in long x $\frac{1}{4}$ in dia (cross blade)
Screwdriver - 3 in long x $\frac{1}{8}$ in dia (electricians)
Pliers - electricians side cutters
Pliers - needle nosed
Pliers - circlip (internal and external)
Cold chisel - $\frac{1}{2}$ inch
Scriber
Scraper
Centre punch
Pin punch
Hacksaw
Valve grinding tool
Steel rule/straight-edge
Allen keys
Selection of files
Wire brush (large)
Axle-stands
Jack (strong scissor or hydraulic type)

Special tools

The tools in this list are those which are not used regularly, are expensive to buy, or which need to be used in accordance with their manufacturers' instructions. Unless relatively difficult mechanical jobs are undertaken frequently, it will not be economic to buy many of these tools. Where this is the case, you could consider clubbing together with friends (or joining a motorists' club) to make a joint purchase, or borrowing the tools against a deposit from a local garage or tool hire specialist.

The following list contains only those tools and instruments freely available to the public, and not those special tools produced by the vehicle manufacturer specifically for its dealer network. You will find occasional references to these manufacturers' special tools in the text of this manual. Generally, an alternative method of doing the job without the use of the vehicle manufacturers' special tool is given. However, sometimes, there is no alternative to using them. Where this is the case and the relevant tool cannot be bought or borrowed, you will have to entrust the work to a franchised garage.

Valve spring compressor (where applicable)
Piston ring compressor
Balljoint separator
Universal hub/bearing puller
Impact screwdriver
Micrometer and/or vernier gauge
Dial gauge
Stroboscopic timing light
Dwell angle meter/tachometer
Universal electrical multi-meter
Cylinder compression gauge
Lifting tackle
Trolley jack
Light with extension lead

Buying tools

For practically all tools, a tool factor is the best source since he will have a very comprehensive range compared with the average garage or accessory shop. Having said that, accessory shops often offer excellent quality tools at discount prices, so it pays to shop around.

There are plenty of good tools around at reasonable prices, but always aim to purchase items which meet the relevant national safety standards. If in doubt, ask the proprietor or manager of the shop for advice before making a purchase.

Care and maintenance of tools

Having purchased a reasonable tool kit, it is necessary to keep the tools in a clean serviceable condition. After use, always wipe off any dirt, grease and metal particles using a clean, dry cloth, before putting the tools away. Never leave them lying around after they have been used. A simple tool rack on the garage or workshop wall, for items such as screwdrivers and pliers is a good idea. Store all normal wrenches and sockets in a metal box. Any measuring instruments, gauges, meters, etc, must be carefully stored where they cannot be damaged or become rusty.

Take a little care when tools are used. Hammer heads inevitably become marked and screwdrivers lose the keen edge on their blades from time to time. A little timely attention with emery cloth or a file will soon restore items like this to a good serviceable finish.

Working facilities

Not to be forgotten when discussing tools, is the workshop itself. If anything more than routine maintenance is to be carried out, some form of suitable working area becomes essential.

It is appreciated that many an owner mechanic is forced by circumstances to remove an engine or similar item, without the benefit of a garage or workshop. Having done this, any repairs should always be done under the cover of a roof.

Wherever possible, any dismantling should be done on a clean, flat workbench or table at a suitable working height.

Any workbench needs a vice: one with a jaw opening of 4 in (100 mm) is suitable for most jobs. As mentioned previously, some clean dry storage space is also required for tools, as well as for lubricants, cleaning fluids, touch-up paints and so on, which become necessary.

Another item which may be required, and which has a much more general usage, is an electric drill with a chuck capacity of at least $\frac{5}{16}$ in (8 mm). This, together with a good range of twist drills, is virtually essential for fitting accessories such as mirrors and reversing lights.

Last, but not least, always keep a supply of old newspapers and clean, lint-free rags available, and try to keep any working area as clean as possible.

Spanner jaw gap comparison table

Jaw gap (in)	Spanner size
0.250	$\frac{1}{4}$ in AF
0.276	7 mm
0.313	$\frac{5}{16}$ in AF
0.315	8 mm
0.344	$\frac{11}{32}$ in AF; $\frac{1}{8}$ in Whitworth
0.354	9 mm
0.375	$\frac{3}{8}$ in AF
0.394	10 mm
0.433	11 mm
0.438	$\frac{7}{16}$ in AF
0.445	$\frac{3}{16}$ in Whitworth; $\frac{1}{4}$ in BSF
0.472	12 mm
0.500	$\frac{1}{2}$ in AF
0.512	13 mm
0.525	$\frac{1}{4}$ in Whitworth; $\frac{5}{16}$ in BSF
0.551	14 mm
0.563	$\frac{9}{16}$ in AF
0.591	15 mm
0.600	$\frac{5}{16}$ in Whitworth; $\frac{3}{8}$ in BSF
0.625	$\frac{5}{8}$ in AF
0.630	16 mm
0.669	17 mm
0.686	$\frac{11}{16}$ in AF
0.709	18 mm
0.710	$\frac{3}{8}$ in Whitworth; $\frac{7}{16}$ in BSF
0.748	19 mm
0.750	$\frac{3}{4}$ in AF
0.813	$\frac{13}{16}$ in AF
0.820	$\frac{7}{16}$ in Whitworth; $\frac{1}{2}$ in BSF
0.866	22 mm
0.875	$\frac{7}{8}$ in AF
0.920	$\frac{1}{2}$ in Whitworth; $\frac{9}{16}$ in BSF
0.938	$\frac{15}{16}$ in AF
0.945	24 mm
1.000	1 in AF
1.010	$\frac{9}{16}$ in Whitworth; $\frac{5}{8}$ in BSF
1.024	26 mm
1.063	$1\frac{1}{16}$ in AF; 27 mm
1.100	$\frac{5}{8}$ in Whitworth; $\frac{11}{16}$ in BSF
1.125	$1\frac{1}{8}$ in AF
1.181	30 mm
1.200	$\frac{11}{16}$ in Whitworth; $\frac{3}{4}$ in BSF
1.250	$1\frac{1}{4}$ in AF
1.260	32 mm
1.300	$\frac{3}{4}$ in Whitworth; $\frac{7}{8}$ in BSF
1.313	$1\frac{5}{16}$ in AF
1.390	$\frac{13}{16}$ in Whitworth; $\frac{15}{16}$ in BSF
1.417	36 mm
1.438	$1\frac{7}{16}$ in AF
1.480	$\frac{7}{8}$ in Whitworth; 1 in BSF
1.500	$1\frac{1}{2}$ in AF
1.575	40 mm; $\frac{15}{16}$ in Whitworth
1.614	41 mm
1.625	$1\frac{5}{8}$ in AF
1.670	1 in Whitworth; $1\frac{1}{8}$ in BSF
1.688	$1\frac{11}{16}$ in AF
1.811	46 mm
1.813	$1\frac{13}{16}$ in AF
1.860	$1\frac{1}{8}$ in Whitworth; $1\frac{1}{4}$ in BSF
1.875	$1\frac{7}{8}$ in AF
1.969	50 mm
2.000	2 in AF
2.050	$1\frac{1}{4}$ in Whitworth; $1\frac{3}{8}$ in BSF
2.165	55 mm
2.362	60 mm

General repair procedures

Whenever servicing, repair or overhaul work is carried out on the car or its components, it is necessary to observe the following procedures and instructions. This will assist in carrying out the operation efficiently and to a professional standard of workmanship.

Joint mating faces and gaskets

Where a gasket is used between the mating faces of two components, ensure that it is renewed on reassembly, and fit it dry unless otherwise stated in the repair procedure. Make sure that the mating faces are clean and dry with all traces of old gasket removed. When cleaning a joint face, use a tool which is not likely to score or damage the face, and remove any burrs or nicks with an oilstone or fine file.

Make sure that tapped holes are cleaned with a pipe cleaner, and keep them free of jointing compound if this is being used unless specifically instructed otherwise.

Ensure that all orifices, channels or pipes are clear and blow through them, preferably using compressed air.

Oil seals

Whenever an oil seal is removed from its working location, either individually or as part of an assembly, it should be renewed.

The very fine sealing lip of the seal is easily damaged and will not seal if the surface it contacts is not completely clean and free from scratches, nicks or grooves. If the original sealing surface of the component cannot be restored, the component should be renewed.

Protect the lips of the seal from any surface which may damage them in the course of fitting. Use tape or a conical sleeve where possible. Lubricate the seal lips with oil before fitting and, on dual lipped seals, fill the space between the lips with grease.

Unless otherwise stated, oil seals must be fitted with their sealing lips toward the lubricant to be sealed.

Use a tubular drift or block of wood of the appropriate size to install the seal and, if the seal housing is shouldered, drive the seal down to the shoulder. If the seal housing is unshouldered, the seal should be fitted with its face flush with the housing top face.

Screw threads and fastenings

Always ensure that a blind tapped hole is completely free from oil, grease, water or other fluid before installing the bolt or stud. Failure to do this could cause the housing to crack due to the hydraulic action of the bolt or stud as it is screwed in.

When tightening a castellated nut to accept a split pin, tighten the nut to the specified torque, where applicable, and then tighten further to the next split pin hole. Never slacken the nut to align a split pin hole unless stated in the repair procedure.

When checking or retightening a nut or bolt to a specified torque setting, slacken the nut or bolt by a quarter of a turn, and then retighten to the specified setting.

Locknuts, locktabs and washers

Any fastening which will rotate against a component or housing in the course of tightening should always have a washer between it and the relevant component or housing.

Spring or split washers should always be renewed when they are used to lock a critical component such as a big-end bearing retaining nut or bolt.

Locktabs which are folded over to retain a nut or bolt should always be renewed.

Self-locking nuts can be reused in non-critical areas, providing resistance can be felt when the locking portion passes over the bolt or stud thread.

Split pins must always be replaced with new ones of the correct size for the hole.

Special tools

Some repair procedures in this manual entail the use of special tools such as a press, two or three-legged pullers, spring compressors etc. Wherever possible, suitable readily available alternatives to the manufacturer's special tools are described, and are shown in use. In some instances, where no alternative is possible, it has been necessary to resort to the use of a manufacturer's tool and this has been done for reasons of safety as well as the efficient completion of the repair operation. Unless you are highly skilled and have a thorough understanding of the procedure described, never attempt to bypass the use of any special tool when the procedure described specifies its use. Not only is there a very great risk of personal injury, but expensive damage could be caused to the components involved.

Jacking and towing

To change a roadwheel, remove the spare wheel and tool kit from the well in the rear compartment. Apply the handbrake and chock the wheel diagonally opposite the one to be changed. Make sure that the car is located on firm level ground. Lever off the wheel trim (photo) and slightly loosen the wheel nuts with the spanner provided. Position the jack under the nearest jacking point to the wheel being removed. Using the handle provided, raise the jack until the wheel is free of the ground. Unscrew the wheel nuts and remove the wheel.

Fit the spare wheel on the studs, then fit and tighten the wheel nuts with their tapered ends towards the wheel. Lower the jack, then finally tighten the wheel nuts and refit the wheel. Remove the chock, and refit the wheel and tool kit in the rear compartment.

When jacking up the car with a trolley jack, position the head of the jack under the jacking bracket/towing hook in the centre of the front crossmember (position 1 in the illustration) (photo) to raise both front wheels.

To raise both rear wheels position the jack head under the rear jacking bracket towing hook (position 2 in the illustration) (photo).

To raise one side of the car at the front or rear, place the jack head between the front or rear jacking points. In all cases make sure the handbrake is firmly applied and the wheels chocked before raising the car. Always position axle stands or suitable supports under a structural member, such as a chassis member or crossmember, to support the car securely when it is raised.

The car may be towed, or tow another vehicle, using the front or rear jacking bracket/towing hooks.

If automatic transmission is fitted the vehicle should only be towed at slow speed for a short distance. If a greater distance must be covered or if there is the possibility of any fault in the transmission, the car must be towed with the front wheels lifted.

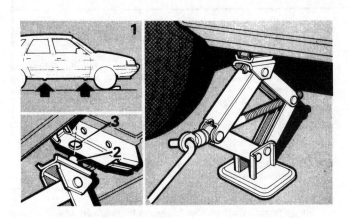

Wheel changing

1 Jacking point locations
2 Peg on jack head
3 Locating hole in jacking point

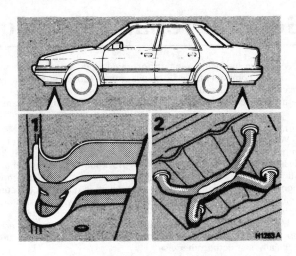

Jacking bracket/towing hook locations

1 Front
2 Rear

Use the tool provided, or a screwdriver, to lever off the wheel trim

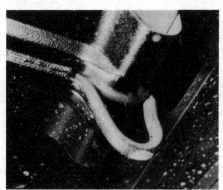

Front jacking bracket/towing hook

Rear jacking bracket/towing hook

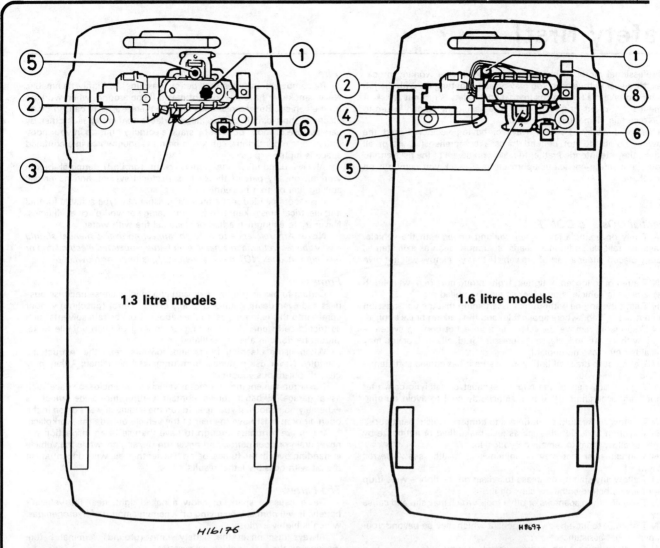

1.3 litre models

1.6 litre models

H16176   H8697

Recommended lubricants and fluids

Component or system	Lubricant type/specification	Duckhams recommendation
1 Engine*	Multigrade engine oil, viscosity SAE 10W/40	Duckhams QXR, Hypergrade, or 10W/40 Motor Oil
2 Manual gearbox: Models up to 1989 1.6 five-speed models, 1989-on*	 Hypoid gear oil, viscosity SAE 80 EP Multigrade engine oil, viscosity SAE 10W/40	 Duckhams Hypoid 80 Duckhams QXR, Hypergrade, or 10W/40 Motor Oil
3 Distributor – 1.3 only	Multigrade engine oil, viscosity SAE 10W/40	Duckhams QXR, Hypergrade, or 10W/40 Motor Oil
4 Final drive (automatic transmission)	Hypoid gear oil, viscosity SAE 90EP	Duckhams Hypoid 90S
5 Carburettor piston damper	Multigrade engine oil, viscosity SAE 10W/40	Duckhams QXR, Hypergrade, or 10W/40 Motor Oil
6 Brake fluid reservoir	Hydraulic fluid to FMVSS 116 DOT 4 or SAE J1703C	Duckhams Universal Brake and Clutch Fluid
7 Automatic transmission	Dexron IID type ATF	Duckhams Uni-Matic or D-Matic
8 Power steering	Dexron IID type ATF	Duckhams Uni-Matic or D-Matic

*Note: *Austin Rover specify a 10W/40 oil to meet warranty requirements for models produced after August 1983. Duckhams QXR and 10W/40 Motor Oil are available to meet these requirements*

Safety first!

Professional motor mechanics are trained in safe working procedures. However enthusiastic you may be about getting on with the job in hand, do take the time to ensure that your safety is not put at risk. A moment's lack of attention can result in an accident, as can failure to observe certain elementary precautions.

There will always be new ways of having accidents, and the following points do not pretend to be a comprehensive list of all dangers; they are intended rather to make you aware of the risks and to encourage a safety-conscious approach to all work you carry out on your vehicle.

Essential DOs and DON'Ts

DON'T rely on a single jack when working underneath the vehicle. Always use reliable additional means of support, such as axle stands, securely placed under a part of the vehicle that you know will not give way.

DON'T attempt to loosen or tighten high-torque nuts (e.g. wheel hub nuts) while the vehicle is on a jack; it may be pulled off.

DON'T start the engine without first ascertaining that the transmission is in neutral (or 'Park' where applicable) and the parking brake applied.

DON'T suddenly remove the filler cap from a hot cooling system – cover it with a cloth and release the pressure gradually first, or you may get scalded by escaping coolant.

DON'T attempt to drain oil until you are sure it has cooled sufficiently to avoid scalding you.

DON'T grasp any part of the engine, exhaust or catalytic converter without first ascertaining that it is sufficiently cool to avoid burning you.

DON'T allow brake fluid or antifreeze to contact vehicle paintwork.

DON'T syphon toxic liquids such as fuel, brake fluid or antifreeze by mouth, or allow them to remain on your skin.

DON'T inhale dust – it may be injurious to health (see *Asbestos* below).

DON'T allow any spilt oil or grease to remain on the floor – wipe it up straight away, before someone slips on it.

DON'T use ill-fitting spanners or other tools which may slip and cause injury.

DON'T attempt to lift a heavy component which may be beyond your capability – get assistance.

DON'T rush to finish a job, or take unverified short cuts.

DON'T allow children or animals in or around an unattended vehicle.

DO wear eye protection when using power tools such as drill, sander, bench grinder etc, and when working under the vehicle.

DO use a barrier cream on your hands prior to undertaking dirty jobs – it will protect your skin from infection as well as making the dirt easier to remove afterwards; but make sure your hands aren't left slippery. Note that long-term contact with used engine oil can be a health hazard.

DO keep loose clothing (cuffs, tie etc) and long hair well out of the way of moving mechanical parts.

DO remove rings, wristwatch etc, before working on the vehicle – especially the electrical system.

DO ensure that any lifting tackle used has a safe working load rating adequate for the job.

DO keep your work area tidy – it is only too easy to fall over articles left lying around.

DO get someone to check periodically that all is well, when working alone on the vehicle.

DO carry out work in a logical sequence and check that everything is correctly assembled and tightened afterwards.

DO remember that your vehicle's safety affects that of yourself and others. If in doubt on any point, get specialist advice.

IF, in spite of following these precautions, you are unfortunate enough to injure yourself, seek medical attention as soon as possible.

Asbestos

Certain friction, insulating, sealing, and other products – such as brake linings, brake bands, clutch linings, torque converters, gaskets, etc – contain asbestos. *Extreme care must be taken to avoid inhalation of dust from such products since it is hazardous to health.* If in doubt, assume that they *do* contain asbestos.

Fire

Remember at all times that petrol (gasoline) is highly flammable. Never smoke, or have any kind of naked flame around, when working on the vehicle. But the risk does not end there – a spark caused by an electrical short-circuit, by two metal surfaces contacting each other, by careless use of tools, or even by static electricity built up in your body under certain conditions, can ignite petrol vapour, which in a confined space is highly explosive.

Always disconnect the battery earth (ground) terminal before working on any part of the fuel or electrical system, and never risk spilling fuel on to a hot engine or exhaust.

It is recommended that a fire extinguisher of a type suitable for fuel and electrical fires is kept handy in the garage or workplace at all times. Never try to extinguish a fuel or electrical fire with water.

Note: *Any reference to a 'torch' appearing in this manual should always be taken to mean a hand-held battery-operated electric lamp or flashlight. It does NOT mean a welding/gas torch or blowlamp.*

Fumes

Certain fumes are highly toxic and can quickly cause unconsciousness and even death if inhaled to any extent. Petrol (gasoline) vapour comes into this category, as do the vapours from certain solvents such as trichloroethylene. Any draining or pouring of such volatile fluids should be done in a well ventilated area.

When using cleaning fluids and solvents, read the instructions carefully. Never use materials from unmarked containers – they may give off poisonous vapours.

Never run the engine of a motor vehicle in an enclosed space such as a garage. Exhaust fumes contain carbon monoxide which is extremely poisonous; if you need to run the engine, always do so in the open air or at least have the rear of the vehicle outside the workplace.

If you are fortunate enough to have the use of an inspection pit, never drain or pour petrol, and never run the engine, while the vehicle is standing over it; the fumes, being heavier than air, will concentrate in the pit with possibly lethal results.

The battery

Never cause a spark, or allow a naked light, near the vehicle's battery. It will normally be giving off a certain amount of hydrogen gas, which is highly explosive.

Always disconnect the battery earth (ground) terminal before working on the fuel or electrical systems.

If possible, loosen the filler plugs or cover when charging the battery from an external source. Do not charge at an excessive rate or the battery may burst.

Take care when topping up and when carrying the battery. The acid electrolyte, even when diluted, is very corrosive and should not be allowed to contact the eyes or skin.

If you ever need to prepare electrolyte yourself, always add the acid slowly to the water, and never the other way round. Protect against splashes by wearing rubber gloves and goggles.

When jump starting a car using a booster battery, for negative earth (ground) vehicles, connect the jump leads in the following sequence: First connect one jump lead between the positive (+) terminals of the two batteries. Then connect the other jump lead first to the negative (–) terminal of the booster battery, and then to a good earthing (ground) point on the vehicle to be started, at least 18 in (45 cm) from the battery if possible. Ensure that hands and jump leads are clear of any moving parts, and that the two vehicles do not touch. Disconnect the leads in the reverse order.

Mains electricity and electrical equipment

When using an electric power tool, inspection light etc, always ensure that the appliance is correctly connected to its plug and that, where necessary, it is properly earthed (grounded). Do not use such appliances in damp conditions and, again, beware of creating a spark or applying excessive heat in the vicinity of fuel or fuel vapour. Also ensure that the appliances meet the relevant national safety standards.

Ignition HT voltage

A severe electric shock can result from touching certain parts of the ignition system, such as the HT leads, when the engine is running or being cranked, particularly if components are damp or the insulation is defective. Where an electronic ignition system is fitted, the HT voltage is much higher and could prove fatal.

Routine maintenance

For modifications, and information applicable to later models, see Supplement at end of manual

Maintenance is essential for ensuring safety, and desirable for the purpose of getting the best in terms of performance and economy from your car. Over the years the need for periodic lubrication has been greatly reduced if not totally eliminated. This has unfortunately tended to lead some owners to think that, because no such action is required, the items either no longer exist, or will last forever. This is certainly not the case; it is essential to carry out regular visual examination as comprehensively as possible in order to spot any possible defects at an early stage before they develop into major expensive repairs.

The following service schedules are a list of the maintenance requirements and the intervals at which they should be carried out, as recommended by the manufacturers. Where applicable these procedures are covered in greater detail throughout this manual, near the beginning of each Chapter.

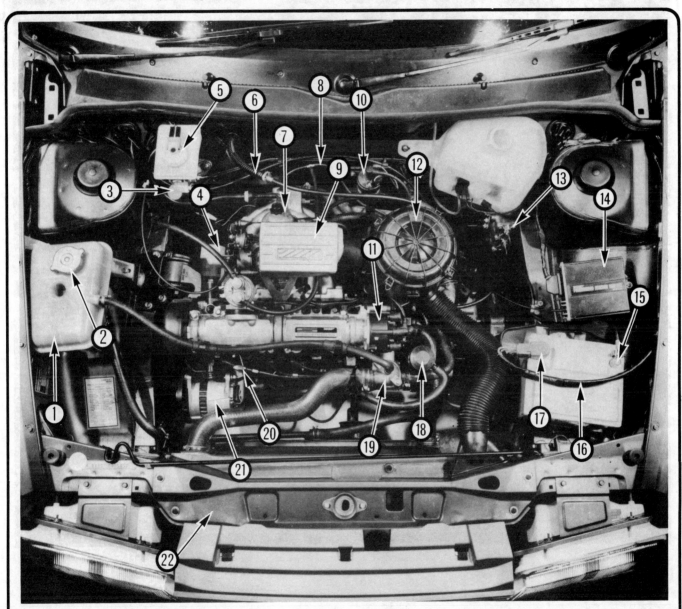

Engine and under-bonnet component locations (1.6 litre models)

1	Cooling system expansion tank	6	Clutch cable self-adjusting mechanism
2	Cooling system filler cap	7	Carburettor
3	Brake master cylinder	8	Main vacuum line
4	Oil filter	9	Air cleaner plenum chamber
5	Master cylinder reservoir filler cap	10	Ignition coil

11	Distributor
12	Air cleaner
13	Braking system twin GP valve
14	Ignition system electronic control unit (ECU)
15	Battery negative terminal

16	Fusible links
17	Battery positive terminal
18	Oil filler/breather cap
19	Water outlet elbow
20	Oil dipstick
21	Alternator
22	Front body panel

Front underbody view (1.6 litre models)

1	Brake caliper	6	Driveshaft inner constant velocity joint	10	Front jacking point
2	Gearbox filler/level plug	7	Suspension crossmember	11	Access panel
3	Steering tie-rod outer balljoint	8	Engine oil drain plug	12	Front snubber cup
4	Suspension lower arm rear mounting	9	Anti-roll bar clamp	13	Front jacking bracket/towing hook
5	Gearbox drain plug				

Rear underbody view (1.6 litre model)

1	Exhaust intermediate silencer	5	Rear jacking bracket/towing hook	9	Rear axle transverse member
2	Handbrake cable adjuster	6	Rear suspension strut lower mounting	10	Fuel tank rear mounting bolts
3	Handbrake cable connectors	7	Rear axle mounting pivot bolt	11	Fuel tank front mounting bolts
4	Exhaust rear silencer	8	Rear brake hose	12	Rear jacking point

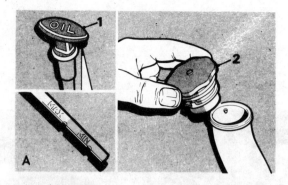

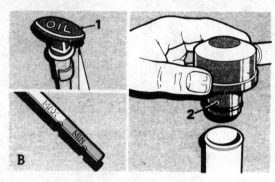

Oil filler and dipstick details

A   1.3 litre models   1   Dipstick
B   1.6 litre models   2   Oil breather/filler cap

Every 250 miles (400 km) or weekly – whichever occurs first

Engine, cooling system and brakes
Check the oil level and top up, if necessary (photos)
Check the coolant level and top up, if necessary (photo)
Check the brake fluid level in the master cylinder and top up, if necessary (photo)

Lights and wipers
Check the operation of all interior and exterior lights, wipers and washers
Check and, if necessary, top up the washer reservoir, adding a screen wash such as Turtle Wax High Tech Screen Wash

Tyres
Check the tyre pressures
Visually examine the tyres for wear or damage

Every 12 000 miles (20 000 km) or 12 months – whichever occurs first

Engine (Chapter 1)
Renew the engine oil and filter (photos)
Check and, if necessary, adjust the valve clearances on 1.3 litre models
Visually check the engine for oil leaks and for the security and condition of all related components and attachments

Cooling system (Chapter 2)
Check the hoses, hose clips and visible joint gaskets for leaks and any signs of corrosion or deterioration
Check and, if necessary, top up the cooling system
Check the condition of the alternator drivebelt and renew if worn. Adjust the belt tension.

Check the engine oil level on the dipstick ...

... and top up if necessary

Top up the cooling system at the expansion tank

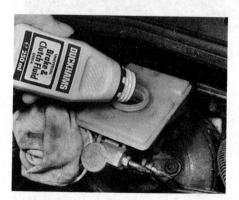

Top up the brake fluid at the master cylinder reservoir

Sump drain plug location on 1.3 litre models ...

... and 1.6 litre models

Oil filter location on 1.3 litre models

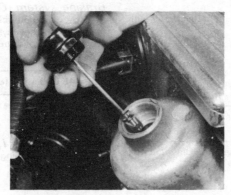

Top up the carburettor piston damper with engine oil

Gearbox filler plug location (5-speed gearbox)

Fuel and exhaust system (Chapter 3)
Renew the air cleaner element
Visually check the fuel pipes and hoses for security, chafing, leaks and corrosion
Check the fuel tank for leaks and any signs of damage or corrosion
Top up the carburettor piston damper (photo)
Check the operation of the accelerator and linkage
Check and, if necessary, adjust the carburettor slow running characteristics
Check the exhaust system for corrosion, leaks and security

Ignition system (Chapter 4)
Renew the spark plugs
Clean the distributor cap, coil tower and HT leads, and check for tracking

Clutch (Chapter 5)
Check the operation of the clutch and clutch pedal

Gearbox (Chapter 6)
Visually check for oil leaks around the gearbox joint faces and oil seals
Check and, if necessary, top up the gearbox oil (photo)
Lubricate the gearchange linkage (Chapter 13)

Automatic transmission (Chapter 7)
Visually check for oil leaks around the transmission joint faces and oil seals
Check and, if necessary, top up the automatic transmission fluid
Check and, if necessary top up the final drive gear oil

Driveshafts (Chapter 8)
Check the driveshaft constant velocity joints for wear or damage and check the rubber gaiters for condition

Braking system (Chapter 9)
Check visually all brake pipes, hoses and unions for corrosion, chafing, leakage and security
Check and, if necessary, top up the brake fluid
Check the operation of the brake warning indicators
Check the brake servo vacuum hose for condition and security
Check the operation of the hand and footbrake
Check the front brake pads for wear, and the discs for condition
Check the rear brake shoes for wear, and the drums for condition

Electrical system (Chapter 10)
Check the condition and security of all accessible wiring connectors, harnesses and retaining clips
Check the operation of all electrical equipment and accessories (lights, indicators, horn, wipers etc)
Check and adjust the operation of the screen washer and, if necessary, top up the reservoir
Clean the battery terminals and smear with petroleum jelly
Have the headlamp alignment checked and, if necessary, adjusted

Suspension, steering, wheels and tyres (Chapter 11)
Check the front and rear suspension struts for fluid leaks
Check the condition and security of the steering gear, steering and suspension joints, and rubber gaiters
Check the front wheel toe setting
Check and adjust the tyre pressures
Check the tyres for damage, tread depth and uneven wear
Inspect the roadwheels for damage
Check the tightness of the wheel nuts
Check the power-assisted steering reservoir fluid level (Chapter 13)

Bodywork (Chapter 12)
Carefully inspect the paintwork for damage and the bodywork for corrosion
Check the condition of the underseal
Oil all hinges, door locks and the bonnet release mechanism with a few drops of light oil (not the steering lock)

Road test
Check the operation of all instruments and electrical equipment
Check the operation of the seat belts
Check for any abnormalities in the steering, suspension, handling or road feel
Check the performance of the engine, clutch and transmission
Check the operation and performance of the braking system

Every 24 000 miles (40 000 km) or 24 months – whichever occurs first

---

In addition to all the items in the annual service, carry out the following:

Engine (Chapter 1)
Renew the engine oil filler cap on 1.3 litre models
Clean the engine breather filter on 1.6 litre models
Check and, if necessary, adjust the valve clearances on 1.6 litre engines

Cooling system (Chapter 2)
Flush the cooling system and renew the antifreeze solution
Renew the alternator drivebelt

Fuel system (Chapter 3)
Check and adjust the carburettor idle speed and mixture settings

Ignition system (Chapter 4)
Check and, if necessary, adjust the ignition timing on 1.3 litre models
Lubricate the distributor centrifugal advance mechanism on 1.3 litre models

Automatic transmission (Chapter 7)
Drain the transmission fluid, clean the oil strainer and refill with fresh fluid
Check the operation of the parking pawl

Braking system (Chapter 9)
Renew the brake fluid

Every 36 000 miles (60 000 km) or 36 months – whichever occurs first

In addition to the items in the annual service, carry out the following:

Braking system (Chapter 9)
Renew the flexible rubber hoses and the rubber seals in the calipers, wheel cylinders and master cylinder
Renew the air filter in the servo unit

Every 48 000 miles (80 000 km) or 4 years – whichever occurs first

In addition to the items listed in the annual service, carry out the following:

Engine (Chapter 1)
Renew the timing belt on 1.6 litre engines

Fault diagnosis

Introduction

The vehicle owner who does his or her own maintenance according to the recommended schedules should not have to use this section of the manual very often. Modern component reliability is such that, provided those items subject to wear or deterioration are inspected or renewed at the specified intervals, sudden failure is comparatively rare. Faults do not usually just happen as a result of sudden failure, but develop over a period of time. Major mechanical failures in particular are usually preceded by characteristic symptoms over hundreds or even thousands of miles. Those components which do occasionally fail without warning are often small and easily carried in the vehicle.

With any fault finding, the first step is to decide where to begin investigations. Sometimes this is obvious, but on other occasions a little detective work will be necessary. The owner who makes half a dozen haphazard adjustments or replacements may be successful in curing a fault (or its symptoms), but he will be none the wiser if the fault recurs and he may well have spent more time and money than was necessary. A calm and logical approach will be found to be more satisfactory in the long run. Always take into account any warning signs or abnormalities that may have been noticed in the period preceding the fault – power loss, high or low gauge readings, unusual noises or smells, etc – and remember that failure of components such as fuses or spark plugs may only be pointers to some underlying fault.

The pages which follow here are intended to help in cases of failure to start or breakdown on the road. There is also a Fault Diagnosis Section at the end of each Chapter which should be consulted if the preliminary checks prove unfruitful. Whatever the fault, certain basic principles apply. These are as follows:

Verify the fault. This is simply a matter of being sure that you know what the symptoms are before starting work. This is particularly important if you are investigating a fault for someone else who may not have described it very accurately.

Don't overlook the obvious. For example, if the vehicle won't start, is there petrol in the tank? (Don't take anyone else's word on this particular point, and don't trust the fuel gauge either!) If an electrical fault is indicated, look for loose or broken wires before digging out the test gear.

Cure the disease, not the symptom. Substituting a flat battery with a fully charged one will get you off the hard shoulder, but if the underlying cause is not attended to, the new battery will go the same way. Similarly, changing oil-fouled spark plugs for a new set will get you moving again, but remember that the reason for the fouling (if it wasn't simply an incorrect grade of plug) will have to be established and corrected.

Don't take anything for granted. Particularly, don't forget that a 'new' component may itself be defective (especially if it's been rattling round in the boot for months), and don't leave components out of a fault diagnosis sequence just because they are new or recently fitted. When you do finally diagnose a difficult fault, you'll probably realise that all the evidence was there from the start.

Electrical faults

Electrical faults can be more puzzling than straightforward mechanical failures, but they are no less susceptible to logical analysis if the basic principles of operation are understood. Vehicle electrical wiring exists in extremely unfavourable conditions – heat, vibration and chemical attack – and the first things to look for are loose or corroded connections and broken or chafed wires, especially where the wires pass through holes in the bodywork or are subject to vibration.

All metal-bodied vehicles in current production have one pole of the battery 'earthed', ie connected to the vehicle bodywork, and in nearly all modern vehicles it is the negative (–) terminal. The various electrical components – motors, bulb holders etc – are also connected to earth, either by means of a lead or directly by their mountings. Electric current flows through the component and then back to the battery via the bodywork. If the component mounting is loose or corroded, or if a good path back to the battery is not available, the circuit will be incomplete and malfunction will result. The engine and/or gearbox are also earthed by means of flexible metal straps to the body or subframe; if these straps are loose or missing, starter motor, generator and ignition trouble may result.

Assuming the earth return to be satisfactory, electrical faults will be due either to component malfunction or to defects in the current supply. Individual components are dealt with in Chapter 10. If supply wires are broken or cracked internally this results in an open-circuit, and the easiest way to check for this is to bypass the suspect wire temporarily with a length of wire having a crocodile clip or suitable connector at each end. Alternatively, a 12V test lamp can be used to verify the presence of supply voltage at various points along the wire and the break can be thus isolated.

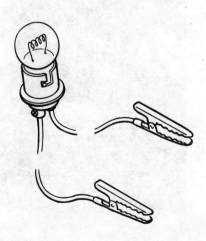

A simple test lamp is useful for checking electrical faults

If a bare portion of a live wire touches the bodywork or other earthed metal part, the electricity will take the low-resistance path thus formed back to the battery: this is known as a short-circuit. Hopefully a short-circuit will blow a fuse, but otherwise it may cause burning of the insulation (and possibly further short-circuits) or even a fire. This is why it is inadvisable to bypass persistently blowing fuses with silver foil or wire.

Spares and tool kit

Most vehicles are supplied only with sufficient tools for wheel changing; the *Maintenance and minor repair* tool kit detailed in *Tools and working facilities,* with the addition of a hammer, is probably sufficient for those repairs that most motorists would consider attempting at the roadside. In addition a few items which can be fitted without too much trouble in the event of a breakdown should be carried. Experience and available space will modify the list below, but the following may save having to call on professional assistance:

Spark plugs, clean and correctly gapped
HT lead and plug cap – long enough to reach the plug furthest from the distributor
Distributor rotor
Drivebelt(s) – emergency type may suffice
Spare fuses
Set of principal light bulbs
Tin of radiator sealer and hose bandage
Exhaust bandage
Roll of insulating tape
Length of soft iron wire
Length of electrical flex
Torch or inspection lamp (can double as test lamp)
Battery jump leads
Tow-rope
Ignition water dispersant aerosol
Litre of engine oil

Sealed can of hydraulic fluid
Emergency windscreen
Worm drive clips

If spare fuel is carried, a can designed for the purpose should be used to minimise risks of leakage and collision damage. A first aid kit and a warning triangle, whilst not at present compulsory in the UK, are obviously sensible items to carry in addition to the above.

When touring abroad it may be advisable to carry additional spares which, even if you cannot fit them yourself, could save having to wait while parts are obtained. The items below may be worth considering:

Clutch and throttle cables
Cylinder head gasket
Alternator brushes
Tyre valve core

One of the motoring organisations will be able to advise on availability of fuel etc in foreign countries.

Engine will not start

Engine fails to turn when starter operated

Flat battery (recharge, use jump leads, or push start)
Battery terminals loose or corroded
Battery earth to body defective
Engine earth strap loose or broken
Starter motor (or solenoid) wiring loose or broken
Automatic transmission selector in wrong position, or inhibitor switch faulty
Ignition/starter switch faulty
Major mechanical failure (seizure)
Starter or solenoid internal fault (see Chapter 10)

Carrying a few spares may save you a long walk!

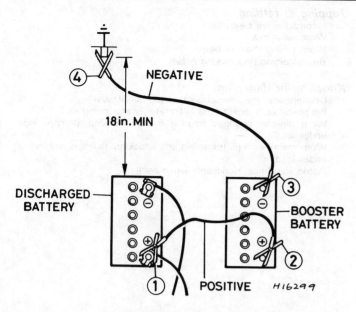

Jump start lead connections for negative earth – connect leads in order shown

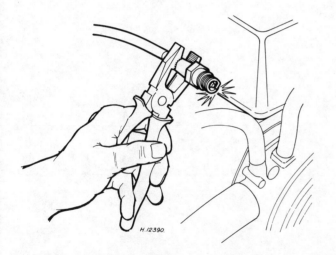

Crank engine and check for spark. Note use of insulated tool to hold plug lead

Starter motor turns engine slowly
Partially discharged battery (recharge, use jump leads, or push start)
Battery terminals loose or corroded
Battery earth to body defective
Engine earth strap loose
Starter motor (or solenoid) wiring loose
Starter motor internal fault (see Chapter 10)

Starter motor spins without turning engine
Flat battery
Starter motor pinion sticking on sleeve
Flywheel gear teeth damaged or worn
Starter motor mounting bolts loose

Engine turns normally but fails to start
Damp or dirty HT leads and distributor cap (crank engine and check for spark), – try moisture dispersant such as Holts Wet Start
No fuel in tank (check for delivery at carburettor)

Fouled or incorrectly gapped spark plugs (remove and regap)
Other ignition system fault (see Chapter 4)
Other fuel system fault (see Chapter 3)
Poor compression (see Chapter 1)
Major mechanical failure (eg camshaft drive)

Engine fires but will not run
Air leaks at carburettor or inlet manifold
Fuel starvation (see Chapter 3)
Other ignition fault (see Chapter 4)

Engine cuts out and will not restart

Engine cuts out suddenly – ignition fault
Loose or disconnected LT wires
Wet HT leads or distributor cap (after traversing water splash)
Coil or condenser failure (check for spark)
Other ignition fault (see Chapter 4)

Engine misfires before cutting out – fuel fault
Fuel tank empty
Fuel pump defective or filter blocked (check for delivery)
Fuel tank filler vent blocked (suction will be evident on releasing cap)
Carburettor needle valve sticking
Other fuel system fault (see Chapter 3)

Engine cuts out – other causes
Serious overheating
Major mechanical failure (eg camshaft drive)

Engine overheats

Ignition (no-charge) warning light illuminated
Slack or broken drivebelt – retension or renew (Chapter 10)

Ignition warning light not illuminated
Coolant loss due to internal or external leakage (see Chapter 2)
Thermostat defective
Low oil level
Brakes binding
Radiator clogged externally or internally
Electric cooling fan not operating correctly
Engine waterways clogged
Ignition timing incorrect or automatic advance malfunctioning
Mixture too weak

Note: *Do not add cold water to an overheated engine or damage may result*

Low engine oil pressure

Gauge reads low or warning light illuminated with engine running
Oil level low or incorrect grade
Defective gauge or sender unit
Wire to sender unit earthed
Engine overheating
Oil filter clogged or bypass valve defective
Oil pressure relief valve defective
Oil pick-up strainer clogged
Oil pump worn or mountings loose
Worn main or big-end bearings

Note: *Low oil pressure in a high-mileage engine at tickover is not necessarily a cause for concern. Sudden pressure loss at speed is far more significant. In any event, check the gauge or warning light sender before condemning the engine.*

Engine noises

Pre-ignition (pinking) on acceleration
Incorrect grade of fuel
Ignition timing incorrect
Distributor faulty or worn
Worn or maladjusted carburettor
Excessive carbon build-up in engine

Whistling or wheezing noises
Leaking vacuum hose
Leaking carburettor or manifold gasket
Blowing head gasket

Tapping or rattling
Incorrect valve clearances
Worn valve gear
Worn timing chain or belt
Broken piston ring (ticking noise)

Knocking or thumping
Unintentional mechanical contact (eg fan blades)
Peripheral component fault (alternator, water pump etc)
Worn big-end bearings (regular heavy knocking, perhaps less under load)
Worn main bearings (rumbling and knocking, perhaps worsening under load)
Piston slap (most noticeable when cold)

Chapter 1 Engine

For modifications, and information applicable to later models, see Supplement at end of manual

Contents

Part A: 1.3 litre engine
General
Type	Four-cylinder in-line, overhead valve (12H – 'A+' series)
Bore	2.780 in (70.61 mm)
Stroke	3.200 in (81.28 mm)
Capacity	1275 cc (77.8 cu in)
Firing order	1–3–4–2 (No 1 cylinder at crankshaft pulley end)
Compression ratio	9.75 : 1

Crankshaft
Main journal diameter	2.0012 to 2.0017 in (50.83 to 50.84 mm)
Main bearing running clearance	0.0003 to 0.0030 in (0.008 to 0.076 mm)
Main journal minimum regrind diameter	1.9811 in (50.32 mm)
Crankpin journal diameter	1.7497 to 1.7504 in (44.44 to 44.46 mm)
Crankpin running clearance	0.0015 to 0.0032 in (0.0381 to 0.0813 mm)
Crankpin minimum regrind diameter	1.7297 in (43.93 mm)
Crankshaft endfloat	0.002 to 0.003 in (0.051 to 0.076 mm)

Connecting rods
Length between centres	5.750 in (146.05 mm)

Pistons
Skirt clearance in cylinder:	
Top	0.0029 to 0.0045 in (0.074 to 0.114 mm)
Bottom	0.0009 to 0.0025 in (0.023 to 0.064 mm)
Oversizes available	0.020 in (0.51 mm)

Piston rings
Clearance in groove (compression rings)	0.0015 to 0.0035 in (0.038 to 0.089 mm)
Fitted end gap:	
Top compression	0.010 to 0.017 in (0.25 to 0.43 mm)
Second compression	0.008 to 0.013 in (0.20 to 0.33 mm)
Oil control	0.015 to 0.041 in (0.38 to 1.04 mm)

Gudgeon pins
Diameter	0.8123 to 0.8125 in (20.63 to 20.64 mm)
Clearance in piston	Hand push fit at 68°F (20°C)
Interference fit in connecting rod	0.0008 to 0.0015 in (0.02 to 0.04 mm)

Camshaft
Journal diameter:	
Front	1.6655 to 1.6660 in (42.304 to 42.316 mm)
Centre	1.62275 to 1.62325 in (41.218 to 41.231 mm)
Rear	1.37275 to 1.3735 in (34.868 to 34.887 mm)
Running clearance in bearings	0.001 to 0.00225 in (0.025 to 0.057 mm)
Endfloat	0.003 to 0.007 in (0.076 to 0.178 mm)
Valve lift	0.318 in (8.08 mm)
Cam followers outside diameter	0.812 in (20.62 mm)

Valves
Seat angle	45°
Head diameter:	
Inlet	1.307 to 1.312 in (33.20 to 33.32 mm)
Exhaust	1.1515 to 1.1565 in (29.25 to 29.38 mm)
Stem diameter:	
Inlet	0.2793 to 0.2798 in (7.094 to 7.107 mm)
Exhaust	0.2788 to 0.2793 in (7.082 to 7.094 mm)
Clearance in guide:	
Inlet	0.0015 to 0.0025 in (0.038 to 0.064 mm)
Exhaust	0.002 to 0.003 in (0.051 to 0.076 mm)
Valve guides:	
Length	1.687 in (42.85 mm)
Outside diameter	0.470 to 0.471 in (11.94 to 11.96 mm)
Inside diameter	0.2813 to 0.2818 in (7.145 to 7.158 mm)
Fitted height above head	0.540 in (13.72 mm)
Valve springs:	
Free length	1.95 in (49.53 mm)
Valve timing (at valve clearance of 0.021 in/0.53 mm):	
Inlet opens	20° BTDC
Inlet closes	52° ABDC
Exhaust opens	55° BBDC
Exhaust closes	17° ATDC
Valve clearances (cold)	0.014 in (0.36 mm)

Lubrication system

Oil type/specification* .... Multigrade engine oil, viscosity SAE 10W/40 (Duckhams QXR, Hypergrade, or 10W/40 Motor Oil)

Oil filter .... Champion B101 with oil cooler or D102 without oil cooler

Oil pump:

Type .... Bi-rotor
Outer rotor endfloat .... 0.005 in (0.127 mm)
Inner rotor endfloat .... 0.005 in (0.127 mm)
Outer rotor-to-body clearance .... 0.010 in (0.254 mm)
Rotor lobe clearance .... 0.006 in (0.152 mm)

System pressure:

Idling .... 15 lbf/in^2 (1.0 bar)
Running .... 60 lbf/in^2 (4.1 bar)
Warning light switch operating pressure .... 6 to 10 lbf/in^2 (0.4 to 0.7 bar)
Pressure relief valve operating pressure .... 60 lbf/in^2 (4.1 bar)
Pressure relief valve spring free length .... 2.86 in (72.64 mm)

*Note: *Austin Rover specify a 10W/40 oil to meet warranty requirements for models produced after August 1983. Duckhams QXR and 10W/40 Motor Oil are available to meet these requirements*

Torque wrench settings

	lbf ft	Nm
Camshaft locating plate bolts	8	11
Camshaft retaining nut	65	88
Connecting rod big-end cap nuts	30	41
Crankshaft pulley bolt	105	142
Cylinder head nuts	55	75
Front plate to main bearing cap	5	7
Front plate to cylinder block	16	22
Gearbox adaptor plate bolts	18	24
Main bearing cap bolts	65	88
Oil pressure switch	18	24
Oil pump retaining bolts	8	11
Oil pressure relief valve cap nut	45	61
Oil separator bolts	16	22
Rocker cover bolts	3	4
Rocker shaft pedestal nuts	24	33
Sump drain plug	28	38
Sump retaining bolts	8	11
Timing cover to front plate:		
$\frac{1}{4}$ in bolts	5	7
$\frac{5}{16}$ in bolts	12	16
Vacuum hose banjo union bolt	37	50
Left-hand engine mounting bracket to body	18	24
Left-hand engine mounting-to-bracket retaining bolts	18	24
Left-hand engine mounting through-bolt	34	46
Left-hand engine mounting to gearbox	30	41
Right-hand engine mounting support strap bolts	18	24
Right-hand engine mounting bracket-to-engine side bolts	18	24
Right-hand engine mounting bracket top and bottom bolts	30	41
Right-hand engine mounting through-bolt	34	46
Tie-rod bracket bolts	34	46
Tie-rod to engine bracket	67	91
Tie-rod to crossmember	34	46

Part B: 1.6 litre engine

General

Type .... Four-cylinder in-line, overhead camshaft (16H – 'S' series)
Bore .... 3.000 in (76.20 mm)
Stroke .... 3.443 in (87.38 mm)
Capacity .... 1598 cc (97.5 cu in)
Firing order .... 1–3–4–2 (No 1 cylinder at crankshaft pulley end)
Compression ratio .... 9.6 : 1

Crankshaft

Main journal diameter .... 2.2515 to 2.2520 in (57.19 to 57.20 mm)
Main bearing running clearance .... 0.002 to 0.0035 in (0.05 to 0.09 mm)
Main journal minimum regrind diameter .... 2.2115 in (56.17 mm)
Crankpin journal diameter .... 1.8759 to 1.8765 in (47.65 to 47.66 mm)
Crankpin running clearance .... 0.0015 to 0.003 in (0.04 to 0.08 mm)
Crankpin minimum regrind diameter .... 1.8359 in (46.63 mm)
Endfloat .... 0.004 to 0.007 in (0.10 to 0.18 mm)

Connecting rods

Length between centres .... 5.830 in (148.08 mm)

Pistons

Skirt clearance in cylinder:

Top .... 0.0028 to 0.0044 in (0.07 to 0.11 mm)
Bottom .... 0.001 to 0.002 in (0.03 to 0.05 mm)
Oversizes available .... 0.020 in (0.51 mm)

Piston rings

Clearance in grooves	0.0015 to 0.0035 in (0.04 to 0.09 mm)
Fitted gap:	
Compression	0.012 to 0.022 in (0.30 to 0.56 mm)
Oil control rails	0.015 to 0.045 in (0.38 to 1.14 mm)

Gudgeon pins

Diameter	0.8123 to 0.8125 in (20.63 to 20.64 mm)
Clearance in piston	Hand push fit at 68°F (20°C)
Interference fit in connecting rod	0.0008 to 0.0015 in (0.02 to 0.04 mm)

Camshaft

Journal diameter:	
Front	1.9355 to 1.9365 in (49.16 to 49.19 mm)
Centre	1.9668 to 1.9678 in (49.96 to 49.98 mm)
Rear	1.998 to 1.999 in (50.75 to 50.77 mm)
Running clearance in bearings	0.001 to 0.0023 in (0.025 to 0.058 mm)
Endfloat	0.002 to 0.007 in (0.05 to 0.18 mm)
Valve lift:	
Inlet	0.342 to 0.344 in (8.69 to 8.74 mm)
Exhaust	0.338 to 0.340 in (8.58 to 8.64 mm)

Tappets

Adjustment	Selective shims
Outside diameter	1.1865 in (30.14 mm)
Adjustment shim number:	**Corresponding thickness**
97 to 49	0.097 in (2.46 mm) to 0.149 in (3.78 mm), in 0.002 in (0.05 mm) steps

Valves

Seat angle	45.25°
Head diameter:	
Inlet	1.500 in (38.1 mm)
Exhaust	1.218 in (30.9 mm)
Stem diameter (standard):	
Inlet	0.3115 to 0.3120 in (7.91 to 7.92 mm)
Exhaust	0.3109 to 0.3114 in (7.90 to 7.91 mm)
Stem diameter (oversize)	0.3331 to 0.3336 in (8.46 to 8.47 mm)
Stem-to-guide clearance	0.0015 in (0.038 mm)
Valve spring free length	1.810 in (45.97 mm)
Valve timing (at tappet clearance of 0.021 in/0.53 mm):	
Inlet opens	17° BTDC
Inlet closes	59° ABDC
Exhaust opens	57° BBDC
Exhaust closes	19° ATDC
Tappet clearances (cold):	
Inlet	0.014 to 0.015 in (0.36 to 0.38 mm)
Exhaust	0.017 to 0.018 in (0.43 to 0.46 mm)
Adjust only if less than	0.012 in (0.30 mm)
Valve sequence (from front of head):	
Inlet	2,3,6,7
Exhaust	1,4,5,8

Timing belt

Tensioning method	Torque wrench engaged with hole in tensioner bracket
Tension:	
Used belt	10 lbf ft (14 Nm)
New belt	15 lbf ft (20 Nm)

Cylinder head

Height:	
New	3.312 to 3.322 in (84.12 to 84.38 mm)
Minimum	3.302 in (83.87 mm)
Maximum clearance under straight-edge across surface	0.002 in (0.05 mm)

Lubrication system

Oil type/specification* .... Multigrade engine oil, viscosity SAE 10W/40 (Duckhams QXR, Hypergrade, or 10W/40 Motor Oil)

Oil filter .... Champion C104
Oil pump:
    Type .... Bi-rotor
    Outer rotor endfloat .... 0.001 to 0.003 in (0.03 to 0.08 mm)
    Outer rotor-to-body clearance .... 0.007 to 0.011 in (0.18 to 0.27 mm)
    Rotor lobe clearance .... 0.001 to 0.006 in (0.025 to 0.15 mm)
System pressure:
    Running .... 55 lbf/in² (3.8 bar)
    Idling .... 10 lbf/in² (0.7 bar)
Pressure relief valve spring free length .... 2.47 in (62.7 mm)
Warning light switch operating pressure .... 6 to 10 lbf/in² (0.4 to 0.7 bar)

***Note:** Austin Rover specify a 10W/40 oil to meet warranty requirements for models produced after August 1983. Duckhams QXR and 10W/40 Motor Oil are available to meet these requirements*

Torque wrench settings

	lbf ft	Nm
Timing belt tensioner bolts	18	24
Camshaft carrier bolts	18	24
Camshaft carrier cover bolts	6	8
Camshaft sprocket bolt	41	56
Connecting rod big-end cap nuts	34	46
Crankshaft pulley bolt	34	46
Cylinder head bolts (lightly oiled):		88
Stage 1	30	40
Stage 2	60	80
Stage 3	Tighten by a further quarter-turn (90°)	Tighten by a further quarter-turn (90°)
Gearbox adaptor plate:		
M8 bolts	18	24
M10 bolts	34	46
M12 bolts	67	91
Knock sensor	9	12
Main bearing cap bolts	67	91
Oil pump-to-cylinder block bolts	19	26
Oil pressure switch	18	24
Sump drain plug	18	24
Sump bolts	6	8
Vacuum hose banjo union bolt	37	50
Left-hand engine mounting to gearbox	34	46
Left-hand mounting body bracket to body	30	41
Left-hand mounting through-bolt	67	91
Right-hand engine mounting to cylinder block and head	34	46
Right-hand mounting support plate-to-body bolts	30	40
Right-hand mounting body bracket to body	34	46
Right-hand mounting through-bolt	67	91
Rear engine mounting support bracket to crossmember	34	46
Rear mounting to support bracket	18	24
Rear mounting to gearbox casing	34	46
Front snubber bracket to engine	38	52
Snubber cup bolts	30	40

PART A: 1.3 LITRE ENGINE

1 General description

The engine is of four-cylinder, in-line overhead valve type, mounted transversely at the front of the car.

The crankshaft is supported in three shell-type main bearings. Thrust washers are fitted at the centre main bearing to control crankshaft endfloat.

The connecting rods are attached to the crankshaft by horizontally split shell-type big-end bearings, and to the pistons by interference fit gudgeon pins. The aluminium alloy pistons are of the slipper type and are fitted with three piston rings; two compression rings and a three piece oil control ring.

The camshaft is chain driven from the crankshaft and operates the rocker arms via pushrods. The inlet and exhaust valves are each closed by a single valve spring and operate in guides pressed into the cylinder head. The valves are actuated directly by the rocker arms.

Engine lubrication is by an eccentric rotor type oil pump. The pump is mounted on the gearbox end of the cylinder block and is driven by the camshaft, as are the distributor and fuel pump.

2 Maintenance and inspection

1 At the intervals given in Routine Maintenance at the beginning of this manual, carry out the following service operations on the engine.
2 Visually inspect the engine joint faces, gaskets and seals for any sign of oil or water leaks. Pay particular attention to the areas around the rocker cover, cylinder head, timing cover and sump joint faces. Rectify any leaks by referring to the appropriate Sections of this Chapter.
3 Place a suitable container beneath the oil drain plug on the right-hand side of the sump. Unscrew the plug using spanner or socket and allow the oil to drain. Inspect the condition of the drain plug sealing washer and renew it, if necessary. Refit and tighten the plug after draining.
4 Move the bowl to the rear of the engine, under the oil filter.
5 Using a strap wrench, or filter removal tool, slacken the filter and then unscrew it from the engine and discard.
6 Wipe the mating face on the cylinder block with a rag and then lubricate the seal of a new filter using clean engine oil.
7 Screw the filter into position and tighten it by hand only, do not use any tools.

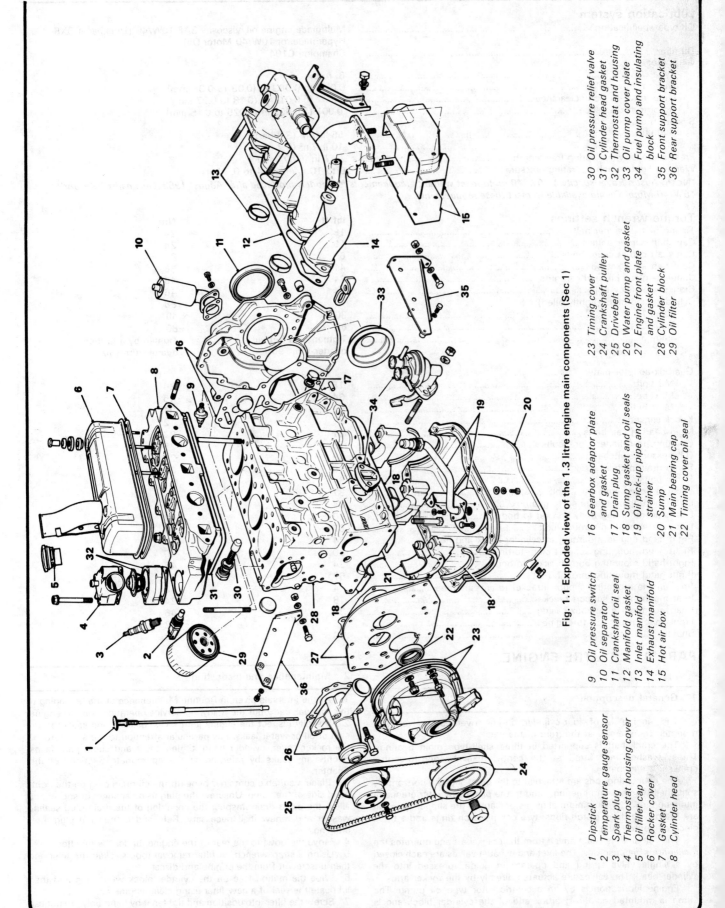

Fig. 1.1 Exploded view of the 1.3 litre engine main components (Sec 1)

1 Dipstick
2 Temperature gauge sensor
3 Spark plug
4 Thermostat housing cover
5 Oil filler cap
6 Rocker cover
7 Gasket
8 Cylinder head
9 Oil pressure switch
10 Oil separator
11 Crankshaft oil seal
12 Manifold gasket
13 Inlet manifold
14 Exhaust manifold
15 Hot air box
16 Gearbox adaptor plate
    and gasket
17 Drain plug
18 Sump gasket and oil seals
19 Oil pick-up pipe and
    strainer
20 Sump
21 Main bearing cap
22 Timing cover oil seal
23 Timing cover
24 Crankshaft pulley
25 Drivebelt
26 Water pump and gasket
27 Engine front plate
    and gasket
28 Cylinder block
29 Oil filter
30 Oil pressure relief valve
31 Cylinder head gasket
32 Thermostat and housing
33 Oil pump cover plate
34 Fuel pump and insulating
    block
35 Front support bracket
36 Rear support bracket

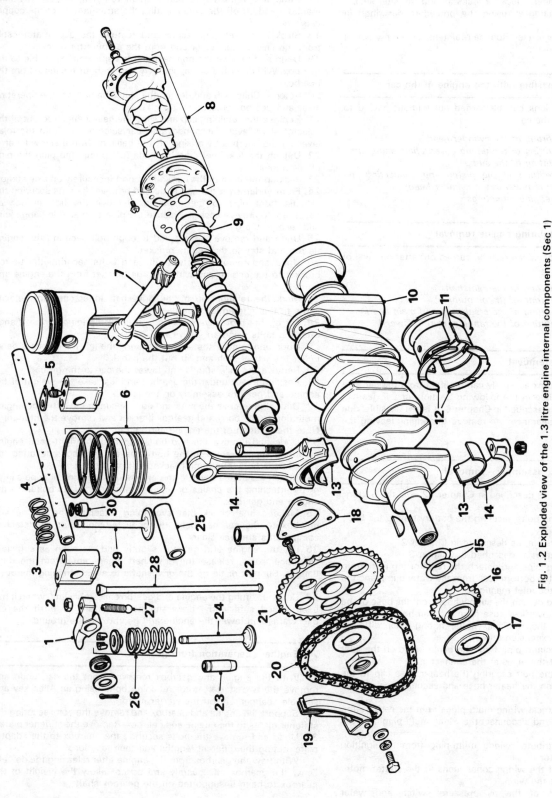

Fig. 1.2 Exploded view of the 1.3 litre engine internal components (Sec 1)

1	Valve rocker	7	Distributor driveshaft	13 Big-end bearing shells	19 Chain tensioner	26 Valve spring, cap and
2	Rocker shaft pedestal	8	Oil pump	14 Connecting rod and cap	20 Timing chain	collets
3	Rocker shaft spring	9	Camshaft	15 Crankshaft sprocket	21 Camshaft sprocket	27 Rocker adjusting screw
4	Rocker shaft	10	Crankshaft	packing shims	22 Tappet	28 Push-rod
5	Rocker shaft locating	11	Main bearing shells	16 Crankshaft sprocket	23 Valve guide	29 Inlet valve
	screw and lockplate	12	Crankshaft thrust washers	17 Crankshaft oil thrower	24 Exhaust valve	30 Valve stem oil seal
6	Piston rings and piston			18 Camshaft locating plate	25 Gudgeon pin	

8 Refill the engine, using the correct grade of oil, through the filler neck on the rocker cover. Fill until the level reaches the MAX mark on the dipstick.
9 With the engine running, check for leaks around the filter seal.
10 Adjust the valve clearances using the procedure described in Section 44.
11 At less frequent intervals (see Routine Maintenance) renew the oil filler cap on the rocker cover.

3 Major operations possible with the engine in the car

The following operations can be carried out without having to remove the engine from the car:

(a) Removal and refitting of the cylinder head
(b) Removal and refitting of the timing cover, chain and gears
(c) Removal and refitting of the sump
(d) Removal and refitting of the piston and connecting rod assemblies (after removal of the cylinder head)
(e) Renewal of the engine mountings

4 Major operations requiring engine removal

The following operations can only be carried out after removal of the engine from the car:

(a) Removal and refitting of the camshaft
(b) Removal and refitting of the oil pump
(c) Removal and refitting of the crankshaft and main bearings
(d) Removal and refitting of the tappets (cam followers)

5 Methods of engine removal

The engine and gearbox assembly can be lifted from the car as a complete unit, as described in the following Section, or the gearbox may first be removed, as described in Chapter 6. It is not possible, due to the limited working clearances, to remove the engine leaving the gearbox in position.

6 Engine and gearbox assembly – removal

1 Remove the bonnet, as described in Chapter 12, and the battery, as described in Chapter 10.
2 Undo and remove the bolts securing the battery tray to the body and remove the tray.
3 Drain the cooling system, as described in Chapter 2.
4 Remove the air cleaner, as described in Chapter 3.
5 Slacken the retaining clips and detach the radiator top hose from the thermostat housing, the bottom hose from the water pump and the two heater hoses from the inlet manifold.
6 Slacken the clips and detach the supply hose from the base of the expansion tank, and the overflow hose from the thermostat housing.
7 Undo and remove the retaining bolts, noting their different lengths, and lift off the expansion tank.
8 Detach the main vacuum pipe from the banjo union on the inlet manifold and disconnect the pipes at the connector.
9 Undo and remove the bolt securing the heater hose clip to the cylinder head and position the heater hose and vacuum pipe clear of the engine.
10 Disconnect the electrical wiring multi-plugs from the carburettor.
11 Spring back the clip and disconnect the wiring multi-plug from the alternator.
12 Disconnect the distributor wiring multi-plug from the ignition amplifier wiring connector.
13 Note the position of the wiring connections at the starter motor solenoid and disconnect the wires.
14 Note the positions of the oil pressure switch and water temperature gauge wires and disconnect them from their sensors. Detach the reversing lamp switch wires from the gearbox.
15 Undo and remove the bolt securing the wiring harness clip to the gearbox, and release the clips securing the harness to the rocker cover brackets. Check that all the wiring has been disconnected and move the harness clear of the engine.

16 Pull the HT leads off the spark plugs and detach the HT lead from the coil and rubber boot. Spring back the clips and remove the distributor cap and leads.
17 Undo and remove the banjo union retaining bolt at the inlet manifold and lift off the union, noting the arrangement of the copper washers.
18 Slacken the retaining screw and remove the accelerator cable from the linkage connector and from the carburettor bracket.
19 Undo and remove the bolt securing the speedometer cable to the gearbox. Withdraw the cable, squarely and without twisting, from the gearbox.
20 Refer to Chapter 5 and detach the clutch cable from the operating lever and gearbox bracket.
21 Remove the retaining clip and slide the gearchange rod out of the selector shaft lever. Disconnect the rear selector rod from the relay lever on the gearbox by prising off the balljoint with a screwdriver.
22 Detach the fuel inlet pipe from the fuel pump and plug the pipe after removal.
23 Jack up the front of the car and support it securely on axle stands.
24 From underneath the car, undo and remove the nuts securing the exhaust front pipe to the manifold. Remove the hot air box as necessary to provide access. Release the pipe-to-manifold flange joint and recover the gasket.
25 Undo and remove the nut and through-bolt securing the engine tie-rod and stay to the engine bracket.
26 Undo and remove the two nuts and bolts securing the tie-rod bracket to the engine and withdraw the bracket from the engine and tie-rod.
27 Mark the relationship of the driveshaft inner constant velocity joints to the differential drive flanges.
28 Lift off the protective caps and unscrew the joint-to-drive flange retaining bolts using an Allen key.
29 Undo and remove the screws securing the left-hand access panel under the wheel arch and lift out the panel.
30 Remove the axle stands and lower the car to the ground.
31 Position a jack under the gearbox and just take the weight of the engine and gearbox assembly on the jack.
32 Undo and remove the nuts and bolts securing the left-hand engine mounting to the body and gearbox bracket and remove the mounting. Note the location of the engine earth strap.
33 Undo and remove the two bolts securing the right-hand engine mounting support strap to the body and the two bolts securing the strap to the body mounting bracket. Lift off the strap.
34 Attach a suitable hoist to the engine using chains or rope slings, or by attaching the chains or ropes to sturdy brackets bolted to the engine and gearbox.
35 Make a final check that everything attaching the engine and gearbox to the car has been disconnected and that all detached components are well clear.
36 Lift the engine and gearbox slightly and, as soon as sufficient clearance exists, release the driveshaft inner joints from the drive flanges. Support or tie up the driveshafts to avoid straining the outer joints.
37 Continue lifting the engine and gearbox assembly and, when it has been raised sufficiently, draw the hoist forward or push the car backwards and lower the engine and gearbox to the ground.

7 Engine – separation from gearbox

1 With the engine and gearbox removed from the car, undo and remove the two starter motor retaining bolts using an Allen key and suitable spanner. Lift off the starter motor.
2 Support the engine and gearbox and remove the bolts securing the stiffener plates to the engine and gearbox. Remove the stiffener plates.
3 Undo and remove the bolts securing the gearbox to the adaptor plate, noting the different lengths and their locations.
4 Withdraw the gearbox from the engine after releasing the dowels. Draw the gearbox off squarely and do not allow the weight of the gearbox to hang unsupported on the gearbox shaft.

8 Engine dismantling – general

1 If possible, mount the engine on a stand for the dismantling procedure, but failing this support it in an upright position with blocks of wood placed under each side of the sump or crankcase.

2 Drain the oil into a suitable container before cleaning the engine or major dismantling.

3 Cleanliness is most important, and if the engine is dirty it should be cleaned with paraffin or a suitable solvent while keeping it in an upright position.

4 Avoid working with the engine directly on a concrete floor, as grit presents a real source of trouble.

5 As parts are removed, clean them in a paraffin bath. However, do not immerse parts with internal oilways in paraffin as it is difficult to remove, usually requiring a high pressure hose. Clean oilways with nylon pipe cleaners.

6 It is advisable to have suitable containers to hold small items, as this will help when reassembling the engine and also prevent possible losses.

7 Always obtain complete sets of gaskets when the engine is being dismantled, but retain the old gaskets with a view to using them as a pattern to make a replacement if a new one is not available.

8 When possible, refit nuts, bolts and washers in their location after being removed, as this helps to protect the threads and will also be helpful when reassembling the engine.

9 Retain unserviceable components in order to compare them with the new parts supplied.

9 Ancillary components – removal

With the engine separated from the gearbox the externally mounted ancillary components, as given in the following list, can be removed. The removal sequence need not necessarily follow the order given:

Inlet and exhaust manifolds and carburettor (Chapter 3)
Fuel pump (Chapter 3)
Alternator (Chapter 10)
Spark plugs (Chapter 4)
Distributor (Chapter 4)
Water pump (Chapter 2)
Thermostat housing and thermostat (Chapter 2)
Oil filter (Section 2 of this Chapter)
Dipstick
Engine right-hand mounting (Section 29 of this Chapter)
Clutch and flywheel (Chapter 5)

10 Cylinder head – removal

Note: *If the engine is still in the car, first carry out the following operations with reference to the relevant Sections and Chapters of this manual:*

(a) *Disconnect the battery negative lead*
(b) *Drain the cooling system*
(c) *Remove the inlet and exhaust manifold, complete with carburettor*
(d) *Disconnect the cooling system top hose, heater hose, and expansion tank hose from the thermostat housing*
(e) *Remove the HT leads and spark plugs*
(f) *Disconnect the lead from the coolant temperature thermistor*
(g) *Release the engine mounting bracket on the thermostat housing*
(h) *Remove the bolt securing the heater hose clip to the cylinder head*

1 Unscrew the two rocker cover retaining bolts and lift off the wiring harness support brackets. Collect the washers and lift off the rocker cover and gasket (photos).

2 Unscrew the rocker shaft pedestal small nuts and remove the washers. Note the lockwasher fitted to the second pedestal from the front.

3 Unscrew the cylinder head nuts half a turn at a time in the reverse order to that shown in Fig. 1.11, and then remove the nuts.

4 Lift the rocker shaft and pedestals from the studs.

5 Shake the pushrods free from the tappets (cam followers), then withdraw them from the cylinder head keeping them in strict order to ensure correct refitting (photo).

6 Lift the cylinder head from the block. If it is stuck, tap it free with a wooden mallet. *Do not insert a lever into the gasket joint* – you may damage the mating surfaces.

7 Remove the cylinder head gasket from the cylinder block.

10.1A Remove the rocker cover washer and seal

10.1B Lift off the rocker cover and gasket

10.5 Withdraw the pushrods, keeping them in order

11.1 With the valve spring compressed, lift out the split collets

11 Cylinder head – dismantling

1 Using a valve spring compressor, compress each valve spring in turn until the split collets can be removed (photo). Release the compressor and remove the cup and spring. If the cups are difficult to release, do not continue to tighten the compressor, but gently tap the top of the tool with a hammer. Always make sure that the compressor is held firmly over the cup.

2 Remove the oil seals from the inlet valve guides (photo). A small seal may also be fitted at the bottom of the collet groove on the valve stems.

3 Remove each valve from the combustion chambers keeping them in their order of removal, together with the respective valve springs and cups (photos). Identify each valve according to the cylinder, remembering that No 1 cylinder is at the thermostat end of the cylinder head.

12 Timing cover, chain and gears – removal

Note: *If the engine is still in the car, carry out the following operations with reference to the relevant Sections and Chapters of this manual:*

   (a) *Disconnect the battery negative lead*
   (b) *Remove the drivebelt and the water pump pulley*
   (c) *Remove the access panel from the right-hand wheel arch*

1 Disconnect the crankcase ventilation hose from the oil separator on the timing cover and release the fuel pipe clips from the timing cover studs.

2 Using a spanner or socket on the crankshaft pulley bolt, rotate the engine until No 4 piston is at top-dead-centre on the compression stroke. This will align the timing gear marks.

3 Undo and remove the crankshaft pulley bolt after releasing the lockwasher. If the engine is removed, lock the crankshaft using a bar or stout screwdriver across two bolts fitted to the crankshaft rear boss. If the engine is in the car, engage top gear and apply the handbrake.

4 Lever the pulley and damper off the front of the crankshaft.

5 Unbolt and remove the timing cover, and remove the gasket (photo).

6 Remove the oil thrower, noting which way round it is fitted (photo).

7 Flatten the lockwasher, then unscrew the camshaft gear retaining nut (photo). Use a screwdriver through one of the gear holes to restrain the gear. Remove the lockwasher.

8 Check that the alignment marks on the timing gears are facing each other, then unbolt and remove the chain tensioner.

11.2 Remove the valve stem oil seals

11.3A Remove the valves ...

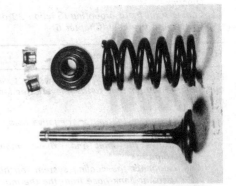

11.3B ... and keep the components together

12.5 Remove the timing cover

12.6 Note the fitted position of the oil thrower

12.7 Remove the camshaft gear retaining nut and lockwasher

Are your plugs trying to tell you something?

Normal.
Grey-brown deposits, lightly coated core nose. Plugs ideally suited to engine, and engine in good condition.

Heavy Deposits.
A build up of crusty deposits, light-grey sandy colour in appearance.
Fault: Often caused by worn valve guides, excessive use of upper cylinder lubricant, or idling for long periods.

Lead Glazing.
Plug insulator firing tip appears yellow or green/yellow and shiny in appearance.
Fault: Often caused by incorrect carburation, excessive idling followed by sharp acceleration. Also check ignition timing.

Carbon fouling.
Dry, black, sooty deposits.
Fault: over-rich fuel mixture.
Check: carburettor mixture settings, float level, choke operation, air filter.

Oil fouling.
Wet, oily deposits. Fault: worn bores/piston rings or valve guides; sometimes occurs (temporarily) during running-in period.

Overheating.
Electrodes have glazed appearance, core nose very white – few deposits. Fault: plug overheating. Check: plug value, ignition timing, fuel octane rating (too low) and fuel mixture (too weak).

Electrode damage.
Electrodes burned away; core nose has burned, glazed appearance. Fault: pre-ignition. Check: for correct heat range and as for 'overheating'.

Split core nose.
(May appear initially as a crack). Fault: detonation or wrong gap-setting technique. Check: ignition timing, cooling system, fuel mixture (too weak).

WHY DOUBLE COPPER IS BETTER FOR YOUR ENGINE.

Unique Trapezoidal Copper Cored Earth Electrode — 50% Larger Spark Area — Copper Cored Centre Electrode

Champion Double Copper plugs are the first in the world to have copper core in both centre _and_ earth electrode. This innovative design means that they run cooler by up to 100°C – giving greater efficiency and longer life. These double copper cores transfer heat away from the tip of the plug faster and more efficiently. Therefore, Double Copper runs at cooler temperatures than conventional plugs giving improved acceleration response and high speed performance with no fear of pre-ignition.

TRAPEZOIDAL COPPER CORED EARTH ELECTRODE
NEW TRAPEZOIDAL COPPER CORED EARTH ELECTRODE — CONVENTIONAL SOLID NICKEL ALLOY EARTH ELECTRODE
50% INCREASE IN SPARK AREA

EARTH ELECTRODE TEMPERATURE VS ENGINE SPEED
SOLID NICKEL EARTH ELECTRODE
COPPER CORED EARTH ELECTRODE
TEMPERATURE
ENGINE SPEED

Champion Double Copper plugs also feature a unique trapezoidal earth electrode giving a 50% increase in spark area. This, together with the double copper cores, offers greatly reduced electrode wear, so the spark stays stronger for longer.

 FASTER COLD STARTING

 FOR UNLEADED OR LEADED FUEL

 ELECTRODES UP TO 100°C COOLER

 BETTER ACCELERATION RESPONSE

 LOWER EMISSIONS

 50% BIGGER SPARK AREA

 THE LONGER LIFE PLUG

Plug Tips/Hot and Cold.
Spark plugs must operate within well-defined temperature limits to avoid cold fouling at one extreme and overheating at the other.
Champion and the car manufacturers work out the best plugs for an engine to give optimum performance under all conditions, from freezing cold starts to sustained high speed motorway cruising.
Plugs are often referred to as hot or cold. With Champion, the higher the number on its body, the hotter the plug, and the lower the number the cooler the plug.

Plug Cleaning
Modern plug design and materials mean that Champion no longer recommends periodic plug cleaning. Certainly don't clean your plugs with a wire brush as this can cause metal conductive paths across the nose of the insulator so impairing its performance and resulting in loss of acceleration and reduced m.p.g.
However, if plugs are removed, always carefully clean the area where the plug seats in the cylinder head as grit and dirt can sometimes cause gas leakage.
Also wipe any traces of oil or grease from plug leads as this may lead to arcing.

CHAMPION

DOUBLE COPPER

1

This photographic sequence shows the steps taken to repair the dent and paintwork damage shown above. In general, the procedure for repairing a hole will be similar; where there are substantial differences, the procedure is clearly described and shown in a separate photograph.

2

First remove any trim around the dent, then hammer out the dent where access is possible. This will minimise filling. Here, after the large dent has been hammered out, the damaged area is being made slightly concave.

3

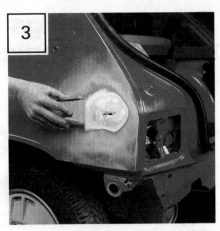

Next, remove all paint from the damaged area by rubbing with coarse abrasive paper or using a power drill fitted with a wire brush or abrasive pad. 'Feather' the edge of the boundary with good paintwork using a finer grade of abrasive paper.

4

Where there are holes or other damage, the sheet metal should be cut away before proceeding further. The damaged area and any signs of rust should be treated with Turtle Wax Hi-Tech Rust Eater, which will also inhibit further rust formation.

5

For a large dent or hole mix Holts Body Plus Resin and Hardener according to the manufacturer's instructions and apply around the edge of the repair. Press Glass Fibre Matting over the repair area and leave for 20-30 minutes to harden. Then ...

5A

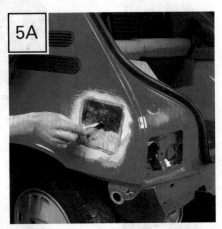

... brush more Holts Body Plus Resin and Hardener onto the matting and leave to harden. Repeat the sequence with two or three layers of matting, checking that the final layer is lower than the surrounding area. Apply Holts Body Plus Filler Paste as shown in Step 5B.

5B

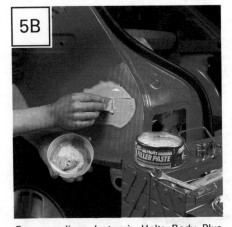

For a medium dent, mix Holts Body Plus Filler Paste and Hardener according to the manufacturer's instructions and apply it with a flexible applicator. Apply thin layers of filler at 20-minute intervals, until the filler surface is slightly proud of the surrounding bodywork.

5C

For small dents and scratches use Holts No Mix Filler Paste straight from the tube. Apply it according to the instructions in thin layers, using the spatula provided. It will harden in minutes if applied outdoors and may then be used as its own knifing putty.

6

Use a plane or file for initial shaping. Then, using progressively finer grades of wet-and-dry paper, wrapped round a sanding block, and copious amounts of clean water, rub down the filler until glass smooth. 'Feather' the edges of adjoining paintwork.

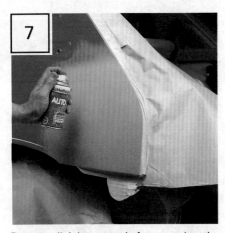

Protect adjoining areas before spraying the whole repair area and at least one inch of the surrounding sound paintwork with Holts Dupli-Color primer.

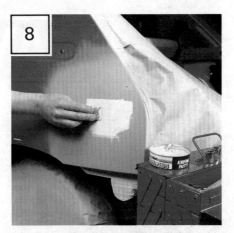

Fill any imperfections in the filler surface with a small amount of Holts Body Plus Knifing Putty. Using plenty of clean water, rub down the surface with a fine grade wet-and-dry paper – 400 grade is recommended – until it is really smooth.

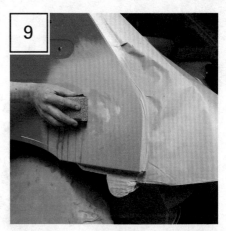

Carefully fill any remaining imperfections with knifing putty before applying the last coat of primer. Then rub down the surface with Holts Body Plus Rubbing Compound to ensure a really smooth surface.

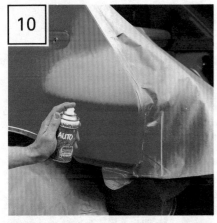

Protect surrounding areas from overspray before applying the topcoat in several thin layers. Agitate Holts Dupli-Color aerosol thoroughly. Start at the repair centre, spraying outwards with a side-to-side motion.

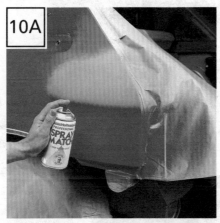

If the exact colour is not available off the shelf, local Holts Professional Spraymatch Centres will custom fill an aerosol to match perfectly.

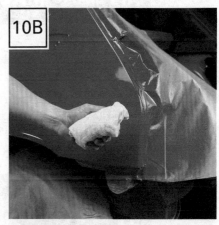

To identify whether a lacquer finish is required, rub a painted unrepaired part of the body with wax and a clean cloth.

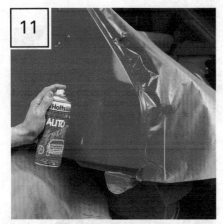

If *no* traces of paint appear on the cloth, spray Holts Dupli-Color clear lacquer over the repaired area to achieve the correct gloss level.

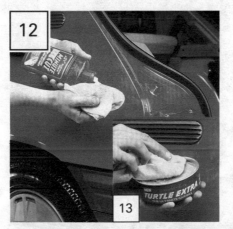

The paint will take about two weeks to harden fully. After this time it can be 'cut' with a mild cutting compound such as Turtle Wax Minute Cut prior to polishing with a final coating of Turtle Wax Extra.

When carrying out bodywork repairs, remember that the quality of the finished job is proportional to the time and effort expended.

9   Using two levers, ease the two gears and chain from the camshaft and crankshaft.

10   Remove the gears from the chain, but identify the outer face of the chain so that it can be refitted in its original position.

13   Sump – removal

Note: *If the engine is still in the car carry out the following operations with reference to the relevant Sections and Chapters of this manual:*

    (a)   Disconnect the battery negative lead
    (b)   Drain the engine oil
    (c)   Unbolt and remove the engine tie-rod

1   Undo and remove the bolts securing the sump to the crankcase and withdraw the sump. If it is initially stuck, give it a tap with a hide or plastic mallet to break the seal. Remove the gaskets and the end seals.

2   To remove the oil pick-up pipe and strainer, unscrew the bolt securing the support bracket to the main bearing cap. Undo the pipe union nut and withdraw the pipe and strainer.

14   Gearbox adaptor plate – removal

1   Remove the engine and gearbox, as described in Section 6, separate the gearbox and remove the clutch assembly, as described in Chapter 5.

2   Remove the sump, as described in the previous Section.

3   Undo and remove the bolts securing the adaptor plate to the engine, noting the location of the three long screws.

4   Withdraw the adaptor plate and recover the gasket.

15   Oil pump – removal

1   Remove the gearbox adaptor plate, as described in the previous Section.

2   Bend back the locktabs and unscrew the bolts securing the oil pump to the cylinder block.

3   Withdraw the oil pump and recover the gasket.

16   Camshaft and tappets – removal

1   Remove the engine and gearbox, as described in Section 6, then separate the gearbox from the engine, as described in Section 7.

2   Remove the fuel pump, as described in Chapter 3, and the distributor, as described in Chapter 4.

3   Remove the timing cover chain and gears, as described in Section 12, and the sump, as described in Section 13.

4   Remove the rocker cover, rocker shaft and the pushrods.

5   Screw a long $\frac{5}{16}$ in bolt into the centre of the distributor driveshaft and withdraw the shaft from its location in the cylinder block (photo).

16.5 Use a bolt to remove the distributor driveshaft

6   Undo and remove the three bolts securing the camshaft locating plate to the front of the cylinder block and lift off the plate (photo).

7   Turn the engine on its side to prevent the tappets fouling the camshaft as it is withdrawn.

8   Turn the camshaft through one complete turn to move all the tappets away from the cam lobes and then withdraw the camshaft from the cylinder block (photo). Take care not to damage the three camshaft bearings as the cam lobes pass through them.

9   Withdraw each of the eight tappets through the crankcase (photo) and keep them in strict order as they must be refitted in their original locations unless they are to be renewed.

17   Pistons and connecting rods – removal

Note: *If the engine is still in the car, carry out the following operations with reference to the relevant Sections and Chapters of this manual:*

    (a)   Remove the sump and oil pick-up pipe
    (b)   Remove the cylinder head (this is not necessary if only the big-end bearings are to be removed)

1   Check the big-end caps for identification marks. If necessary, use a centre-punch on the caps and connecting rods to identify them; mark them on the camshaft side to ensure correct refitting.

2   Turn the crankshaft so that No 1 crankpin is at its lowest point. Using a suitable socket, undo and remove the nuts securing the connecting rod cap to the rod.

16.6 Remove the camshaft locating plate

16.8 Withdraw the camshaft from the front of the engine

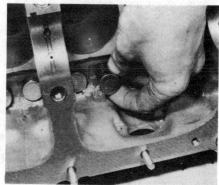

16.9 Remove the tappets from their bores

3   Withdraw the cap complete with the bearing shell (photo).
4   Using the handle of a hammer, tap the piston and connecting rod from the bore and withdraw it from the top of the cylinder block.
5   Loosely refit the cap to the connecting rod.
6   Repeat the procedure given in paragraphs 1 to 5 on No 4 piston and connecting rod, then turn the crankshaft through half a turn and repeat the procedure on No 2 and No 3 pistons.

18.5 Engine front plate removal

17.3 Remove the connecting rod cap and bearing shell

18.7 Check the crankshaft endfloat with feeler gauges

18 Crankshaft and main bearings – removal

1   Remove the engine and gearbox from the car, as described in Section 6, and separate the gearbox, as described in Section 7.
2   Remove the timing cover gears and chain, as described in Section 12.
3   Remove the gearbox adaptor plate, as described in Section 14.
4   Follow the procedure for removing the pistons and connecting rods described in Section 17, but it is not necessary to completely withdraw them from the cylinder block.
5   Unbolt the front plate from the engine, and remove the gasket (photo). Invert the engine.
6   Check the main bearing caps for identification marks, and if necessary use a centre-punch to identify them.
7   Before removing the crankshaft, check that the endfloat is within the specified limits by inserting a feeler blade between the centre crankshaft web and the thrust washers (photo). This will indicate whether new thrust washers are required or not.
8   Unscrew the bolts and remove the main bearing caps, complete with bearing shells. Recover the thrust washers from the centre main bearing cap.
9   Lift the crankshaft from the crankcase and remove the remaining centre bearing thrust washers.
10 Extract the bearing shells from the crankcase recesses and the caps, and identify them for location (photo).

19 Crankcase ventilation system – description

The crankcase ventilation system consists of hoses from the crankcase area linked and connected to a port on the carburettor.
One hose is attached to an oil separator bolted to the gearbox adaptor plate and the other is attached to an oil separator on the timing cover.
Periodically the hoses should be examined for security and condition. Cleaning them will not normally be necessary except when the engine is well worn and sludge has accumulated.

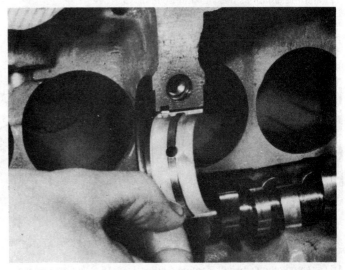

18.10 Remove the main bearing shells

20 Examination and renovation – general

With the engine completely stripped, clean all the components and examine them for wear. Each part should be checked and, where necessary, renewed or renovated as described in the following Sections. Renew main and big-end shell bearings as a matter of course, unless you know that they have had little wear and are in perfect condition.

21 Oil pump – examination

1   Remove the retaining screw(s) and withdraw the cover (photo).
2   Lift the two rotors from the pump body.
3   Clean the components with paraffin and wipe dry.
4   Refit the rotors to the pump body, making sure that the chamfer on the outer rotor enters the body first.
5   Using a feeler blade, and where necessary a straight-edge, check that the rotor clearances are as given in the Specifications (photos). If any clearance is outside that specified, or if damage is evident on any component, renew the complete oil pump.
6   If the oil pump is serviceable, refit the cover and tighten the retaining screw. Operate the pump in clean engine oil to prime it.

22 Crankshaft and main bearings – examination and renovation

1   Examine the bearing surfaces of the crankshaft for scratches or scoring and, using a micrometer, check each journal and crankpin for ovality. Where this is found to be in excess of 0.001 in (0.0254 mm) the crankshaft will have to be reground and undersize bearings fitted.
2   Crankshaft regrinding should be carried out by a suitable engineer-ing works, who will normally supply the matching undersize main and big-end shell bearings.
3   If the crankshaft endfloat is more than the maximum specified amount, new thrust washers should be fitted to the centre main bearing; these are usually supplied, together with the main and big-end bearings, with a reground crankshaft.

23 Cylinder block and crankcase – examination and renovation

1   The cylinder bores must be examined for taper, ovality, scoring and scratches. Start by examining the top of the bores; if these are worn, a slight ridge will be found which marks the top of the piston ring travel. If the wear is excessive, the engine will have had a high oil consumption rate accompanied by blue smoke from the exhaust.
2   If available, use an inside dial gauge to measure the bore diameter just below the ridge and compare it with the diameter at the bottom of the bore, which is not subject to wear. If the difference is more than 0.006 in (0.152 mm), the cylinders will normally require boring with new oversize pistons fitted.
3   Provided the cylinder bore wear does not exceed 0.008 in (0.203 mm), however, special oil control rings and pistons can be fitted to restore compression and stop the engine burning oil.
4   If new pistons are being fitted to old bores, it is essential to roughen the bore walls slightly with fine glasspaper to enable the new piston rings to bed in properly.
5   Thoroughly examine the crankcase and cylinder block for cracks and damage, and use a piece of wire to probe all oilways and waterways to ensure they are unobstructed.
6   Check the tappet bores for wear and scoring; if excessive, they can be reamed and oversize tappets fitted.
7   Unscrew the oil pressure relief valve cap and remove the valve and spring. Check the valve seating for excessive wear and check that the spring free length is as specified. Renew the valve and spring as necessary, and refit them to the cylinder block.

21.1 Remove the screw and lift off the oil pump cover

21.5A Checking the oil pump inner rotor endfloat ...

21.5B ... and outer rotor endfloat ...

21.5C ... outer rotor clearance ...

21.5D ... and rotor lobe clearance ...

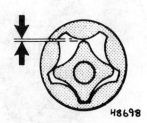

Fig. 1.3 Oil pump rotor lobe clearance checking point (Sec 21)

24.1 Piston and connecting rod components

24.3 Piston ring removal

24.5 Checking the piston ring-to-groove clearance

24 Pistons and connecting rods – examination and renovation

1   Examine the pistons for ovality, scoring and scratches. Check the connecting rods for wear and damage (photo).
2   If the pistons or connecting rods are to be renewed it is recommended that this work is carried out by a BL garage, who willl have the necessary tooling to extract the gudgeon pins frrom the connecting rods.
3   If new rings are to be fitted to the original pistons, expand the old rings over the top of the pistons. The use of two or three old feeler blades will be helpful in preventing the rings from dropping into empty grooves (photo). Note that the oil control ring is in three sections.
4   Before fitting the new rings to the piston, insert them into the cylinder bore and use a feeler gauge to check that the end gaps are within the specified limits.
5   After fitting the rings, check the compression rings for groove clearances using a feeler blade (photo). Make sure that the word 'Top', where marked on the compression rings, is toward the top of the piston. Arrange the compression ring gaps at 90 degrees to each other on the camshaft side of the piston.

25 Camshaft and tappets – examination

1   Examine the camshaft bearing surfaces, cam lobes and skew gear for wear. If excessive, renew the shaft.
2   Check the locating plate for wear, and renew it if necessary.
3   Check the camshaft bearings for wear and if necessary remove them with a suitable diameter length of tubing. Fit the new prefinished bearings with their oil holes aligned with the oilways in the cylinder block.
4   Examine the tappets for wear, and renew them if necessary.

26 Timing cover, chain and gears – examination

1   Examine all the teeth on the camshaft and crankshaft sprockets. If these are 'hooked' in appearance, renew the sprockets.
2   Examine the chain tensioner for wear, and renew it if necessary.
3   Examine the timing chain for wear. If it has been in operation for a considerable time, or if when held horizontally (rollers vertical) it takes on a deeply bowed appearance, renew it.
4   Check the timing cover for damage, and renew it if necessary. It is good practice to renew the timing cover oil seal whenever the timing cover is removed. To do this, drive out the old seal with a suitable drift, and install the new seal using a block of wood to make sure that it enters squarely (photo).

27 Gearbox adaptor plate – examination and renovation

1   The oil seal in the gearbox adaptor plate should be renewed as a matter of course if the engine is being overhauled.
2   Remove the old seal by prising or drifting it out of its location (photo).
3   Lubricate a new seal in engine oil and place it in position on the adaptor plate, with its sealing lip towards the engine (photo).
4   Using a block of wood and a hammer, drive the seal into its location until it is flush with the adaptor plate face.

28 Cylinder head – decarbonising, valve grinding and renovation

1   The operation will normally only be required at comparatively high mileages. However, if persistent pinking occurs and performance has

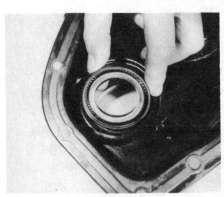

26.4 Fitting the timing cover oil seal

27.2 Removing the crankshaft rear oil seal in the adaptor plate

27.3 Fit the crankshaft rear oil seal with its sealing lip towards the engine

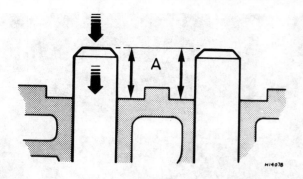

Fig. 1.4 Valve guide fitted height dimension 'A' (Sec 28)

Arrow indicates direction of both removal and refitting

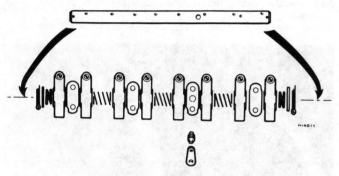

Fig. 1.5 Rocker shaft components (Sec 28)

deteriorated even though the engine adjustments are correct, decarbonising and valve grinding may be required.

2 With the cylinder head removed, use a scraper to remove the carbon from the combustion chambers and ports. Remove all traces of gasket from the cylinder head surface, then wash it thoroughly with paraffin.

3 Use a straight-edge and feeler blade to check that the cylinder head surface is not distorted. If it is, it must be resurfaced by a suitably equipped engineering works.

4 If the engine is still in the car, clean the piston crowns and cylinder bore upper edges, but make sure that no carbon drops between the pistons and bores. To do this, locate two of the pistons at the top of their bores and seal off the remaining bores with paper and masking tape. Press a little grease between the two pistons and their bores to collect any carbon dust; this can be wiped away when the piston is lowered. To prevent carbon build-up, polish the piston crown with metal polish, but remove all traces of polish afterwards.

5 Examine the heads of the valves for pitting and burning, especially the exhaust valve heads. Renew any valve which is badly burnt. Examine the valve seats at the same time. If the pitting is very slight, it can be removed by grinding the valve heads and seats together with coarse, then fine, grinding paste.

6 Where excessive pitting has occurred, the valve seats must be recut or renewed by a suitably equipped engineering works.

7 Valve grinding is carried out as follows. Place the cylinder head upside down on a bench with a block of wood at each end to give clearance for the valve stems.

8 Smear a trace of coarse carborundum paste on the seat face and press a suction grinding tool onto the valve head. With a semi-rotary action grind the valve head to its seat, lifting the valve occasionally to redistribute the grinding paste. When a dull matt even surface is produced on both the valve seat and the valve, wipe off the paste and repeat the process with fine carborundum paste. A light spring placed under the valve head will greatly ease this operation. When a smooth unbroken ring of light grey matt finish is produced on both the valve and seat, the grinding operation is complete.

9 Scrape away all carbon from the valve head and stem, and clean away all traces of grinding compound. Clean the valves and seats with a paraffin-soaked rag, then wipe with a clean rag.

10 If the valve guides are worn, indicated by a side-to-side motion of the valve, new guides must be fitted. To do this, use a suitable mandrel to press the worn guides downwards and out through the combustion chamber. Press the new guides into the cylinder head in the same direction until they are at the specified fitted height.

11 If the original valve springs have been in use for 20 000 miles (32 000 km) or more renew them. Where fitted, the inlet valve oil seals should also be renewed whenever the cylinder head is dismantled.

12 Examine the pushrods and rocker shaft assembly for wear, and renew them as necessary. Dismantling and reassembly of the rocker components is straightforward if reference is made to Figs. 1.2 and 1.5.

29 Engine mountings – removal and refitting

Left-hand mounting

1 Place a jack beneath the gearbox, with a block of wood between the jack head and the gearbox. Raise the jack and just take the weight of the gearbox.

2 Undo and remove the through-bolt securing the engine mounting to the gearbox bracket. Lift off the upper rebound washer, lower the jack slightly and slide out the lower rebound washer.

3 Undo and remove the two bolts and withdraw the mounting from the body bracket.

4 Refitting is the reverse sequence to removal.

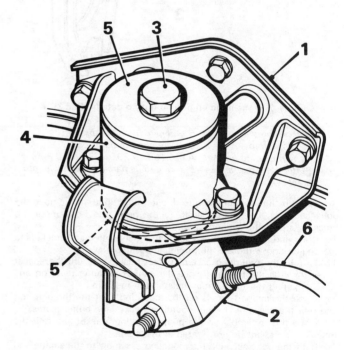

Fig. 1.6 Left-hand engine mounting details (Sec 29)

1 Engine mounting-to-body bracket	4 Mounting
2 Gearbox mounting bracket	5 Rebound washers
3 Through-bolt	6 Earth cable

Right-hand mounting

5 Jack up the front of the car so that the front wheels are just clear of the ground and support it securely on axle stands.

6 Undo and remove the securing screws and withdraw the access panel from under the right-hand wheel arch.

7 Undo and remove the bolts securing the cooling system expansion tank to the body and move the tank to one side.

8 Position a jack beneath the right-hand side of the sump with a block of wood placed between the jack head and sump. Raise the jack and just take the weight of the engine.

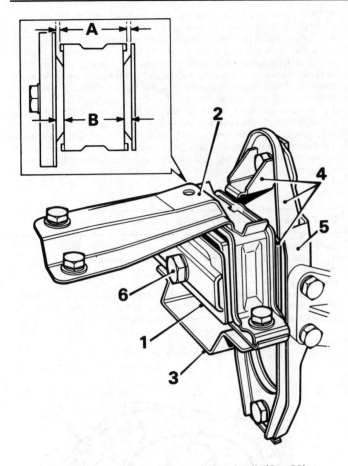

Fig. 1.7 Right-hand engine mounting details (Sec 29)

1 Buffer plates	*5 Engine mounting bracket*
2 Support strap	*6 Mounting through-bolt*
3 Body mounting bracket	*A = 0.16 to 0.20 in (4.0 to 5.0 mm)*
4 Buffer plate restraint	*B = Gap to be parallel over full length*
* bracket*	

9   Slacken the mounting centre through-bolt, undo and remove the bolts securing the support strap to the body and body mounting bracket. Lift off the support strap.
10   Undo and remove the bolts securing the mounting and brackets to the engine and withdraw the mounting assembly.
11   To refit the mounting assembly, first position it with the spacer offset towards the top, and place the reinforced buffer plate on one side, engaging the top location hole on the mounting.
12   Place the support straps on the mounting, refit the through-bolt and refit the assembly to the mounting bracket and buffer plates.
13   Place this assembly on the body mounting bracket and refit the retaining bolts, but do not tighten at this stage.
14   Refit the bolts securing the mounting assembly to the engine and tighten fully. Tighten the through-bolt.
15   Refer to Fig. 1.7 and move the support straps as necessary to give the specified clearances, then tighten the support strap retaining bolts.
16   Refit the access panel, remove the axle stands and lower the car to the ground.

30   Engine tie-rod – removal and refitting

1   Jack up the front of the car and support it securely on axle stands.
2   Undo and remove the through-bolts securing the tie-rod to the engine support bracket (photo) and the crossmember bracket. Withdraw the tie-rod from its location.
3   If necessary the tie-rod engine support bracket can be removed after unscrewing the two nuts and bolts securing the bracket to the engine (photo).
4   Refitting is the reverse sequence to removal.

30.2 Tie-rod-to-engine and crossmember bracket through-bolts (arrowed)

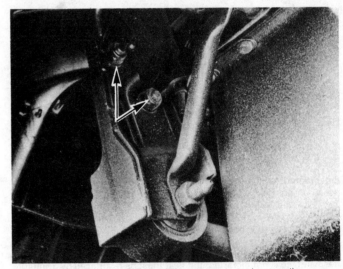

30.3 Tie-rod engine support bracket retaining bolts (arrowed)

31   Engine reassembly – general

1   To ensure maximum life with minimum trouble from a rebuilt engine, not only must everything be correctly assembled, but it must also be spotlessly clean. All oilways must be clear, and locking washers and spring washers must be fitted where indicated. Oil all bearings and other working surfaces thoroughly with engine oil during assembly.
2   Before assembly begins, renew any bolts or studs with damaged threads.
3   Gather together a torque wrench, oil can, clean rags and a set of engine gaskets and oil seals, together with a new oil filter.

32   Crankshaft and main bearings – refitting

1   Clean the backs of the bearing shells and the bearing recesses in both the cylinder block and main bearing caps.
2   Press the main bearing shells into the cylinder block and caps and oil them liberally.
3   Using a little grease, stick the thrust washers to each side of the centre main bearings with their oilways facing away from the bearing

32.3A Installing the centre main bearing thrust washers

32.3B Installing the centre main bearing cap and thrust washers

32.5 Tightening the main bearing bolts with a torque wrench

(photo). Similarly fit the thrust washers to the centre main bearing cap (photo).
4   Lower the crankshaft into position, then fit the main bearing caps in their previously noted locations.
5   Insert and evenly tighten the main bearing cap bolts to the specified torque (photo). Check that the crankshaft rotates freely, then check that the endfloat is within the specified limits by inserting a feeler blade between the centre crankshaft web and the thrust washers.
6   Smear the front plate gasket with sealing compound and locate it on the front of the cylinder block. Fit the engine front plate and tighten the two lower retaining boltss.
7   With the engine upright, refit the timing cover, chain and gears, as described in Section 38.
8   Refit the pistons and connecting rods as described in Section 33.

33 Pistons and connecting rods – refitting

1   Clean the backs of the bearing shells and the recesses in the connecting rods and big-end caps.
2   Press the big-end bearing shells into the connecting rods and caps in their correct position and oil them liberally.
3   Fit a ring compressor to No 1 piston, then insert the piston and connecting rod into No 1 cylinder. With No 1 crankpin at its lowest point, drive the piston carefully into the cylinder with the wooden handle of a hammer, and at the same time guide the connecting rod onto the crankpin. Make sure that the 'Front' mark on the piston crown is facing the timing chain end of the engine, and that the connecting rod offset is as shown in Fig. 1.8.
4   Fit the big-end bearing cap in its previously noted position, then tighten the nuts evenly to the specified torque.
5   Check that the crankshaft turns freely.
6   Repeat the procedure given in paragraphs 3 to 5 for No 4 piston and connecting rod, then turn the crankshaft through half a turn and repeat the procedure on No 2 and No 3 pistons.
7   Refit the gearbox adaptor plate, as described in Section 36.
8   Refit the sump, as described in Section 37.
9   Refit the cylinder head, as described in Section 39.
10   Attach the gearbox to the engine (Section 41) and then refit the engine and gearbox assembly, as described in Section 42.

34 Oil pump – refitting

Note: *Prime the pump with clean engine oil before fitting*
1   Make sure that the mating faces of the oil pump and cylinder block are clean then fit the oil pump, together with a new gasket, and tighten the retaining bolts evenly to the specified torque. Make sure that the cut-outs in the gasket are correctly aligned with the pump, and, if the camshaft is already in position, make sure that the pump spindle engages the slot in the camshaft.
2   Bend the lockwashers to lock the bolts.
3   If the engine and gearbox were removed purposely to remove the

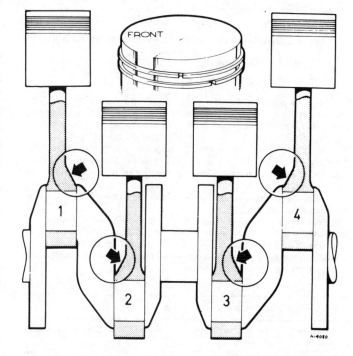

Fig. 1.8 Connecting rod offset positioning – arrowed (Sec 33)

oil pump, refit the gearbox adaptor plate, with reference to Section 36, and attach the gearbox. Refit the engine and gearbox, as described in Section 42.

35 Camshaft and tappets – refitting

1   Lubricate the tappets (cam followers) with engine oil and insert them into their bores. If the original tappets are being refitted, insert them in their original locations.
2   Oil the camshaft bearings and carefully insert the camshaft from the timing chain end of the cylinder block. Make sure that the oil pump spindle engages the slot in the camshaft.
3   Fit the locating plate to the front plate and tighten the bolts evenly.
4   Using a dial gauge, vernier calipers or feeler blade and bridging piece, check that the camshaft endfloat is within the specified limits. If not, renew the locating plate.
5   With the engine upright, refit the timing cover, chain and gears, as described in Section 38.

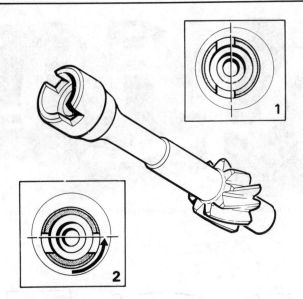

Fig. 1.9 Distributor driveshaft fitting procedure (Sec 35)

1   Initial fitting position   2   Fitted position

6   Turn the engine until No 1 piston is at top-dead-centre (TDC) on the compression stroke. If the cylinder head is not yet fitted, use two pushrods to determine the point when No 4 cylinder valves are rocking – in this position No 1 piston is at TDC compression.
7   Using the $\frac{5}{16}$ in bolt used for removal, insert the distributor driveshaft into the cylinder block with the slot in the vertical position and the larger offset segment towards the crankshaft pulley end of the engine (see Fig. 1.9). As the driveshaft engages the skew gear on the camshaft, it will turn anti-clockwise so that the slot is in the horizontal position. Remove the bolt.
8   Refit the distributor (Chapter 4) and the fuel pump (Chapter 3).
9   Refit the pushrods and rocker shaft and adjust the valve clearances, as described in Section 44. Refit the rocker cover.
10   Attach the gearbox to the engine (Section 41) and refit the engine and gearbox, as described in Section 42.

36 Gearbox adaptor plate – refitting

1   Ensure that the mating faces are perfectly clean and place a new gasket in position on the cylinder block.
2   Lubricate the crankshaft oil seal with engine oil and refit the adaptor plate to the cylinder block.
3   Refit the retaining bolts and tighten them evenly to the specified torque.
4   Refit the sump, as described in Section 37.

5   Refit the clutch assembly, as described in Chapter 5, attach the gearbox to the engine (Section 41) and refit the engine and gearbox as described in Section 42.

37 Sump – refitting

1   If the oil pick-up pipe and strainer were removed, refit the pipe to the crankcase and engage the union nut.
2   Refit and tighten the bolt securing the support bracket to the main bearing cap and then tighten the union nut.
3   Ensure that the sump and crankcase mating faces are clean and free from all traces of old gasket.
4   Apply jointing compound to both sides of the new side gaskets and position them on the sump.
5   Locate the new end seals in the sump recesses so that when they are pushed fully into place, an equal amount of gasket at each side protrudes above the sump face.
6   Refit the sump to the crankcase and secure with the retaining bolts. Tighten the bolts in a diagonal sequence to the specified torque in two or three stages.
7   If the engine is in the car, refit the tie-rod, as described in Section 30, fill the engine with oil and reconnect the battery.

38 Timing cover, chain and gears – refitting

1   Locate the timing gears on the crankshaft and camshaft without the chain, and check their alignment using a straight-edge.
2   Remove the gears, and if necessary extract the Woodruff key and fit shims to the crankshaft to obtain the alignment (photo). Refit the key.
3   Turn the crankshaft so that the Woodruff key is at top-dead-centre, and turn the camshaft so that the key is at 2 o'clock. In this position No 4 cylinder is at TDC compression.
4   Loop the timing chain over the two gears so that the timing marks are facing each other on the centre line (see Fig. 1.10).
5   Locate the two gears on the crankshaft and camshaft and press them firmly home. Using a straight-edge check that the timing marks are still on the centre line.
6   Fit the camshaft gear retaining nut and lockwasher, and tighten the nut while using a screwdriver through one of the gear holes to restrain the gear (photo). Bend the lockwasher to lock the nut (photo).
7   Fit the chain tensioner and tighten the bolts, while keeping firm thumb pressure against the top of the bracket to provide the preload (photo).
8   Locate the oil thrower on the crankshaft with the side marked F facing outwards.
9   Stick the timing cover gasket to the front plate, then fit the timing cover and retain it with two upper bolts inserted loosely.
10   Oil the timing cover oil seal, then temporarily fit the crankshaft pulley to centralise the timing cover. Insert and tighten evenly the upper retaining bolts, then remove the pulley and fit the lower bolts.
11   Fit the crankshaft pulley on the crankshaft followed by the lockwasher and bolt. Tighten the bolt to the specified torque (photo) and bend over the lockwasher tab. Adopt the same procedure as used

38.2 Crankshaft gear shim location (arrowed)

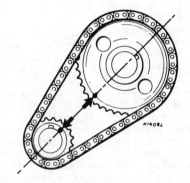

Fig. 1.10 Timing gear alignment marks (arrowed) and centre line (Sec 38)

38.6A Tightening the camshaft gear retaining nut

38.6B Camshaft gear retaining nut locked and timing marks aligned

38.7 Fitting the timing chain tensioner

38.11 Tightening the crankshaft pulley bolt

during removal to lock the crankshaft as the bolt is tightened (Sec 12).

12 Connect the crankcase ventilation hose to the oil separator and attach the fuel pipe clips to the timing cover studs.

13 If the engine is in the car, refit the access panel to the wheel arch, refit the water pump pulley and drivebelt, with reference to Chapter 2, and reconnect the battery.

39 Cylinder head – reassembly and refitting

1 Fit the valves in their original sequence or, if new valves have been obtained, in the seat to which they have been ground.

2 Oil the valve stems liberally and fit the oil seals to the inlet valve guides and collet grooves, where applicable.

3 Working on one valve, fit the spring and cup, then compress the spring with the compressor and insert the split collets. Release the compressor and remove it.

4 Repeat the procedure given in paragraph 3 on the remaining valves. Tap the end of each valve stem with a non-metallic mallet to settle the collets.

5 Make sure that the faces of the cylinder head and block are perfectly clean, then fit the new gasket over the studs with the words TOP and FRONT correctly positioned. FRONT being the crankshaft pulley end of the engine. Do not use jointing compound.

6 Lower the cylinder head over the studs and onto the gasket.

7 Insert the pushrods in their original locations, then lower the rocker shaft and pedestals over the studs, at the same time guiding the adjusting screws into the pushrods (photo).

8 Locate the coil and bracket on the stud furthest from the crankshaft pulley, and fit the rocker shaft lockwasher to the second pedestal from the front (crankshaft pulley end).

9 Fit the cylinder head nuts and tighten them to half the specified torque in the order shown in Fig. 1.11 (photo). After several minutes, tighten the nuts to the final torque, again in the order recommended.

10 Fit the rocker shaft pedestal washers and nuts and tighten them evenly to the specified torque (photo).

11 Adjust the valve clearances, as described in Section 44.

12 Fit the rocker cover with a new gasket, place the wiring harness support brackets in position and refit the rocker cover retaining bolts and washers. Tighten the bolts to the specified torque.

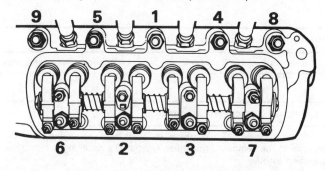

Fig. 1.11 Cylinder head nut tightening sequence (Sec 39)

39.7 Fit the rocker shaft assembly over the cylinder head studs

39.9 Tightening the cylinder head ...

39.10 ... and rocker pedestal nuts using a torque wrench

13 If the engine is in the car, reverse the introductory procedure given in Section 10, and refill the cooling system, with reference to Chapter 2.

14 Drive the car for five to ten miles then allow the engine to cool, and remove the rocker cover. Working in the order shown in Fig. 1.11, slacken half a turn, then immediately tighten each cylinder head nut to the specified torque. Readjust the valve clearances then refit the rocker cover.

40 Ancillary components – refitting

Refer to Section 9 and refit the listed components with reference to the Chapters indicated.

41 Engine – attachment to gearbox

Refer to Section 7 and attach the gearbox to the engine using the reverse of the removal procedure. Tighten the retaining bolts to the specified torque.

42 Engine and gearbox assembly – refitting

Refer to Section 6 and refit the engine and gearbox using the reverse of the removal procedure, noting the following additional points:

   (a) Align the marks on the driveshaft joints and drive flanges made during removal
   (b) Adjust the accelerator cable, as described in Chapter 3, and the clutch cable, as described in Chapter 5
   (c) Refill the engine with oil
   (d) Refill the cooling system, as described in Chapter 2

43 Engine – adjustments after major overhaul

1 With the engine and gearbox refitted to the car, make a final check to ensure that everything has been reconnected and that no rags or tools have been left in the engine compartment.

2 Make sure that the oil and water levels are topped up and then start the engine; this may take a little longer than usual as the fuel pump and carburettor float chamber may be empty.

3 As soon as the engine starts, watch for the oil pressure light to go out and check for any oil, fuel or water leaks. Don't be alarmed if there are some odd smells and smoke from parts getting hot and burning off oil deposits.

4 Drive the car until normal operating temperature is reached and then allow it to cool. Re-torque the cylinder head nuts, as described in Section 39, and adjust the valve clearances, as described in Section 44.

5 If new pistons, ring or crankshaft bearings have been fitted the engine must be run-in for the first 500 miles (800 km). Do not exceed 45 mph (72 kph), operate the engine at full throttle or allow it to labour in any gear.

44 Valve clearances – adjustment

1 The valve clearances must be adjusted with the engine cold.

2 Remove the rocker cover and gasket.

3 Turn the engine with a spanner on the crankshaft pulley bolt until No 8 valve (No 4 cylinder exhaust) is fully open.

4 Insert a feeler blade of the correct thickness between the rocker arm and valve stem of No 1 valve (No 1 cylinder exhaust). If the blade is not a firm sliding fit, loosen the locknut on the rocker arm with a ring spanner and turn the adjusting screw with a screwdriver (photo). Tighten the locknut whilst holding the adjusting screw stationary, then recheck the adjustment.

5 Repeat the procedure given in paragraphs 3 and 4 on the remaining valves using the 'rule of nine' method as given below:

Valve open	Adjust valve
8 exhaust	1 exhaust
6 inlet	3 inlet
4 exhaust	5 exhaust
7 inlet	2 inlet
1 exhaust	8 exhaust
3 inlet	6 inlet
5 exhaust	4 exhaust
2 inlet	7 inlet

6 Check the rocker cover gasket for damage, and renew it if necessary. Refit the rocker cover and gasket with the filler cap towards the crankshaft pulley end of the engine. Tighten the nuts to the specified torque.

44.4 Valve clearance adjustment

PART B: 1.6 LITRE ENGINE

45 General description

The engine is of four-cylinder, in-line overhead camshaft type, mounted transversely at the front of the car.

The crankshaft is supported in five shell-type main bearings. Thrust washers are fitted to the centre main bearing in the crankcase to control crankshaft endfloat.

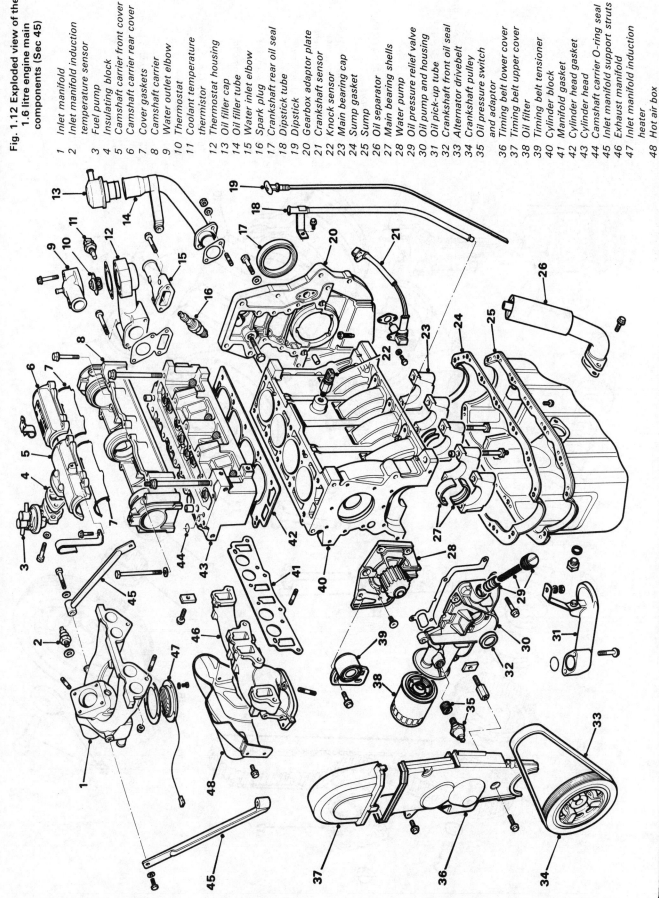

Fig. 1.12 Exploded view of the 1.6 litre engine main components (Sec 45)

1 Inlet manifold
2 Inlet manifold induction temperature sensor
3 Fuel pump
4 Insulating block
5 Camshaft carrier front cover
6 Camshaft carrier rear cover
7 Cover gaskets
8 Camshaft carrier
9 Water outlet elbow
10 Thermostat
11 Coolant temperature thermistor
12 Thermostat housing
13 Oil filler cap
14 Oil filler tube
15 Water inlet elbow
16 Spark plug
17 Crankshaft rear oil seal
18 Dipstick tube
19 Dipstick
20 Gearbox adaptor plate
21 Crankshaft sensor
22 Knock sensor
23 Main bearing cap
24 Sump gasket
25 Sump
26 Oil separator
27 Main bearing shells
28 Water pump
29 Oil pressure relief valve
30 Oil pump and housing
31 Oil pick-up tube
32 Crankshaft front oil seal
33 Alternator drivebelt
34 Crankshaft pulley
35 Oil pressure switch and adaptor
36 Timing belt lower cover
37 Timing belt upper cover
38 Oil filter
39 Timing belt tensioner
40 Cylinder block
41 Manifold gasket
42 Cylinder head gasket
43 Cylinder head
44 Camshaft carrier O-ring seal
45 Inlet manifold support struts
46 Exhaust manifold
47 Inlet manifold induction heater
48 Hot air box

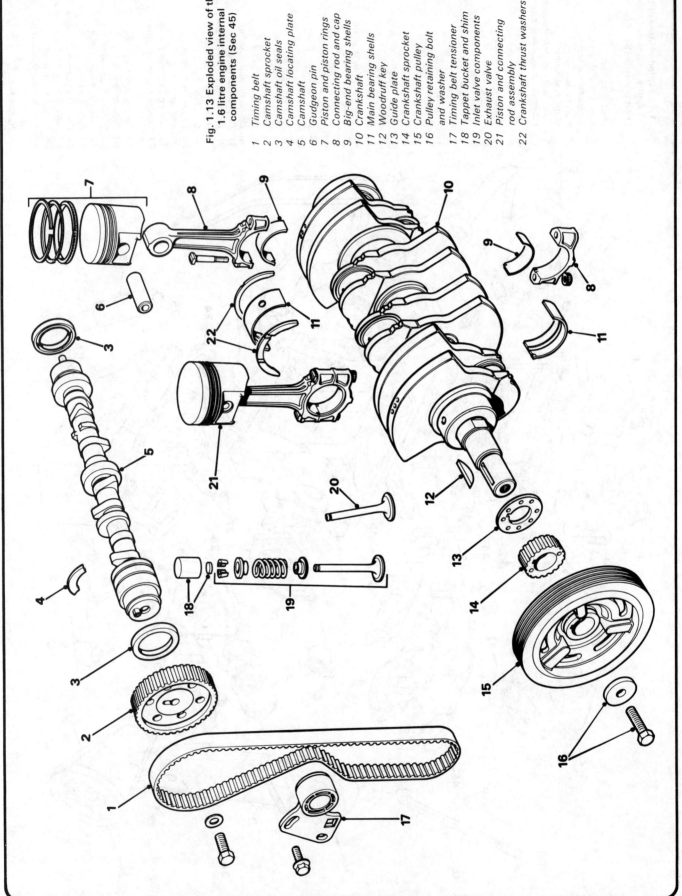

Fig. 1.13 Exploded view of the 1.6 litre engine internal components (Sec 45)

1 Timing belt
2 Camshaft sprocket
3 Camshaft oil seals
4 Camshaft locating plate
5 Camshaft
6 Gudgeon pin
7 Piston and piston rings
8 Connecting rod and cap
9 Big-end bearing shells
10 Crankshaft
11 Main bearing shells
12 Woodruff key
13 Guide plate
14 Crankshaft sprocket
15 Crankshaft pulley
16 Pulley retaining bolt and washer
17 Timing belt tensioner
18 Tappet bucket and shim
19 Inlet valve components
20 Exhaust valve
21 Piston and connecting rod assembly
22 Crankshaft thrust washers

The connecting rods are attached to the crankshaft by horizontally split shell-type big-end bearings, and to the pistons by interference fit gudgeon pins. The aluminium alloy pistons are of the slipper type and have their gudgeon pins offset to the thrust side to reduce piston slap. Two compression rings and a three-piece oil control ring are fitted to each piston.

The overhead camshaft is mounted in a carrier attached to the cylinder head and is driven via a toothed belt by the crankshaft. The camshaft operates the valves through inverted bucket type tappets which are also housed in the camshaft carrier. Tappet adjustment is by shims fitted between the valve stems and the tappet buckets.

The inlet and exhaust valves are mounted at an angle in the cylinder head and are each closed by a single valve spring.

The oil pump, pressure relief valve and full-flow oil filter are located in a housing attached to the front of the cylinder block. The rotor type oil pump is driven directly by the crankshaft.

The distributor rotor is driven directly by the camshaft whereas the fuel pump is operated by an eccentric camshaft lobe. The toothed timing belt also drives the water pump and a separate drivebelt is used for the alternator, driven by a sprocket on the crankshaft.

46 Maintenance and inspection

1   At the intervals given in Routine Maintenance at the beginning of this manual, carry out the following service operations on the engine.
2   Visually inspect the engine joint faces, gaskets and seals for any sign of oil or water leaks. Pay particular attention to the areas around the camshaft cover, cylinder head, crankshaft front oil seal and sump joint faces. Rectify any leaks by referring to the appropriate Sections of this Chapter.
3   Place a suitable container beneath the oil drain plug located on the rear facing side of the sump. Unscrew the plug using a spanner or socket and allow the oil to drain. Inspect the condition of the drain plug sealing washer, and renew it if necessary. Refit and tighten the plug after draining.
4   Move the bowl to the right-hand side of the engine under the oil filter.
5   Using a strap wrench or filter removal tool, slacken the filter and unscrew it from the engine and discard.
6   Wipe the filter mating face on the sump with a rag and then lubricate the seal of a new filter using clean engine oil.
7   Screw the filter into position and tighten it by hand only, do not use any tools.
8   Refill the engine using the correct grade of oil through the filler on the front of the engine. Fill until the level reaches the MAX mark on the dipstick – 0.5 litres will raise the level from MIN to MAX.
9   With the engine running, check for leaks around the filter seal.
10 At less frequent intervals (see Routine Mainenance), check and if necessary, adjust the tappet clearances, as described in Section 86.
11 At the same interval clean the filter gauze in the oil filler cap using paraffin or a suitable solvent and dry using compressed air.
12 When specified, renew the timing belt using the procedure given in Sections 55 and 85.

47 Major operations possible with the engine in the car

The following operations can be carried out without having to remove the engine from the car:

(a) *Removal, refitting and adjustment of the timing belt and tensioner*
(b) *Removal and refitting of the camshaft and tappets*
(c) *Removal and refitting of the cylinder head*
(d) *Removal and refitting of the sump*
(e) *Removal and refitting of the big-end bearings*
(f) *Removal and refitting of the piston and connecting rod assemblies*
(g) *Removal and refitting of the oil pump*
(h) *Removal and refitting of the gearbox adaptor plate (after removal of the gearbox or automatic transmission)*
(i) *Removal and refitting of the engine mountings*

48 Major operations requiring engine removal

Strictly speaking it is only necessary to remove the engine if the crankshaft or main bearings require attention. However, due to the possibility of dirt entry, and to allow greater working access, it is preferable to remove the engine if working on the piston and connecting rod assemblies, or when carrying out any major engine overhaul or repair.

49 Methods of engine removal

The engine, complete with gearbox or automatic transmission, is removed by lifting it upwards and out of the engine compartment as a complete assembly. The gearbox or automatic transmission is then separated from the engine after removal. Due to the limited working clearances it is not possible to remove the engine on its own.

50 Engine – removal with manual gearbox

1   Remove the bonnet, as described in Chapter 12, and the battery, as described in Chapter 10.
2   Undo and remove the bolts securing the battery tray to the body and take out the tray. Note the arrangement of earth wires on the front retaining bolt (photos).
3   Drain the cooling system, as described in Chapter 2.
4   Remove the air cleaner assembly and the plenum chamber, as described in Chapter 3.
5   Slide back the protective cover, then disconnect the two reversing lamp wires at the gearbox switch (photo).
6   Undo the nut and bolt securing the earth strap and cable retaining clip to the gearbox casing (photo).

50.2A Earth wire arrangement on battery tray front retaining bolt

50.2B Remove the battery tray

50.5 Disconnect the reversing lamp switch wires (arrowed)

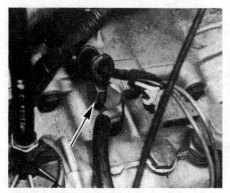

50.6 Remove the earth strap and cable clip retaining bolt (arrowed)

50.8 Disconnect the crankshaft sensor wiring plug ...

50.9 ... and knock sensor wiring plug

7   Make a note of the wiring connections at the starter motor solenoid and disconnect the wires.

8   Disconnect the crankshaft sensor wiring plug (photo) then, using a spanner and suitable Allen key, undo the two starter motor retaining bolts. Remove the starter, crankshaft sensor bracket and engine front snubber.

9   Disconnect the wiring plug at the knock sensor on the front face of the cylinder block (photo).

10 Spring back the clip and disconnect the wiring plug at the rear of the alternator.

11 Disconnect the wiring plug from the coolant temperature thermistor in the thermostat housing.

12 Disconnect the inlet manifold induction heater lead at the wiring connector and the two wires at the inlet manifold induction temperature sensor.

13 Disconnect the wire at the oil pressure switch adjacent to the crankshaft pulley (photo).

14 Disconnect the wiring plug at the carburettor stepping motor and the two leads at the fuel shut-off valve solenoid (photos).

15 Release the retaining clip and remove the fuel inlet hose from the fuel pump (photo). Plug the hose after removal.

16 Pull the float chamber vent hose off the carburettor outlet and remove the hose and pipe from the engine (photo).

17 Detach the crankcase breather hoses at the carburettor, oil filler cap and oil separator, then remove the hose assembly.

18 Disconnect the two ignition vacuum advance hose connectors at the carburettor (photo).

19 Disconnect the vacuum hose from the inlet manifold banjo union, undo the union bolt and recover the two washers (photo). Place the servo vacuum hose to one side.

50.13 Disconnect the oil pressure switch wire

50.14A Disconnect the wiring at the carburettor stepping motor ...

50.14B ... and fuel shut-off valve solenoid

50.15 Remove the fuel inlet hose from the pump

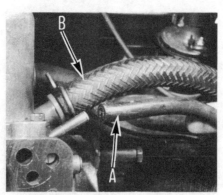

50.16 Detach the float chamber vent hose (A) and crankcase breather hose (B)

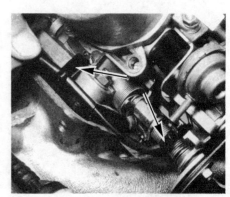

50.18 Detach the two ignition vacuum advance hose connectors (arrowed)

50.19 Remove the banjo union bolt and washers from the inlet manifold

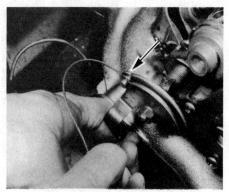

50.20A Slip the accelerator cable end (arrowed) out of the throttle lever ...

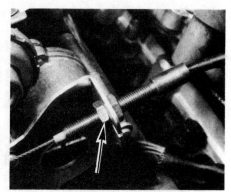

50.20B ... unscrew the lower accelerator cable locknut (arrowed) and remove the cable

20 Open the throttle linkage by hand and slip the accelerator cable end out of the slot on the throttle lever (photo). Slacken the outer cable locknuts, unscrew the lower locknut fully and release the cable from the support bracket (photo). Place the cable to one side.

21 Slacken the retaining clips and disconnect the heater hoses at the inlet manifold.

22 Slacken the retaining clips and disconnect the heater hose and radiator bottom hose from the water inlet elbow (photo), followed by the expansion tank hose, radiator top hose and heater hose from the water outlet elbow and thermostat housing (photos).

23 Refer to Chapter 5 and detach the clutch cable from the operating lever and gearbox bracket.

24 Undo and remove the bolt securing the speedometer cable to the gearbox. Withdraw the cable and pinion assembly and place them aside (photo).

25 Remove the retaining clip and slide the gearchange rod out of the selector shaft lever (photo). Disconnect the rear selector rod from the relay lever on the gearbox by prising off the balljoint socket with a screwdriver (photo).

26 Jack up the front of the car and support it securely on axle stands.

27 From underneath the car, undo the nuts securing the exhaust front pipes to the manifold. Separate the joint flange and recover the gasket.

28 Mark the relationship of the driveshaft inner constant velocity joints to the differential drive flanges. Lift off the protective caps and unscrew the joint-to-drive flange retaining bolts using a suitable Allen key.

29 Undo the two nuts securing the rear engine mounting to the front crossmember (photo).

30 Undo the retaining screws and remove the left-hand and right-hand access panels from under the wheel arches.

50.22A Disconnect the hoses at the water inlet elbow ...

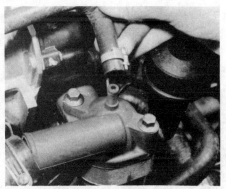

50.22B ... water outlet elbow ...

50.22C ... and thermostat housing

50.24 Remove the speedometer cable and pinion assembly

50.25A Extract the retaining clip and release the gearchange rod ...

50.25B ... then prise the selector rod balljoint socket (arrowed) off the relay lever

50.29 Undo and remove the rear mounting-to-crossmember nuts (arrowed)

50.33 Remove the right-hand mounting support bracket

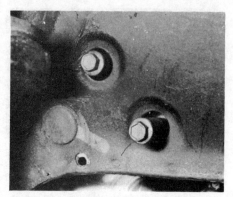

50.34 Undo the right-hand mounting-to-body bolts from under the wheel arch

50.35 Remove the right-hand mounting from the engine

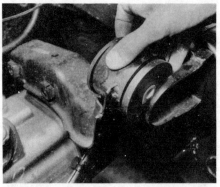

50.36 Remove the left-hand mounting from the gearbox and body bracket ...

50.37 ... then undo the three bolts and remove the body bracket

50.40 Removing the engine and gearbox from the car

31 Remove the axle stands and lower the car to the ground.
32 Attach a suitable hoist to the engine using chains or rope slings, or by attaching the chains or ropes to sturdy brackets bolted to the engine and gearbox. Raise the hoist to just take the weight of the engine.
33 Undo the screw securing the expansion tank to the support bracket and the two bolts securing the bracket to the body. Withdraw the bracket (photo).
34 From under the right-hand wheel arch, undo the two bolts securing the right-hand engine mounting to the body, noting the distance spacer on each bolt (photo).
35 Undo the bolts securing the right-hand engine mounting to the engine and remove the complete mounting assembly (photo).
36 Undo the two bolts securing the left-hand engine mounting to the

gearbox, and the through-bolt and nut securing the mounting to the body bracket. Remove the mounting (photo).
37 Undo the three bolts securing the left-hand engine mounting bracket to the body and remove the bracket (photo).
38 Make a final check that everything attaching the engine and gearbox to the car has been disconnected and that all detached components are well clear.
39 Lift the engine and gearbox slightly and, as soon as sufficient clearance exists, release the driveshaft inner joints from the drive flanges. Support or tie up the driveshafts to avoid straining the outer joints.
40 Continue raising the engine and gearbox assembly (photo), and, when it has been raised sufficiently, draw the hoist forward or push the car backwards and lower the engine and gearbox to the ground.

51 Engine – removal with automatic transmission

1   Remove the bonnet, as described in Chapter 12, and the battery, as described in Chapter 10.

2   Undo and remove the bolts securing the battery tray to the body and take out the tray.

3   Drain the cooling system, as described in Chapter 2.

4   Remove the air cleaner assembly and the plenum chamber, as described in Chapter 3.

5   Make a note of the wiring connections at the starter motor solenoid and disconnect the wires.

6   Disconnect the crankshaft sensor wiring plug then, using a spanner and suitable Allen key, undo the two starter motor retaining bolts.

7   Remove the starter motor, crankshaft sensor bracket and the engine front snubber bracket.

8   Undo the nut and release the selector cable trunnion from the selector lever on the transmission.

9   Slip the kickdown cable end out of its operating lever on the transmission. Release the clevis pin and disconnect the other end of the kickdown cable from the carburettor linkage and support bracket.

10 Undo the retaining bolts securing the selector cable and kickdown cable support bracket to the transmission. Move the bracket and cables clear of the engine and transmission.

11 Disconnect the speedometer cable from its attachment on the transmission.

12 Refer to Section 50 and carry out the operations described in paragraphs 9 to 22 inclusive.

13 Jack up the front of the car and support it securely on axle stands.

14 From underneath the car, undo the nuts securing the exhaust front pipes to the manifold. Separate the joint flange and recover the gasket.

15 Mark the relationship of the driveshaft inner constant velocity joints to the differential drive flanges. Lift off the protective caps and unscrew the joint-to-drive flange retaining bolts using a suitable Allen key.

16 Undo the nuts securing the rear engine/transmission mounting to the crossmember.

17 Undo the retaining screws and remove the left-hand and right-hand access panels from under the wheel arches.

18 Remove the axle stands and lower the car to the ground.

19 Attach a suitable hoist to the engine using chains or rope slings, or by attaching the chains or ropes to sturdy brackets bolted to the engine and transmission. Raise the hoist to just take the weight of the engine.

20 Undo the screw securing the expansion tank to the support bracket and the two bolts securing the bracket to the body. Withdraw the bracket.

21 From under the right-hand wheel arch, undo the two bolts securing the right-hand engine mounting to the body, noting the distance spacer on each bolt.

22 Undo the bolts securing the right-hand mounting to the engine and remove the mounting.

23 Undo the bolt securing the left-hand mounting to the transmission and body then remove the complete mounting assembly.

24 Make a final check that everything attaching the engine and transmission to the car has been disconnected and that all attached components are well clear.

25 Lift the engine and transmission slightly and, as soon as sufficient clearance exists, release the driveshaft inner joints from the drive flanges. Support or tie up the driveshafts to avoid straining the outer joints.

26 Continue raising the engine and transmission assembly and, when it has been raised sufficiently, draw the hoist forward or push the car backwards and lower the engine and transmission to the ground.

52 Engine – separation from manual gearbox or automatic transmission

Manual gearbox

1   With the engine and gearbox removed from the car, support the two units and undo all the bolts securing the gearbox to the engine adaptor plate. Note the locations of the different lengths of bolts and also the arrangement of retaining nut and captive nut on the bolt that secures the inlet manifold support strut (photo).

2   When all the bolts have been removed, ease the gearbox off its locating dowels and withdraw it squarely from the engine (photo). Do not allow the weight of the gearbox to hang unsupported on the gearbox shaft.

52.1 Gearbox-to-adaptor plate bolt, nut and support strut arrangement

52.2 Removing the gearbox from the engine

Automatic transmission

3   Turn the crankshaft as necessary until each of the torque converter retaining bolts becomes accessible through the starter motor aperture. Undo the bolts using a socket and extension bar.

4   Undo the bolts securing the transmission to the engine adaptor plate noting the different bolt lengths and the arrangement of support brackets, where applicable.

5   Ease the transmission off the adaptor plate dowels and withdraw it from the engine, ensuring that the torque converter remains on the transmission shaft.

53 Engine dismantling – general

Refer to Section 8.

54 Ancillary components – removal

1 If the engine has been removed from the car for major overhaul or repair the externally-mounted ancillary components, as given in the following list, should first be removed. Removal is straighforward, but where necessary reference should be made to the relevant Chapters of this manual as indicated. The removal sequence need not necessarily follow the order given:

Alternator (Chapter 10)
Timing belt covers – photos
Fuel pump (Chapter 3)
Distributor cap and rotor arm (Chapter 4) – photos
Oil filler tube
Thermostat, housing and water inlet elbow (Chapter 2)
Alternator mounting bracket – photo
Knock sensor – photo
Crankshaft sensor (Chapter 4) – photo
Spark plugs (Chapter 4)
Inlet and exhaust manifolds and carburettor (Chapter 3)
Clutch and flywheel assembly (Chapter 5)
Water pump (Chapter 2) – after removal of the timing belt
Oil filter (Section 46 of this Chapter)
Dipstick

55 Timing belt – removal

1 Jack up the front of the car and support it on axle stands.
2 Remove the right-hand front roadwheel and the access panel under the wheel arch.
3 Disconnect the battery negative terminal.
4 Slacken the alternator pivot mounting bolt and the adjustment arm bolts. Move the alternator towards the engine and remove the drivebelt from the pulleys (photo).
5 Lift off the timing belt upper cover then undo the screws and remove the lower cover (photos).

6 Using a socket or spanner on the crankshaft pulley bolt, turn the engine over until the dimple on the rear face of the camshaft sprocket is aligned with the notch on the camshaft carrier (photo). Check also that the notch on the crankshaft pulley is aligned with the timing mark on the oil pump housing (photo).
7 Slacken the two timing belt tensioner retaining bolts and move the tensioner away from the engine (photo).
8 Slip the timing belt off the three sprockets and remove it from the engine (photo).
9 If the original belt is to be reused, mark it with chalk to indicate its direction of rotation and also its outer facing edge. Store the belt on its edge while off the engine.

56 Cylinder head removal – engine in car

1 Refer to Section 55 and remove the timing belt.
2 Drain the cooling system, as described in Chapter 2.
3 Remove the air cleaner and plenum chamber, as described in Chapter 3.
4 Remove the alternator, as described in Chapter 10.
5 Undo the three bolts securing the alternator mounting bracket to the cylinder head and block, then remove the bracket (photo).
6 Disconnect the fuel inlet hose at the carburettor and plug the hose after removal.
7 Undo the two bolts securing the fuel pump to the cam cover. Withdraw the pump, insulating block and gaskets, and place it to one side with hoses still attached (photo).
8 Disconnect the vacuum hose from the inlet manifold banjo union, undo the union bolt and recover the two copper washers. Place the servo vacuum hose to one side.
9 Undo the small bolt and release the oil dipstick tube from the camshaft cover.
10 Slacken the retaining clips and disconnect the expansion tank hose, radiator top hose and heater hose from the water outlet elbow and thermostat housing (photo).
11 Slacken the retaining clips and disconnect the heater hose and radiator bottom hose from the water inlet elbow (photo).

54.1A Lift off the timing belt upper cover ...

54.1B ... then undo the screws (arrowed) and remove the lower cover

54.1C Remove the distributor cap and HT leads ...

54.1D ... followed by the rotor arm and shield

54.1E Undo the bolts (arrowed) and remove the alternator mounting bracket

54.1F Unscrew the knock sensor

54.1G Remove the crankshaft sensor

55.4 Remove the alternator drivebelt

55.5A Remove the timing belt upper ...

55.5B ... and lower covers

55.6A Align the dimple on the camshaft sprocket with the notch on the carrier ...

55.6B ... and the notch on the crankshaft pulley with the mark on the oil pump housing

55.7 Slacken the two tensioner retaining bolts (arrowed)

55.8 Remove the timing belt from the sprockets

56.5 Undo the bolts (arrowed) and remove the alternator mounting bracket

56.7 Remove the pump with the hoses still attached

56.10 Disconnect the hoses at the water outlet elbow and thermostat housing ...

56.11 ... and at the water inlet elbow

56.16 Disconnect the vacuum hoses at the carburettor connectors

56.17 Disconnect the induction temperature sensor leads at the manifold

56.18 Disconnect the coolant temperature thermistor wiring plug

56.19 Undo the manifold support strut retaining bolts

56.26 Removing the cylinder head, complete with manifolds and carburettor

12 Slacken the retaining clips and disconnect the heater hoses at the inlet manifold.

13 Disconnect the wiring plug at the carburettor stepping motor and the two wires at the fuel shut-off valve solenoid.

14 Open the throttle linkage by hand and slip the accelerator cable end out of the slot on the throttle lever. Slacken the outer cable locknuts, unscrew the inner cable locknut and release the cable from the support bracket. Place the cable to one side.

15 On vehicles equipped with automatic transmission, remove the clevis pin securing the kickdown cable to the throttle lever and remove the cable from the support bracket.

16 Disconnect the two ignition vacuum advance hose connectors at the carburettor (photo).

17 Disconnect the inlet manifold induction heater lead at the wiring connector and the two wires at the inlet manifold induction temperature sensor (photo).

18 Disconnect the wiring plug from the coolant temperature thermistor in the thermostat housing (photo).

19 Undo the two bolts securing the support struts to the inlet manifold (photo).

20 Undo the nuts securing the exhaust front pipes to the manifold. Separate the joint flange and recover the gasket.

21 Undo the two bolts securing the oil filler tube to the adaptor plate and remove the tube and gasket.

22 Disconnect the spark plug HT leads from the plugs then undo the two distributor cap retaining screws. Remove the cap and leads and place them to one side.

23 Place a suitable jack beneath the right-hand side of the engine with a block of wood between the jack head and the sump. Raise the jack and just take the weight of the engine.

24 Undo the bolts securing the right-hand engine mounting to the cylinder head. Slacken the mounting through-bolt and swivel the mounting as far away from the cylinder head as possible.

25 Gradually slacken the cylinder head retaining bolts, half a turn at a time, in the reverse order to that shown in Fig. 1.18. When all the

bolts have been slackened, remove them from their locations noting where the bolts having upper thread extensions are fitted.

26 Lift the cylinder head and manifold assembly from the engine (photo). If it is stuck, tap it free with a soft-faced mallet, **do not** attempt to prise it free using a lever between the head and cylinder block, or the mating faces may be damaged.

27 Remove the cylinder head gasket from the cylinder block.

57 Cylinder head removal – engine on bench

The procedure for removing the cylinder head with the engine on the bench is similar to that for removal when the engine is in the car with the exception of disconnecting the controls and services. Refer to Section 56 and follow the procedure given as applicable.

58 Camshaft and tappets – removal

Note: *If the engine is in the car, carry out the following operations with reference to the relevant Sections and Chapters of this manual:*

(a) *Disconnect the battery negative terminal*
(b) *Remove the air cleaner*
(c) *Remove the timing belt*
(d) *Remove the fuel pump*
(e) *Remove the distributor cap, rotor arm and shield*

1 Undo the retaining bolts and lift off the two camshaft carrier covers (photo). Note the position of the cable and hose clips on the cover bolts.

2 Using a suitable socket, or spanner, undo the camshaft sprocket retaining bolt. Engage a stout screwdriver or bar through the sprocket holes and in contact with the carrier to prevent the camshaft turning.

58.1 Remove the camshaft carrier covers

58.8 Carefully withdraw the camshaft from the carrier

Remove the sprocket retaining bolt and washer then withdraw the sprocket from the camshaft. Carefully ease it off using two levers if it is tight.

3   Using pliers, withdraw the camshaft locating plate from the carrier (photo).

4   Progressively slacken the camshaft carrier retaining bolts and, when all tension on the bolts has been relieved, remove them.

5   Raise the camshaft carrier slightly and push down the tappet buckets until the cam lobes are clear.

6   Move the camshaft towards the rotor arm end until sufficient clearance exists to enable the oil seal at the sprocket end to be removed. Hook the seal out using a screwdriver.

7   Now move the camshaft towards the sprocket end and remove the remaining oil seal in the same way.

8   The camshaft can now be carefully removed from the rotor arm end of the carrier (photo).

9   Lift out each of the tappet buckets in turn and keep them in strict order (photo). Make sure that the small adjustment shim has remained in place inside the bucket.

10 Withdraw the camshaft carrier from the cylinder head (photo) noting the location of the O-ring oil seal.

58.9 Lift out the tappet buckets, complete with shims

58.3 Withdraw the camshaft locating plate

58.10 Remove the camshaft carrier from the cylinder head

59.1 Compress the valve springs and lift out the split collets

59.3A Remove the valve ...

59.3B ... then prise off the valve stem oil seals

59 Cylinder head – dismantling

Note: *If the cylinder head was removed with the engine in the car, refer to the relevant Sections and Chapters of this manual and remove the following components:*

    (a) Inlet and exhaust manifolds and carburettor
    (b) Distributor cap, rotor arm and shield
    (c) Thermostat housing and water inlet elbow
    (d) Camshaft and tappets

1   Using a valve spring compressor, compress each valve spring in turn until the split collets can be removed (photo). Release the compressor and lift off the cap and spring.
2   If, when the valve spring compressor is screwed down, the valve spring cap refuses to free and expose the split collets, gently tap the top of the tool directly over the cap with a light hammer. This will free the cap.
3   Remove the valve through the combustion chamber then prise the valve stem oil seals off the valve guides using a screwdriver (photos).
4   It is essential that the valves are kept in their correct sequence unless they are so badly worn that they are to be renewed. If they are going to be kept and used again, place them in a sheet of card having eight holes numbered 1 to 8 – corresponding to the relative fitted positions of the valves. **Note** that No 1 valve is nearest to the crankshaft pulley end of the engine.

60 Sump – removal

Note: *If the engine is still in the car, carry out the following operations:*

    (a) Drain the engine oil
    (b) Disconnect the breather hose from the oil separator on the front face of the sump

1   Undo the bolts securing the sump to the gearbox adaptor plate.
2   Progressively slacken the sump retaining bolts and then remove them. Make a note of the locations of the different lengths of bolts and of the ones which also secure cable clips and support brackets.
3   Tap the sump with a soft-faced mallet to free the joint face, then remove the sump from the crankcase. Recover the one-piece rubber gasket.
4   If necessary undo the two bolts and one nut securing the oil pick-up tube to the oil pump housing and main bearing cap and remove the tube. Recover the O-ring oil seal at the base of the pick-up tube.

61 Oil pump and housing – removal

Note: *If the engine is in the car, carry out the following operations with reference to the relevant Sections and Chapters of this manual:*

    (a) Remove the timing belt
    (b) Remove the sump and oil pick-up tube
    (c) Disconnect the oil pressure switch wire

1   Undo the bolt securing the crankshaft pulley to the crankshaft. If the engine is in the car, engage 1st gear (manual transmission) and firmly apply the footbrake to prevent crankshaft rotation as the bolt is undone. On automatic transmission models it is necessary to remove the starter motor (Chapter 10) and prevent the crankshaft from turning by inserting a suitable bar into the driveplate ring gear. If the engine is on the bench refit two of the clutch pressure plate retaining bolts to the crankshaft and engage a stout bar between them (photo).
2   Withdraw the crankshaft pulley followed by the crankshaft sprocket and guide plate.
3   Undo the pump retaining bolts, noting their lengths and locations, and also the retaining plate fitted under the top centre bolt.
4   Withdraw the oil pump housing and recover the gasket.

61.1 Using two bolts and a stout bar to lock the crankshaft

62 Gearbox adaptor plate – removal

Note: *If the engine is in the car, carry out the following operations with reference to the relevant Sections and Chapters of this manual:*

    (a) Remove the manual gearbox or automatic transmission
    (b) Remove the clutch assembly and flywheel
    (c) Remove the sump
    (d) Remove the oil filler tube
    (e) Remove the crankshaft sensor

1   Undo the bolts securing the adaptor plate to the cylinder block.
2   Tap the adaptor plate using a soft-faced mallet to free it from the locating dowels and remove it from the engine. Remove the two sump retaining bolts and gearbox retaining bolt from their captive recesses in the adaptor plate (photo).

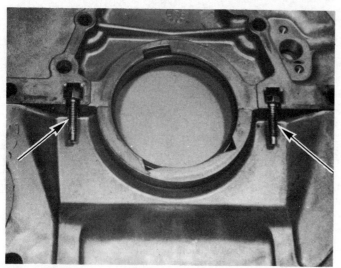

62.2 Sump retaining bolt locations (arrowed) in the gearbox adaptor plate

1   Check the big-end caps for identification marks. If necessary use a centre punch on the caps and rods to identify them; mark them 1 to 4 on the dipstick tube side to ensure correct refitting (photo). **Note** that No 1 is nearest ot the crankshaft pulley end of the engine.
2   Turn the crankshaft so that No 1 crankpin is at its lowest point. Using a suitable socket, undo the two nuts securing the connecting rod cap to the rod.
3   Withdraw the cap, complete with bearing shell.
4   Using the handle of a hammer, carefully push the piston and connecting rod up through the bore and withdraw it from the top of the cylinder block.
5   Loosely refit the cap to the connecting rod.
6   Repeat the procedure given in paragraphs 1 to 5 on No 4 piston and connecting rod, then turn the crankshaft through half a turn and repeat the procedure on No 2 and No 3 piston and connecting rod assemblies.

63 Pistons and connecting rods – removal

Note: *If the engine is in the car, carry out the following operations with reference to the relevant Sections and Chapters of this manual:*

(a)   Remove the cylinder head (this is not necessary if only the big-end bearings are to be removed)

(b)   Remove the sump and oil pick-up tube

64 Crankshaft and main bearings – removal

1   With the engine removed from the car and dismantled as described in the previous Sections of this Chapter, the crankshaft and main bearings can now be removed.
2   Check the crankcase and main bearing caps for identification marks (photo) and if no marks are present use a centre punch to mark them.
3   Undo the bolts securing the main bearing caps and remove the caps, complete with bearing shells.
4   Before removing the crankshaft, check that the endfloat is within the specified limits by inserting feeler blades between No 4 crankshaft web and the thrust washers (photo). If the clearance is not as specified, new thrust washers will be required for reassembly.
5   Lift the crankshaft out of the crankcase (photo) then remove the main bearing shell upper halves and the thrust washers (photo). Keep the main bearing shells in order with their respective caps.

63.1 Identify the connecting rod and cap with centre dots if manufacturer's marks are not apparent

64.2 Identification marks on the main bearing cap and crankcase

64.4 Checking crankshaft endfloat

64.5A Lift out the crankshaft ...

64.5B ... and remove the thrust washers and bearing shells

65 Crankcase ventilation system – description

The crankcase ventilation system consists of an oil separator attached to the engine sump, a breather filter located in the oil filler cap and a hose connecting the oil separator and filler cap to the carburettor.

Air is drawn into the crankcase through the filter in the filler cap. Inlet manifold depression draws the crankcase fumes from the oil separator and oil filler tube through the hose into the carburettor airstream where they mix with the incoming fuel/air mixture passing to the cylinders for combustion.

The breather filter should be cleaned at the specified service intervals (see Routine Maintenance) and the hoses periodically checked for condition and security.

66 Examination and renovation – general

With the engine completely dismantled, clean all the components and examine them for wear. Each part should be checked and where necessary renewed or renovated, as described in the following Sections. All oil seals, gaskets and O-rings should be renewed as a matter of course and also main and big-end shell bearings unless they have had little wear and are in perfect condition.

67 Cylinder block and crankcase – examination and renovation

Refer to Section 23, paragraphs 1 to 5.

68 Crankshaft and main bearings – examination and renovation

Refer to Section 22.

69 Pistons and connecting rods – examination and renovation

Refer to Section 24, but note that, when arranging the piston rings, the compression ring gaps should be positioned at 90° to each other towards the dipstick tube side of the cylinder block.

70 Gearbox adaptor plate – examination and renovation

Refer to Section 27.

71 Oil pump and housing – examination and renovation

1 Remove the oil filter from the pump housing if not already done.
2 Undo the six screws and lift off the pump backplate (photo).
3 Lift out the two rotors from the pump body (photo).
4 Unscrew the pressure relief valve cap using a wide-bladed tool and remove the spring and plunger (photos).

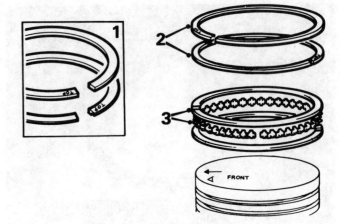

Fig. 1.14 Piston ring details (Sec 69)

1 *Compression ring upper faces (marked TOP)*
2 *Compression ring gaps positioned at 90° to each other towards the dipstick tube side of the engine*
3 *Oil control ring rails and spreader spring*

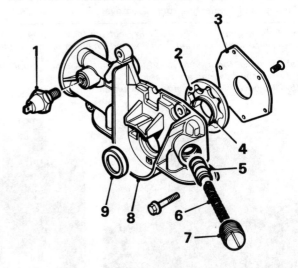

Fig. 1.15 Oil pump housing and pump components (Sec 71)

1 *Oil pressure switch*	6 *Pressure relief valve spring*
2 *Outer rotor*	7 *Pressure relief valve cap*
3 *Pump backplate*	8 *Oil pump housing*
4 *Inner rotor*	9 *Crankshaft front oil seal*
5 *Pressure relief valve plunger*	

71.2 Undo the screws and remove the oil pump backplate

71.3 Lift out the oil pump rotors

71.4A Unscrew the pressure relief valve cap ...

71.4B ... and withdraw the relief valve components

71.6A Checking the oil pump rotor lobe clearance ...

71.6B ... outer rotor-to-body clearance ...

71.6C ... and the outer rotor endfloat

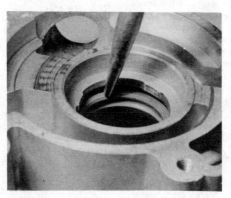

71.7A Remove the crankshaft front oil seal from the pump housing ...

71.7B ... and fit a new seal using a block of wood or socket

5   Clean all the parts in paraffin or a suitable solvent and dry with a lint-free cloth. Examine the components for signs of scoring, wear ridges or other damage and renew the pump as a complete assembly if any of these conditions are apparent.

6   If the pump is satisfactory so far, refit the rotors to the pump body and check the rotor lobe clearance, outer rotor-to-body clearance and the outer rotor endfloat using feeler gauges and a straight-edge (photo). Renew the pump if any of the clearances are outside the figures given in the Specificationss.

7   If the pump is serviceable, renew the crankshaft front oil seal in the pump housing. Tap the old seal out from the inside using a punch (photo) and fit a new seal using a tube, block of wood or socket by tapping it squarely into the housing (photo).

8   Lubricate the pressure relief valve plunger with clean engine oil and refit plunger followed by the spring and cap. Tighten the cap securely.

9   Liberally lubricate the two rotors with clean engine oil and place them in position in the pump housing.

10   Refit the pump backplate and secure with the six screws.

72 Camshaft and tappets – examination and renovation

1   The camshaft itself should show no signs of wear, but if very slight score marks on the cams are noticed, they can be removed by gently rubbing down with very fine emery cloth or an oilstone. *The greatest care must be taken to keep the cam profiles smooth.*

2   Carefully examine the camshaft bearing surfaces for wear and, if evident, the camshaft must be renewed.

3   Check the fit of the camshaft in the carrier and if excessive bearing journal clearance is apparent a new carrier must be obtained. The camshaft bearings run directly in the machined journals of the carrier; renewable bearings are not used.

4   The faces of the tappet buckets which bear on the camshaft lobes should exhibit no signs of pitting, scoring, cracks or other forms of

wear and should be a smooth sliding fit in the carrier. Slight scuffing and blackening of the tappet bucket sides is normal, providing this is not accompanied by scoring or wear ridges.

5   The small shims found inside the tappet bucket should show no signs of indentation from contact with the valve stem. Renew the shim with one of an identical size if wear has taken place. Make sure that each shim is kept with its tappet bucket and not interchanged.

73 Timing belt, sprockets and tensioner – examination and renovation

1   Carefully examine the belt for any sign of cracking, particularly at the root of the teeth, fraying, oil contamination or any other sign of deterioration. Renew the belt if any of these conditions are found, or as a matter of course if the belt is nearing the end of its recommended service life (see Routine Maintenance).

2   Check the sprockets for signs of cracked or chipped teeth and the tensioner for roughness of its bearings or excessive endfloat. Renew the sprockets or tensioner as necessary. Note that the water pump sprocket is an integral part of the pump and cannot be renewed separately. If this sprocket is damaged or if there is any play in the pump spindle, a complete water pump must be obtained (see Chapter 2).

74 Cylinder head – decarbonising, valve grinding and renovation

1   Decarbonising can be carried out with the engine either in or out of the car. With the cylinder head off, carefully remove, with a wire brush and blunt scraper, all traces of carbon deposits from the combustion spaces and the ports. The valve stems and valve guides should also be free from any carbon deposits. Wash the combustion spaces and ports down with paraffin and scrape the cylinder head surface free of any foreign matter with the side of a steel rule or similar article. Take care not to scratch the surfaces.

2   Clean the pistons and top of the cylinder bores. If the pistons are still in the cylinder bores, it is essential that great care is taken to ensure that no carbon gets into the bores as this could scratch the cylinder walls or cause damage to the piston and rings. To ensure that this does not happen, first turn the crankshaft so that two of the pistons are at the top of the bores. Place clean lint-free rags into the two bores, or seal them off with paper and masking tape. The water and oil ways should also be covered with a small piece of masking tape to prevent particles of carbon entering the cooling system and damaging the water pump, or entering the lubrication system and causing damage to a bearing surface.

3   Before starting, press a little grease into the gap between the cylinder walls and the two pistons which are to be worked on. With a blunt scraper carefully scrape away the carbon from the piston crowns, taking care not to scratch the aluminium. Also scrape away the carbon from the surrounding lip of the cylinder wall. When all carbon has been removed, scrape away the grease which will now be contaminated with carbon particles, taking care not to press any into the bore. To assist prevention of carbon build-up, the piston crown can be polished with metal polish. Remove the rags or masking tape from the other two cylinders and turn the crankshaft so that the two pistons which were at the bottom are now at the top. Place rag into the other two bores, or seal them with paper and masking tape. Do not forget the waterways and oilways as well. Proceed as previously described.

4   With the valves removed from the cylinder head, examine the heads for signs of cracking, burning away and pitting of the edges where they seat in the ports. The seats of the valves in the cylinder head should also be examined for the same signs. Usually it is the valve that deteriorates first, but if a bad valve is not rectified the seat will suffer and this is more difficult to repair.

5   Provided there are no obvious signs of serious pitting, the valve should be ground to its seat. This may be done by placing a smear of carborundum paste on the edge of the valve head and using a suction type valve holder, grinding the valve *in situ*. Use a semi-rotary action, rotating the handle of the valve holder between the hands and lifting it occasionally to re-distribute the traces of paste. Start with a coarse paste and finish with a fine paste.

6   As soon as a matt grey unbroken line appears on both the valve and seat, the valve is 'ground in'. All traces of carbon should also be cleaned from the head and neck of the valve stem. A wire brush mounted in a power drill is a quick and effective way of doing this.

7   If the valve requires renewal, the new one should be ground into the seat in the same way as the old valve.

8   Another form of valve wear can occur on the stem where it runs in the guide in the cylinder head. This can be detected by trying to rock the valve from side-to-side. If there is any movement at all it is an indication that the valve stem or guide is worn. Check the stem first with a micrometer at points along and around its length, and if they are not within the specified tolerance new valves will probably solve the problem. If the guides are worn, however, they will need reboring for oversize valves to be fitted. The valve seats will also need recutting to ensure they are concentric with the stems. This work should be given to your local BL garage or engineering works.

9   When the valve seats are badly burnt or pitted, requiring renovation, inserts may be fitted – or renewed if previously fitted – and once again this is a specialist task to be carried out by a suitable engineering firm.

10   When all valve grinding is completed it is essential that every trace of grinding paste is removed from the valves and ports in the cylinder head. This should be done by thorough washing in paraffin and blowing out with a jet of air. If particles of carborundum paste should work their way into the engine this would cause havoc with bearings or cylinder walls.

75 Engine mountings – removal and refitting

Left-hand mounting

1   Jack up the front of the car and support it on axle stands.

2   Remove the left-hand front roadwheel and the access panel under the wheel arch.

3   Remove the air cleaner cold air intake tube from the front body panel and air cleaner body.

4   Disconnect the two reversing lamp wires from the switch on top of the gearbox casing.

5   Position a jack beneath the gearbox with an interposed block of wood, and just take the weight of the engine and gearbox assembly.

6   Undo the two bolts securing the mounting to the gearbox and the through-bolt securing the mounting to the body bracket.

7   Lower the gearbox slightly and manipulate the mounting out of its location.

8   Refitting is the reverse sequence to removal. Tighten the retaining bolts to the specified torque and centralize the front snubber, as described in paragraph 22.

Right-hand mounting

9   Position a jack beneath the engine sump with an interposed block of wood, and just take the weight of the engine.

10   Lift off the timing belt upper cover then undo the two upper bolts securing the lower cover.

11   Undo the bolts securing the right-hand mounting to the cylinder block and head.

12   Undo the through-bolt securing the mounting to the body bracket.

13   Undo the four bolts securing the support plate to the body and body bracket.

14   Ease the mounting out of the body bracket and manipulate it out of its location.

15   Refitting is the reverse sequence to removal. Tighten the retaining bolts to the specified torque and centralize the front snubber, as described in paragraph 22.

Rear mounting

16   Jack up the front of the car and support it on axle stands.

17   Undo the two bolts securing the rear mounting support bracket to the crossmember. Recover the thread plate from the mounting bracket.

18   Undo the two nuts securing the mounting to the gearbox casing, slide the mounting off the casing studs and remove the assembly from under the car.

19   Refitting is the reverse sequence to removal. Tighten the retaining bolts to the specified torque and centralize the front snubber, as described in paragraph 22.

Front snubber

20   Undo the nuts from the starter motor retaining bolts, remove the crankshaft sensor wiring plug bracket and lift off the front snubber bracket and snubber.

21   Refit the snubber using the reverse of this procedure, tightening the starter motor retaining bolts to the specified torque.

22   Slacken the snubber cup retaining bolts (photo), centralize the snubber cup around the snubber rubber then tighten the bolts. This should be done whenever **any** of the engine mountings are renewed or in any way disturbed.

75.22 Engine front snubber cup retaining bolts (arrowed)

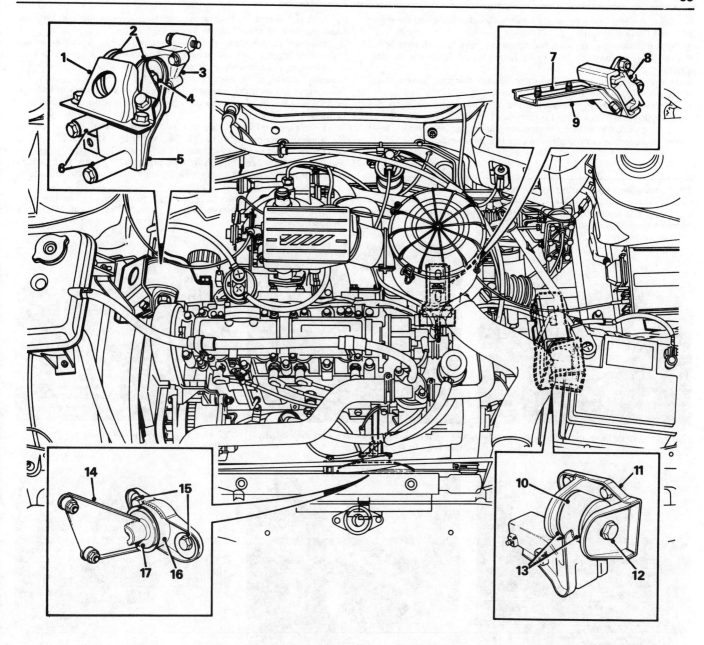

Fig. 1.16 Engine mounting component details (Sec 75)

1	Right-hand mounting support plate	6	Right-hand mounting distance spacers
2	Washers	7	Rear mounting thread plate
3	Right-hand mounting	8	Rear mounting
4	Through-bolt	9	Rear mounting support bracket
5	Right-hand mounting body bracket		

10	Left-hand mounting	14	Front snubber bracket
11	Left-hand mounting body bracket	15	Snubber cup retaining bolts
12	Through-bolt	16	Snubber cup
13	Washers	17	Snubber rubber

76 Engine reassembly – general

1   To ensure maximum life with minimum trouble from a rebuilt engine, not only must everything be correctly assembled, but it must also be spotlessly clean. All oilways must be clear, and locking washers and spring washers must be fitted where indicated. Oil all bearings and other working surfaces thoroughly with engine oil during assembly.

2   Before assembly begins, renew any bolts or studs with damaged threads.

3   Gather together a torque wrench, oil can, clean rags and a set of engine gaskets and oil seals, together with a new oil filter.

4   A tube of Loctite 574 sealant will be required for the camshaft carrier to cylinder head joint face and an RTV silicone sealant for the remainder of the joint faces that do not have gaskets. These compounds, together with conventional gasket jointing compound, are available from BL dealers or motor factors.

77 Crankshaft and main bearings – refitting

1   Clean the backs of the bearing shells and the bearing recesses in both the cylinder block and main bearing caps.

2   Press the main bearing shells into the cylinder block and caps and oil them liberally (photo).

3   Using a little grease, stick the thrust washers to each side of No 4 main bearing with their oilways facing away from the bearing (photo).

4   Lower the crankshaft into position, then fit the main bearing caps in their previously noted locations (photos).

5   Insert and tighten evenly the main bearing cap bolts to the specified torque. Check that the crankshaft rotates freely, then check that the endfloat is within the specified limits by inserting a feeler blade between the crankshaft web and the thrust washers.

78 Pistons and connecting rods – refitting

1   Clean the backs of the bearing shells and the recesses in the connecting rods and big-end caps.

2   Fit the big-end bearing shells into their connecting rod and cap locations and oil them liberally.

3   Position a piston ring compressor around No 1 piston then insert the connecting rod and piston into No 1 cylinder. With No 1 crankpin at its lowest point, drive the piston carefully into the cylinder with the wooden handle of a hammer, and at the same time guide the connecting rod onto the crankpin. Make sure that the mark FRONT, A or an arrow on the piston crown is towards the crankshaft pulley end of the engine (photos).

4   Fit the big-end cap in its previously noted position, then screw on the nuts and tighten them to the specified torque (photos).

5   Check that the crankshaft turns freely.

6   Repeat the procedures given in paragraphs 3 to 5 for No 4 piston, then turn the crankshaft through half a turn and repeat the procedure on No 2 and No 3 piston.

7   If the engine is in the car, refit the sump and oil pick-up tube, and the cylinder head with reference to the relevant Sections and Chapters of this manual.

77.2 Fit the main bearing shells to the cylinder block

77.3 Fit the thrust washers with their oilways facing away from the bearing

77.4A Lower the crankshaft into the crankcase ...

77.4B ... then fit the main bearing caps

78.3A With the rings compressed, insert the piston and connecting rod into its cylinder ...

78.3B ... then tap it down into the bore

78.3C Ensure that the word FRONT is towards the crankshaft pulley end of the engine

78.4A Fit the big-end cap ...

78.4B ... and tighten the nuts to the specified torque

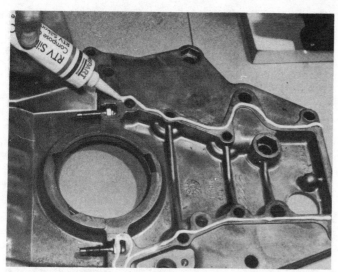

79.1 Apply RTV sealant to the adaptor plate mating face

adaptor plate then carefully fit the adaptor plate to the cylinder block (photo).
4 Refit the retaining bolts and tighten them evenly to the specified toque.
5 If the engine is in the car, refer to Section 62 and refit the items listed, in the reverse sequence to removal, with reference to the relevant Sections and Chapters of this manual.

80 Oil pump and housing – refitting

1 Make sure that the mating faces of the pump and cylinder block are clean then place a new gasket in position on the pump housing.
2 Wrap some insulating tape around the end of the crankshaft to protect the oil seal as the pump is fitted.
3 Liberally lubricate the lips of the oil seal and the insulating tape with engine oil.
4 Position the flats of the pump inner rotor to correspond with the flats on the crankshaft and carefully fit the pump to the cylinder block (photo). Remove the tape.
5 Refit the pump retaining bolts (photo) with the exception of the top centre bolt which also secures the water pump. This bolt is fitted later unless the water pump is already in place.
6 Tighten the bolts evenly to the specified torque.
7 If the engine is in the car, refer to Section 61 and refit the items listed, in the reverse sequence to removal, with reference to the relevant Sections and Chapters of this manual.

79 Gearbox adaptor plate – refitting

1 Ensure that the mating faces of the cylinder block and adaptor plate are thoroughly clean then apply a bead of RTV sealant to the adaptor plate face (photo).
2 Make sure that the two sump retaining bolts and the gearbox retaining bolt are fitted in their adaptor plate locations and apply additional sealant to their bolt heads (photo).
3 Liberally lubricate the lips of the crankshaft rear oil seal in the

81 Sump – refitting

1 Insert a new O-ring seal into the groove in the oil pick-up tube then refit the tube to the oil pump housing and main bearing cap (photos). Refit the retaining bolts and nut and tighten them securely.
2 Ensure that all traces of old sealant are removed from the sump

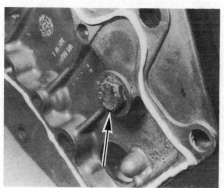

79.2 Apply additional sealant to the sump bolt heads and gearbox retaining bolt head (arrowed)

79.3 Fit the adaptor plate to the cylinder block

80.4 Fit the oil pump housing to the cylinder block using tape on the crankshaft to protect the seal

80.5 Refit and tighten the oil pump housing retaining bolts (arrowed)

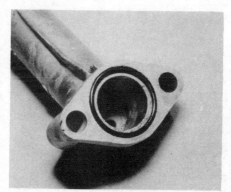

81.1A Insert a new O-ring into the oil pick-up tube groove ...

81.1B ... then secure the tube to the pump housing and bearing cap

and cylinder block mating faces and from around the one-piece rubber gasket if this in intact and to be reused.

3   Apply a thick bead of RTV sealant to the semi-circular joint faces of the sump, oil pump housing and gearbox adaptor plate. Extend the bead of sealant about 0.5 in (12 mm) beyond the ends of the semi-circular joint faces (photo).

4   Place the rubber gasket on the sump then position the sump on the engine (photo).

5   Refit the sump retaining bolts and nuts with the longer bolts, cable clips and brackets in the positions noted during removal (photo). Progressively tighten the bolts in a diagonal sequence to the specified torque.

6   With a new gasket in place, fit the oil separator and secure with the two nuts (photo).

7   Refit the bolts securing the sump to the gearbox adaptor plate (photo).

8   If the engine is in the car, refer to Section 60 and refit the items listed, in the reverse sequence to removal, with reference to the relevant Sections and Chapters of this manual.

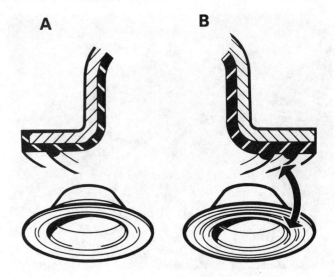

Fig. 1.17 Valve stem oil seals (Sec 82)

A   Standard size stem oil seal with one flange ring
B   Oversize stem oil seal with two flange rings

82 Cylinder head – reassembly

1   Place the new valve stem oil seals in place over the valve guides and push them fully into place using a small tube or socket (photos). Note that, in service, new oil seals are only supplied for the inlet valves and if valves with oversize stems are being fitted, oil seals with two rings on their flange must be used (Fig. 1.17).

2   Oil the valve stems liberally and fit each valve to its original guide, or if new valves have been obtained, to the seat in which they have been ground (photo).

3   Working on one valve at a time, fit the spring and cap then compress the spring with the compressor and insert the split collets (photos). Release the compressor and remove it.

4   Repeat the procedure given in paragraph 3 on the remaining valves. After fitting, tap the end of each valve stem with a mallet to settle the collets.

83 Camshaft and tappets – refitting

1   Smear the tappet shims with petroleum jelly and then locate the shims in the recesses of their respective tappet buckets.

2   Place the camshaft carrier in position on the cylinder head and fit the tappet buckets to their locations in the carrier.

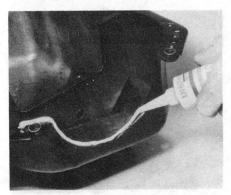

81.3 Apply RTV sealant to the semi-circular joint faces

81.4 Position the sump on the engine ...

81.5 ... and secure with the nuts. Note the cable clip locations (arrowed)

81.6 Refit the oil separator to the sump

81.7 Refit the sump-to-adaptor plate bolts

82.1A Place new valve stem oil seals over the valve guides ...

82.1B ... and push them fully into place

82.2 Fit the valves to their original guides

82.3A Fit the valve spring and cap ...

82.3B ... then compress the spring and insert the split collets

83.5 Camshaft square protrusion (arrowed) for spanner engagement

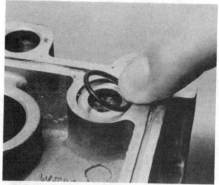

83.7A Locate a new O-ring in the camshaft carrier recess ...

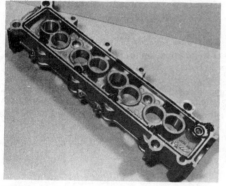

83.7B ... then fill the carrier groove with the special sealant

83.8A Place the carrier in position on the cylinder head ...

83.8B ... then fit the tappet buckets to their original bores

3   Lift the carrier slightly, push the tappet buckets down and carefully insert the camshaft. Fit the camshaft locating plate to the carrier front bearing journal.

4   Fit the camshaft carrier retaining bolts and progressively tighten them to the specified torque.

5   Before proceeding further the tappet clearances should be checked and adjusted using the procedure described in Section 86, paragraphs 2 to 6. For the purposes of tappet clearance checking the camshaft may be turned using an adjustable wrench on the square protrusion between No 6 and 7 camshaft lobe (photo).

6   With the tappet clearance checked and the correct new shims obtained as necessary, remove the camshaft, camshaft carrier and the tappet buckets (if not already done).

7   Locate a new O-ring seal in the carrier recess then fill the carrier groove with Loctite 574 sealant (photos).

8   Place the carrier in position on the cylinder head once more and insert the tappet buckets in their respective locations (photos).

9   Lift the carrier slightly, push down the tappet buckets and slide the camshaft into the carrier. Refit the camshaft locating plate.

10 Fit the camshaft carrier retaining bolts and progressively tighten them to the specified torque.

11 Thoroughly lubricate the lips of new camshaft front and rear oil seals and carefully locate them over the camshaft journals and into their positions in the carrier (photos). Tap the seals squarely into the carrier.

12 Place the camshaft sprocket on the camshaft and use the retaining bolt and washer to draw the sprocket fully home (photo). Tighten the retaining bolt to the specified torque.

13 Locate the one-piece rubber gaskets in the camshaft carrier covers (photo) using new gaskets if necessary.

14 Apply a continuous bead of RTV sealant, 3 mm wide, to the cover mating faces in the camshaft carrier (photo).

15 Fit the two covers, retaining bolts and brackets, where applicable (photo). Tighten the cover bolts progressively to the specified torque.

83.11A Fit new camshaft front ...

83.11B ... and rear oil seals

83.12 Refit the camshaft sprocket and retaining bolt

83.13 Locate the rubber gaskets in the camshaft carrier covers

83.14 Apply RTV sealant to the camshaft carrier ...

83.15 ... then fit the covers, noting the cable clip locations under the retaining bolts

16 If the engine is in the car refer to Section 58 and refit the items listed, in the reverse sequence to removal, with reference to the relevant Sections and Chapters of this manual.

84 Cylinder head – refitting

1 Ensure that the cylinder block and head mating faces are perfectly clean and free from any traces of oil, grease or water.
2 Place a new head gasket in position over the cylinder block dowels (photo). The gasket is pre-coated and jointing compound **should not** be used.
3 Lower the cylinder head into position (photo) and, with their threads lightly oiled, refit the retaining bolts.
4 Tighten the cylinder head bolts in the sequence shown in Fig. 1.18, to the specified torque.
5 If the engine is in the car reverse the procedure given in Section 56, paragraphs 2 to 23 then refit the timing belt, as described in Section 85.

84.2 Lay a new cylinder head gasket over the cylinder block dowels ...

84.3 ... then lower the head onto the gasket

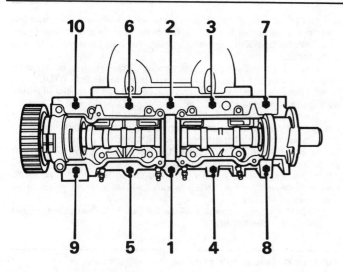

Fig. 1.18 Cylinder head bolt tightening sequence (Sec 84)

85.2B ... followed by the sprocket ...

6   After running the engine, retorque the cylinder head as described in Section 90.

85 Timing belt – refitting and adjustment

1   If the engine is being reassembled after major overhaul, refer to Chapter 2 and refit the water pump.
2   Place the crankshaft sprocket guide plate on the crankshaft followed by the sprocket and crankshaft pulley (photos). Fit the pulley retaining bolt and washer and tighten the pulley to the specified torque (photo).
3   Turn the camshaft as necessary, using a spanner on the sprocket bolt, until the dimple on the rear face of the sprocket is aligned with the notch on the camshaft carrier (photo). Turn the crankshaft until the notch on the pulley is aligned with the timing mark on the oil pump housing.
4   Refit the timing belt tensioner, but do not tighten the two retaining bolts at this stage.
5   Slip the belt over the sprockets and around the tensioner so that it is taut on the straight (driving) side (photo).
6   To tension the belt, engage a torque wrench of the type having a

85.2C ... and crankshaft pulley

85.2A Place the crankshaft sprocket guide plate on the crankshaft ...

85.2D Secure the pulley with the retaining bolt and washer

dial gauge scale or sliding pointer scale and $\frac{3}{8}$ in square drive into the hole in the tensioner bracket. Tension the belt to the torque figure given in the Specifications and tighten the two tensioner retaining bolts (photo).

7   Turn the crankshaft clockwise through three quarters of a turn so that the crankshaft pulley notch is at approximately 90° BTDC.

8   Slacken the tensioner retaining bolts again and re-tension the belt, as described in paragraph 6.

9   Turn the crankshaft clockwise through one and a quarter turns and realign the crankshaft pulley timing notch with the mark on the oil pump housing. Check that the camshaft sprocket and carrier timing marks are aligned. If not, repeat the belt refitting and adjustment procedure, starting at paragraph 3.

10   If the engine is in the car, refit the timing belt upper and lower covers, followed by the alternator drivebelt. Adjust the drivebelt as described in Chapter 2.

11   Refit the access cover and roadwheel, lower the car to the ground and reconnect the battery.

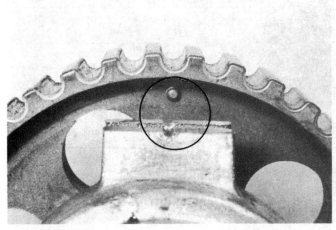

85.3 Camshaft timing marks aligned

86 Tappet clearances – checking and adjustment

1   To check the tappet clearances, disconnect the battery negative terminal then undo the retaining bolts and lift off the camshaft carrier covers. Note which bolts also secure cable and hose retaining clips.

2   Using a feeler gauge, check the clearance between the cam lobe and the tappet bucket of each valve (photo) in the order given in the following table and record each clearance. The engine may be turned using a spanner or socket on the crankshaft pulley bolt. If necessary remove the access panel from under the right-hand wheel arch to provide greater access to the pulley bolt.

Check No 1 tappet with No 8 valve fully open
Check No 3 tappet with No 6 valve fully open
Check No 5 tappet with No 4 valve fully open
Check No 2 tappet with No 7 valve fully open
Check No 8 tappet with No 1 valve fully open
Check No 6 tappet with No 3 valve fully open
Check No 4 tappet with No 5 valve fully open
Check No 7 tappet with No 2 valve fully open

85.5 Slip the timing belt over the sprockets and tensioner

86.2 Checking tappet clearance with a feeler blade

85.6 Using a torque wrench to adjust the timing belt tension

3   Once the readings have been tabulated for all valves it should be noted that, unless new parts have been fitted or the valve seats reground, adjustment of the valve tappet clearance to the standard setting is only necessary if the clearance of either inlet or exhaust is less than 0.012 in (0.30 mm).

4   If adjustment is necessary, refer to Section 58 and remove the camshaft and tappets.

5   Remove the adjusting shim from each maladjusted tappet bucket in turn and note its thickness. The shim reference number is stamped

on the face of the shim (photo) – see Specifications. By using the following calculation, determine the thickness of the new shim required to give the correct tappet clearance:

A = clearance measured in paragraph 2
B = thickness of existing shim
C = correct clearance
New shim thickness required = A + B − C

6   With new shims obtained as necessary, refer to Section 83, paragraphs 7 to 16, and refit the camshaft and tappets.

86.5 Tappet adjusting shim with reference number stamped on its face

87 Ancillary components – refitting

Refer to Section 54 and refit the listed components with reference to the Sections and Chapters specified, where applicable.

88 Engine – attachment to manual gearbox or automatic transmission

Refer to Section 52 and attach the gearbox or transmission to the engine using the reverse of the removal procedure. Tighten the retaining bolts to the specified torque.

89 Engine – refitting with manual gearbox or automatic transmission

Refer to Section 50 or 51, as applicable, and refit the engine and manual gearbox or automatic transmission using the reverse of the removal procedure, noting the following additional points:

(a)   Do not tighten any of the engine mountings fully until all have been fitted, then tighten them in this order: right-hand mounting, left-hand mounting, rear mounting, front snubber. Centralize the front snubber cap as described in Section 75, after tightening the other mountings

(b)   Align the marks on the driveshaft joints and drive flanges made during removal

(c)   On manual gearbox models, refit the clutch cable with reference to Chapter 5

(d)   On automatic transmission models, adjust the selector cable and kickdown cable if necessary, as described in Chapter 7

(e)   Adjust the accelerator cable to give a small amount of free play in the cable with the throttle closed

(f)   Refill the cooling system, as described in Chapter 2

(g)   Refill the engine with oil

90 Engine – adjustments after major overhaul

1   With the engine refitted to the car, make a final check to ensure that everything has been reconnected and that no rags or tools have been left in the engine compartment.

2   Make sure that the oil and water levels are topped up and then start the engine; this may take a little longer than usual as the fuel pump and carburettor float chamber may be empty.

3   As soon as the engine starts, watch for the oil pressure light to go out and check for any oil, fuel or water leaks. Don't be alarmed if there are some odd smells and smoke from parts getting hot and burning off oil deposits.

4   Run the engine for at least 15 minutes or drive the car for approximately 5 miles then switch it off and allow it to cool. Working in the sequence shown in Fig. 1.18, slacken each cylinder head bolt one at a time by half a turn and then tighten it to the specified torque.

5   If new pistons, rings or crankshaft bearings have been fitted the engine must be run-in for the first 500 miles (800 km). Do not exceed 45 mph (72 kph), operate the engine at full throttle or allow it to labour in any gear.

Fault diagnosis appears overleaf

PART C: FAULT DIAGNOSIS

91 Fault diagnosis – engine

Symptom	Reason(s)
Engine fails to start	Discharged battery Loose battery connection Loose or broken ignition leads Moisture on spark plugs, distributor cap, or HT leads Incorrect spark plug gaps Cracked distributor cap or rotor Other ignition system fault Dirt or water in carburettor Empty fuel tank Faulty fuel pump Other fuel system fault Faulty starter motor Low cylinder compressions
Engine idles erratically	Inlet manifold air leak Leaking cylinder head gasket Worn rocker arms, timing chain, timing belt or gears Worn camshaft lobes Faulty fuel pump Incorrect valve clearances Loose crankcase ventilation hoses Carburettor adjustment incorrect Uneven cylinder compressions
Engine misfires	Spark plugs worn or incorrectly gapped Dirt or water in carburettor Carburettor adjustment incorrect Burnt out valve Leaking cylinder head gasket Distributor cap cracked Incorrect valve clearances Uneven cylinder compressions Worn carburettor
Engine stalls	Carburettor adjustment incorrect Inlet manifold air leak Ignition timing incorrect (where applicable)
Excessive oil consumption	Worn pistons, cylinder bores or piston rings Valve guides and valve stem seals worn Major gasket or oil seal leakage
Engine backfires	Carburettor adjustment incorrect Ignition timing incorrect (where applicable) Incorrect valve clearances Inlet manifold air leak Sticking valve

Chapter 2 Cooling system

For modifications, and information applicable to later models, see Supplement at end of manual

Contents

Specifications

System type .... Pressurized, pump-assisted thermosyphon with front-mounted radiator and electric cooling fan

Expansion tank cap pressure .... 15 lbf/in² (1.0 bar)

Thermostat
Opening temperature .... 168° to 176°F (76° to 80°C)
Fully open .... 190°F (88°C)
Lift height .... 0.32 in (8.1 mm)

System capacity (including heater)
1.3 litre models .... 11 Imp pints (6.6 litres)
1.6 litre models .... 15 Imp pints (8.5 litres)

Antifreeze type/specification .... Ethylene glycol-based antifreeze with non-phosphate corrosion inhibitors, suitable for mixed-metal engines, containing no methanol and meeting specifications BS6580 and BS5117 (Duckhams Universal Antifreeze and Summer Coolant)

Antifreeze properties and quantities
33% antifreeze (by volume):

Commences freezing	−2°F (−19°C)	
Frozen solid	−33°F (−36°C)	
Quantities (system refill):	**Antifreeze**	**Water**
1.3 litre models	3.5 Imp pints (2.0 litres)	7.5 Imp pints (4.3 litres)
1.6 litre models	5.0 Imp pints (2.8 litres)	10.0 Imp pints (5.7 litres)

50% antifreeze (by volume):

Commences freezing	−33°F (−36°C)	
Frozen solid	−53°F (−48°C)	
Quantities (system refill):	**Antifreeze**	**Water**
1.3 litre models	5.75 Imp pints (3.3 litres)	5.75 Imp pints (3.3 litres)
1.6 litre models	7.5 Imp pints (4.25 litres)	7.5 Imp pints (4.25 litres)

Torque wrench settings

1.3 litre models

	lbf ft	Nm
Coolant temperature thermistor	40	54
Water outlet elbow bolts	16	22
Water pump bolts	16	22
Water pump pulley	8	11
Cylinder block drain plug	27	37

1.6 litre models

	lbf ft	Nm
Coolant temperature thermistor	11	15
Thermostat housing bolts	13	18
Water inlet elbow to cylinder head	18	24
Water outlet elbow to thermostat housing	18	24
Water pump to cylinder block	9	12
Timing belt tensioner bolts	18	24

1   General description

The cooling system is of the pressurised, pump-assisted thermo-syphon type. The system consists of the radiator, water pump, thermostat, electric cooling fan, expansion tank and associated hoses.

The system functions as follows. Cold coolant in the bottom of the radiator left-hand tank passes through the hose to the water pump on 1.3 litre models, or water inlet elbow on 1.6 litre models, where it is pumped around the cylinder block and head passages. After cooling the cylinder bores, combustion surfaces and valve seats the coolant reaches the underside of the thermostat, which is initially closed, and is diverted through the heater inlet hose to the heater. After passing through the heater the coolant travels through the water jacket of the inlet manifold before returning to the water pump inlet hose or water outlet elbow. When the engine is cold the thermostat remains closed and the coolant only circulates around the engine, heater and inlet manifold. When the coolant reaches a predetermined temperature the thermostat opens and the coolant passes through the top hose to the radiator right-hand tank. As the coolant circulates around the radiator it is cooled by the inrush of air when the car is in forward motion. Airflow is supplemented by the action of the electric cooling fan when necessary. Upon reaching the left-hand side of the radiator, the coolant is now cooled and the cycle is repeated.

When the engine is at normal operating temperature the coolant expands and some of it is displaced into the expansion tank. This coolant collects in the tank and is returned to the radiator when the system cools.

The electric cooling fan mounted in front of the radiator is controlled by a thermostatic switch located in the radiator side tank. At a predetermined coolant temperature the switch contacts close, thus actuating the fan.

On 1.3 litre models the water pump is driven by the crankshaft pulley in the conventional way via a drivebelt. On 1.6 litre models the pump is driven by the toothed camshaft timing belt.

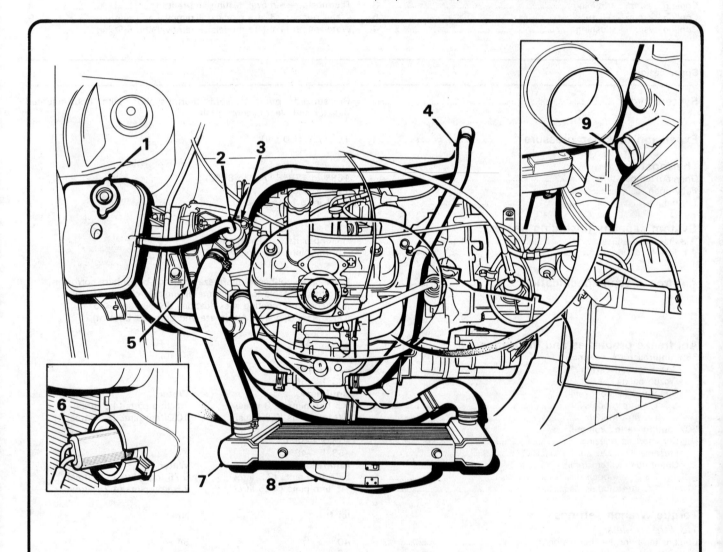

Fig. 2.1 Layout of the cooling system and components – 1.3 litre models (Sec 1)

1 Expansion tank filler cap	3 Thermostat housing	6 Cooling fan thermostatic switch	8 Cooling fan assembly
2 Water outlet elbow	4 Heater hoses	7 Radiator	9 Cylinder block drain plug
	5 Water pump		

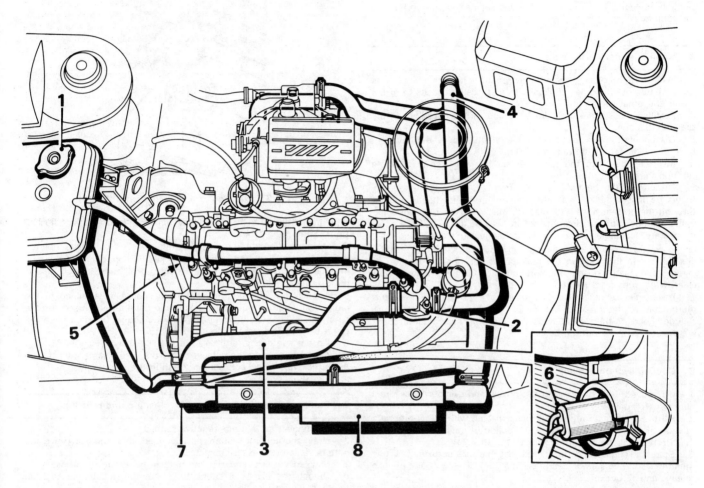

Fig. 2.2 Layout of the cooling system and components – 1.6 litre models (Sec 1)

1	Expansion tank filler cap	3	Radiator top hose	6	Cooling fan thermostatic switch	7	Radiator
2	Thermostat housing	4	Heater hoses			8	Cooling fan assembly
		5	Water pump				

2 Maintenance and inspection

1   Check the coolant level in the system weekly and, if necessary, top up with a water and antifreeze mixture until the level is up to the indicator in the expansion tank. With a sealed type cooling system , topping-up is required, it is likely there is a leak in the system. Check all hoses and joint faces for any staining, or actual wetness, and rectify if necessary. If no leaks can be found it is advisable to have the system pressure tested, as the leak could possibly be internal. It is a good idea to keep a check on the engine oil level as a serious internal leak can often cause the level in the sump to rise, thus confirming suspicions.
2   At the service intervals given in Routine Maintneance at the beginning of this manual carefully inspect all the hoses, hose clips and visible joint gaskets for cracks, corrosion, deterioration or leakage. Renew any hoses and clips that are suspect and also renew any gaskets, if necessary.
3   At the same service interval check the condition of the water pump and alternator drivebelt and renew it if there is any sign of cracking or fraying. Check and adjust the tension of the belt, as described in Section 13.
4   At the less frequent service intervals indicated drain, flush and refill the cooling system using fresh antifreeze, as described in Sections 3, 4, and 5 respectively.
5   Also, at this service interval, the drivebelt should be renewed, using the procedure described in Section 13.

3 Cooling system – draining

Note: *It is not possible to completely drain the engine on 1.6 litre models due to the absence of a cylinder block drain plug. To drain all coolant from the engine or to bring the coolant level to below that of the water inlet elbow, it will be necessary to remove the water pump.*
1   It is preferable to drain the cooling system when the engine is cold. *If the engine is hot the pressure in the cooling system must be released before attempting to drain the system.* Place the cloth over the pressure cap of the expansion tank and turn the cap anti-clockwise until it reaches its stop.
2   Wait until the pressure has escaped, then press the cap downwards and turn it further in an anti-clockwise direction. Release the downward pressure on the cap very slowly and, after making sure that all the pressure in the system has been relieved, remove the cap.
3   Place a suitable container beneath the left-hand side of the radiator. Slacken the hose clip and carefully ease the bottom hose off the radiator outlet. Allow the coolant to drain into the container. Be prepared for some spillage as the chassis member directly below the radiator outlet blocks the directional flow of the coolant as it flows from the outlet.
4   On 1.3 litre models, place a second container beneath the cylinder block drain plug located on the front left-hand side of the engine. Unscrew the plug and drain the remainder of the coolant into the container.

5   If the system is to be flushed after draining, proceed to the next Section otherwise refit the drain plug and secure the bottom hose to the radiator.

4   Cooling system – flushing

1   With time the cooling system may gradually lose its efficiency as the radiator core becomes choked with rust, scale deposits from the water and other sediment.
2   To flush the system, first drain the coolant, as described in the previous Section.
3   Disconnect the top hose at the outlet elbow and leave the bottom hose disconnected at the radiator outlet.
4   Insert a hose into the top hose and allow water to circulate through the radiator until it runs clear from the outlet.
5   Disconnect the heater inlet hose from the thermostat housing. Insert the hose and allow water to circulate through the heater and out through the bottom hose until clear. If, after a reasonable period the water still does not run clear, the radiator can be flushed with a good proprietary cleaning agent such as Holts Radflush or Holts Speedflush.
6   In severe cases of contamination the system should be reverse flushed. To do this remove the radiator, as described in Section 7, invert it and insert a hose in the bottom outlet. Continue flushing until clean water runs from the top hose outlet.
7   The engine should also be flushed. To do this remove the thermostat, as described in Section 8, and insert the hose into the cylinder head or thermostat housing. Flush the system until clean water runs from the bottom hose.

5   Cooling system – filling

1   If removed, refit the thermostat, hoses, and cylinder block drain plug (where applicable).
2   Fill the system through the expansion tank with the appropriate mixture of water and antifreeze (see Section 6) until the tank is half full. Do not refit the expansion tank cap at this stage.
3   Start the engine and run it at a fast idle for approximately one minute. During this time compress the top hose several times to release any air pockets in the system.
4   Stop the engine, top up the expansion tank to the indicated level then refit the filler cap.

6   Antifreeze mixture

1   The antifreeze should be renewed at regular intervals (see Routine Maintenance). This is necessary not only to maintain the antifreeze properties, but also to prevent corrosion which would otherwise occur as the corrosion inhibitors become progressively less effective.
2   Use an antifreeze, in the specified concentration, which conforms to the standard given in the Specifications at the beginning of this Chapter. Note that Austin Rover recommend the use of Unipart Superplus antifreeze, and state that no other antifreeze should be mixed with this type, such as when topping up is required. If another

Fig. 2.3 Expansion tank level indicator (Sec 5)

type of antifreeze is to be used, the system must be drained and flushed prior to its use.
3   Before adding antifreeze the cooling system should be completely drained and flushed, and all hoses checked for condition and security.
4   The quantity of antifreeze and levels of protection are indicated in the Specifications.
5   After filling with antifreeze, a label should be attached to the radiator stating the type and concentration of antifreeze used and the date installed. Any subsequent topping-up should be made with the same type and concentration of antifreeze.
6   Do not use engine antifreeze in the screen washer system, as it will cause damage to the vehicle paintwork. Use a screen wash such as Turtle Wax High Tech Screen Wash in the washer system.

7   Radiator – removal, inspection, cleaning and refitting

1   Disconnect the battery negative terminal.
2   Drain the cooling system, as described in Section 3. Leave the bottom radiator hose disconnected.
3   Slacken the retaining clip and detach the radiator top hose.
4   Detach the electrical connectors from the radiator cooling fan thermostatic switch (photo).
5   Release the upper retaining clips, tip the radiator grille forward at the top and lift it to release the lower mounting lugs from their locations.
6   Undo and remove the three bolts each side securing the front body panel in position. Release the air cleaner cold air intake hose from the panel then withdraw the panel and place it to one side, leaving the bonnet release cable still attached (photos).
7   Release the two plastic panel support clips from the left-hand side of the radiator (photo).
8   Disconnect the cooling fan motor wiring multi-plug (photo) and carefully lift out the radiator.
9   Radiator repair is best left to a specialist, but minor leaks may be sealed using a proprietary coolant additive such as Holts Radweld with the radiator *in situ*. Clean the radiator matrix of flies and small leaves with a soft brush, or by hosing.

7.4 Disconnect the wiring at the cooling fan thermostatic switch

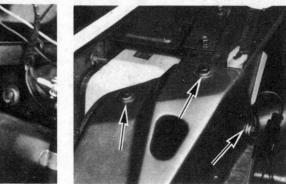

7.6A Front body panel retaining bolts (arrowed)

7.6B Air cleaner cold air intake hose

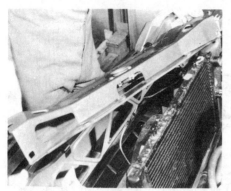
7.6C Front body panel removal

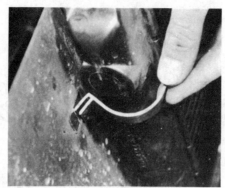
7.7 Release the two panel support clips

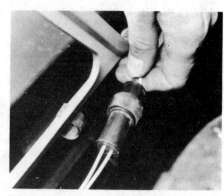
7.8 Cooling fan motor wiring multi-plug

10 Reverse flush the radiator, as described in Section 4. renew the top and bottom hoses and clips if they are damaged or have deteriorated.

11 Refitting the radiator is the reverse sequence to removal, but ensure that the lower mounting lugs engage in the rubber grommets (photo). Fill the cooling system, as described in Section 5, after fitting.

7.11 Ensure proper location of radiator mounting lugs in the rubber grommets

8 Thermostat – removal, testing and refitting

1 Remove the expansion tank filler cap. *If the engine is hot, place a cloth over the cap and turn it slowly anti-clockwise until the first stop is reached.* Wait until all the pressure has been released and then remove the cap completely.

2 Place a suitable container beneath the radiator bottom hose outlet. Disconnect the bottom hose and drain approximately 4 pints (2.3 litres) of the coolant. Reconnect the bottom hose and tighten the clip.

3 Slacken the clips and detach the radiator top hose, heater hose and expansion tank hose from the water outlet elbow.

4 Undo and remove the bolts securing the water outlet elbow to the cylinder head or thermostat housing (photo), and lift off the elbow. On 1.3 litre models it will be necessary to remove the upper retaining bolt and lift away the engine mounting support bracket to facilitate removal of the elbow. After removal, recover the gasket.

5 Withdraw the thermostat from its seat in the thermostat housing.

6 To test whether the unit is serviceable, suspend it on a string in a saucepan of cold water together with a thermometer. Heat the water and note the temperature at which the thermostat begins to open. Continue heating the water until the thermostat is fully open and then remove it from the water.

7 The temperature at which the thermostat should start to open is given in the Specifications and the fully open temperature is stamped on the base of the unit. If the thermostat does not start to open or fully open at the specified temperatures, does not attain the specified lift height or does not fully close when cold, then it must be discarded and a new unit fitted. Under no circumstances should the car be used without a thermostat, as uneven cooling of the cylinder walls and head passages will occur, causing distortion and possible seizure of the engine internal components.

8 Refitting the thermostat is the reverse sequence to removal. Ensure that all traces of old gasket are removed from the mating faces of the water outlet elbow and thermostat housing. Use a new gasket lightly smeared with jointing compound and tighten the retaining bolts to the specified torque. On completion top up the cooling system, with reference to Section 5.

8.4 Water outlet elbow removal (1.6 litre model shown)

9 Water pump (1.3 litre models) – removal and refitting

Note: *Water pump failure is indicated by coolant leaking from the gland at the front of the pump or by rough and noisy operation. This is usually accompanied by excessive play of the pump spindle which can be checked by moving the pulley from side to side. Repair or*

overhaul of a faulty pump is not possible, as internal parts are not available separately. In the event of failure a replacement pump must be obtained.

1   Disconnect the battery negative terminal.
2   Refer to Section 3 and drain the cooling system.
3   Slacken the alternator adjusting arm and pivot bolts, move the alternator towards the engine and slip the drivebelt off the three pulleys.
4   Undo and remove the four water pump pulley retaining bolts.
5   Slacken the retaining clip and disconnect the water hose from the pump.
6   Undo and remove the water pump retaining bolts then tap the pump body with a soft-faced mallet to release it from the locating dowels. Manoeuvre the pump and pulley out from between the cylinder block and engine mounting bracket.
7   Refitting is the reverse sequence to removal, bearing in mind the following points:

(a)   *Remove all traces of old gasket from the cylinder block and pump faces, and ensure that both mating surfaces are clean and dry*
(b)   *Use a new gasket lightly smeared with jointing compound*
(c)   *Adjust the drivebelt tension, as described in Section 13, and refill the cooling system, as described in Section 5*

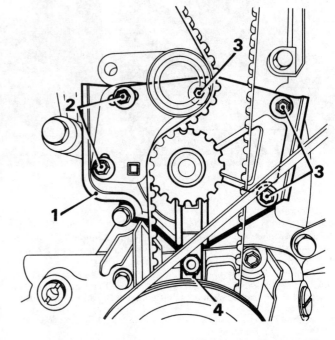

Fig. 2.5 Water pump removal – 1.6 litre models (Sec 10)

1   Water pump
2   Timing belt tensioner retaining bolts
3   Water pump retaining bolts
4   Clamp plate

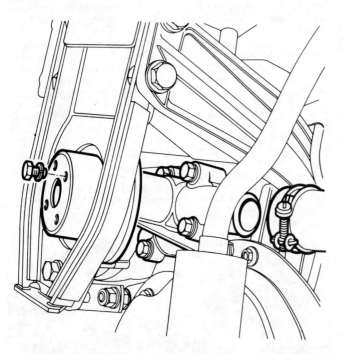

Fig. 2.4 Water pump removal – 1.3 litre models (Sec 9)

10 Water pump (1.6 litre models) – removal and refitting

1   Refer to the introductory note at the beginning of Section 9 before proceeding.
2   Disconnect the battery negative terminal.
3   Refer to Section 3 and drain the cooling system.
4   Refer to Chapter 1 and remove the timing belt.
5   Undo the two bolts and lift off the timing belt tensioner (photo).
6   Place a suitable container beneath the water pump to collect the coolant remaining in the cylinder block.
7   Using an Allen key, undo the water pump upper retaining bolt then undo the remaining three bolts using a spanner or socket (photo). Recover the clamp plate on the lower bolt.
8   Ease the water pump off the engine using a screwdriver to lever between the block and pump side flanges if necessary.

10.5 Timing belt tensioner retaining bolts (arrowed)

10.7 Water pump retaining bolts (arrowed)

10.9 Apply RTV sealant to the pump mating face before fitting

9   Refitting the pump is the reverse sequence to removal, bearing in mind the following points:

(a) *Remove all traces of old sealant from the cylinder block and pump faces, and ensure that both mating surfaces are clean and dry*

(b) *Apply a bead of RTV sealant around the pump mating face (photo) and, with the pump in position, tighten the retaining bolts to the specified torque.*

(c) *Refit and adjust the timing belt, as described in Chapter 1*

(d) *Refill the cooling system, as described in Section 5.*

11   Cooling fan assembly – removal and refitting

1   Disconnect the battery negative terminal.
2   Release the upper retaining clips, tip the radiator grille forward at the top and lift it to release the lower mounting lugs from their locations.
3   Undo and remove the three bolts each side securing the front body panel in position. Release the air cleaner cold air intake hose from the panel then lift the panel upwards and place it to one side, leaving the bonnet release cable still attached (photos 7.6a, 7.6b and 7.6c).
4   Disconnect the fan motor wiring multi-plug.
5   Undo the three nuts securing the fan cowl to the radiator and withdraw the assembly from its location (photo).
6   If required the fan may be removed by pulling it off the motor shaft. Mark the outer face to ensure correct refitment. To remove the fan motor, drill off the rivet heads, tap out the rivets securing the motor to the cowl and lift away the motor.
7   Refitting is the reverse sequence to removal.

11.5 Fan cowl-to-radiator lower retaining nuts (arrowed)

12   Cooling fan thermostatic switch – testing, removal and refitting

1   The radiator cooling fan is operated by a thermostatic switch located on the right-hand side of the radiator. When the coolant exceeds a predetermined temperature the switch contacts close and the fan is activated.
2   If the operation of the fan or switch is suspect, run the engine until normal operating temperature is reached and then allow it to idle. If the fan does not cut in within a few minutes, switch off the engine and disconnect the wiring plug from the thermostic switch. Bridge the two terminals in the wiring plug with a length of wire and switch on the ignition. If the fan now operates, the thermostatic switch is faulty and must be renewed. If the fan still fails to operate check that battery voltage is present at the plug terminals. If not, check for a blown fuse or wiring fault. If voltage is present the fan motor is faulty.
3   To remove the switch, disconnect the battery negative terminal then drain the cooling system, as described in Section 3.

4   Disconnect the wiring plug, release the retaining plate and withdraw the switch and seal from the radiator (photo).
5   Refitting is the reverse sequence to removal, but use a new seal on the switch and refill the cooling system, as described in Section 5.

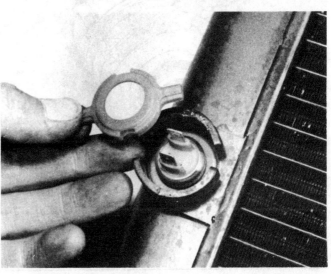

12.4 Cooling fan thermostatic switch removal

13   Drivebelt – renewal and adjustment

Note: *On 1.3 litre models the drivebelt runs in the crankshaft, water pump and alternator pulleys. On 1.6 litre models the belt only runs in the crankshaft and alternator pulleys as the water pump is driven by the toothed timing belt. However, the renewal and adjustment procedures on all models are similar.*
1   To remove the drivebelt, slacken the alternator pivot mounting bolt and the adjusting arm bolts then move the alternator towards the engine. Slip the drivebelt off the pulleys and remove it from the engine.
2   Fit the new drivebelt over the pulleys then lever the alternator away from the engine until moderate tension is obtained. The alternator must only be levered at the drive end bracket.
3   Hold the alternator in this position and tighten the adjusting arm bolts followed by the pivot mounting bolt.
4   Start the engine and run it for five minutes at a fast idle.
5   Switch off the engine and readjust the drivebelt tension so that the belt deflects by 0.25 to 0.50 in (6.0 to 12.0 mm) at the point shown in Fig. 2.6 or 2.7 under moderate finger pressure.

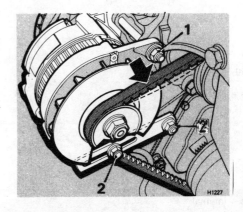

Fig. 2.6 Drivebelt adjustment points – 1.3 litre models (Sec 13)

Arrow indicates tension checking point

1   *Alternator pivot mounting bolt*

2   *Adjusting arm bolts*

Fig. 2.7 Drivebelt adjustment points – 1.6 litre models (Sec 13)

Arrow indicates tension checking point

1   Alternator pivot   2   Adjusting arm bolts
     mounting bolt

---

14 Coolant temperature thermistor – removal and refitting

1   The coolant temperature thermistor incorporates a temperature-sensitive element, the resistance of which alters according to coolant temperature. The unit controls the operation of the temperature gauge and also influences the settings of the fuel and ignition system electronic control units.

2   Place a suitable container beneath the radiator bottom hose outlet. Slacken the retaining clip, disconnect the bottom hose and drain approximately 4 pints (2.3 litres) of coolant. Refit the hose and tighten the clip.

3   Disconnect the wiring plug from the thermistor which is located at the rear right-hand side of the cylinder head on 1.3 litre models and in the thermostat housing on 1.6 litre models (photo).

4   Unscrew the thermistor and remove it from the engine.

5   Refitting is the reverse sequence to removal, but top up the cooling system with reference to Section 5.

14.3 Coolant temperature thermistor location (1.6 litre model shown)

---

15 Fault diagnosis – cooling system

Symptom	Reason(s)
Overheating	Low coolant level (this may be the result of overheating for other reasons)
	Drivebelt slipping or broken (1.3 litre models only)
	Radiator blockage (internal or external), or grille restricted
	Thermostat defective
	Ignition timing incorrect or distributor defective – automatic advance inoperative (1.3 litre models only)
	Carburettor maladjustment
	Faulty cooling fan thermostatic switch
	Blown cylinder head gasket (combustion gases in coolant)
	Water pump defective
	Expansion tank pressure cap faulty
	Brakes binding
Overcooling	Thermostat missing, defective or wrong heat range
Water loss – external	Loose hose clips
	Perished or cracked hoses
	Radiator core leaking
	Heater matrix leaking
	Expansion tank pressure cap leaking
	Boiling due to overheating
	Water pump or thermostat housing leaking
	Core plug leaking
Water loss – internal	Cylinder head gasket blown
	Cylinder head cracked or warped
	Cylinder block cracked
Corrosion	Infrequent draining and flushing
	Incorrect antifreeze mixture or inappropriate type
	Combustion gases contaminating coolant

Chapter 3 Fuel and exhaust systems

For modifications, and information applicable to later models, see Supplement at end of manual

Contents

Specifications

Air cleaner

Type	Automatic air temperature control type, with renewable paper element
Application:	
1.3 litre engine	Champion W135
1.6 litre engine	Champion W114

Fuel pump

Type	Mechanical, operated by eccentric on camshaft
Make	SU AUF 800
Delivery pressure	4.0 lbf/in^2 (0.3 bar)

Carburettor (general)

Type	SU HIF 44 variable choke with electronic mixture control
Identification:	
1.3 litre models	FZX 1422 or 1468
1.6 litre models:	
Manual gearbox	FZX 1424, 1463 or 1476
Automatic transmission	FZX 1425
Piston damper	LZX 1511
Piston spring colour	Red
Jet size	0.100 in (2.54 mm)
Needle identification:	
1.3 litre models	BCZ
1.6 litre models	BFM
Piston damper oil type/specification	Multigrade engine oil, viscosity SAE 10W/40 (Duckhams QXR, Hypergrade, or 10W/40 Motor Oil)

Adjustment data

Fast idle rod minimum clearance	0.005 in (0.13 mm)
Throttle lever lost motion gap	0.060 to 0.080 in (1.5 to 2.0 mm)
Float height (see text)	0.040 to 0.060 in (1.0 to 1.5 mm)
Idling speed:	
1.3 litre models	700 to 800 rpm
1.6 litre models:	
Manual gearbox	700 to 800 rpm
Automatic transmission	775 to 925 rpm
Fast idle speed:	
1.3 litre models	1150 to 1225 rpm
1.6 litre models:	
Manual gearbox	1150 to 1250 rpm
Automatic transmission	1300 to 1400 rpm
Exhaust gas CO content:	
1.3 litre models	1.5 to 3.5%
1.6 litre models	2.0 to 3.5%

Fuel tank capacity	11.25 Imp gals (51 litres)		
Fuel octane rating	see Supplement		
Torque wrench settings	lbf ft		Nm
Carburettor retaining nuts	16		22
Fuel pump retaining nuts or bolts:			
1.3 litre models	16		22
1.6 litre models	11		15
Manifold retaining nuts and bolts	21		28

1 General description

The fuel system consists of a centrally-mounted fuel tank, camshaft-operated mechanical fuel pump and an SU variable choke carburettor with electronic mixture control system.

The air cleaner contains a disposable paper filter element and incorporates an automatic air temperature control system. The system is controlled by a flap valve located at the junction of the air cleaner hot and cold air intakes. The flap is operated by inlet manifold vacuum acting on a thermac unit in conjunction with a temperature-sensitive thermac switch. The system allows hot or cold air to be delivered to the carburettor, depending on the position of the flap valve which varies according to engine temperature and load.

The SU variable choke carburettor incorporates a computer-operated electronic mixture control system working in conjunction with the electronic ignition system to provide a sophisticated engine management package. Further details of this system will be found in Section 11.

The exhaust system is in three sections; the front section incorporates the twin downpipes and is fitted with ball and socket type flexible joints, the intermediate section incorporates the single or twin front silencers and the rear section incorporates the rear silencer and tailpipe. The system is supported on rubber mountings at the centre and rear and is bolted to the exhaust manifold at the front.

The inlet and exhaust manifolds are of cast aluminium and cast iron construction respectively. The inlet manifold is heated by engine coolant to improve fuel vaporisation and additionally, on 1.6 litre models, an electrically-operated manifold heater is used. This unit warms the manifold during initial start-up conditions to improve cold start driveability. The unit is switched off by a thermostatic switch when a predetermined temperature is reached.

Warning: *Many of the procedures in this Chapter entail the removal of fuel pipes and connections which may result in some fuel spillage. Before carrying out any operation on the fuel system refer to the precautions given in Safety First! at the beginning of this manual and follow them implicity. Petrol is a highly dangerous and volatile liquid and the precautions necessary when handling it cannot be overstressed.*

2 Maintenance and inspection

1   At the service intervals given in Routine Maintenance at the beginning of this manual the following checks and adjustments should be carried out on the fuel and exhaust system components.
2   With the car over a pit, raised on a vehicle lift or securely supported on axle stands, carefully inspect the fuel pipes, hoses and unions for chafing, leaks and corrosion. Renew any pipes that are severely pitted with corrosion or in any way damaged. Renew any hoses that show signs of cracking or other deterioration.
3   Examine the fuel tank for leaks, particularly around the fuel gauge sender unit, and for signs of corrosion or damage.
4   Check the condition of the exhaust system, described in Section 17.
5   From within the engine compartment, check the security of all fuel hose attachments and inspect the fuel hoses and vacuum hoses for kinks, chafing or deterioration.
6   Renew the air cleaner element and check the operation of the air cleaner automatic temperature control, described in Sections 3 and 4.
7   Check the operation of the accelerator linkage and lubricate the linkage, cable and pedal pivot with a few drops of engine oil.
8   Top up the carburettor piston damper to the top of the hollow

piston rod with engine oil. Note that although this operation is only considered necessary by the manufacturers at the annual major service interval, practical experience has shown that the damper oil may require topping-up more frequently.
9   Check and, if necessary, adjust the carburettor idle, fast idle and mixture settings, as described in Section 12.

3 Air cleaner and element (1.3 litre models) – removal and refitting

1   To remove the air cleaner element unscrew the three large screws and one small screw securing the air cleaner top cover and body. Withdraw the screws, release the cover from the locating tags and withdraw it from the air cleaner body.
2   The paper element can now be removed.
3   If the air cleaner body is to be removed, detach the cold air intake hose from its location in the front body panel.

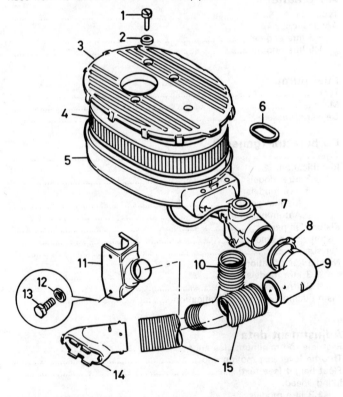

Fig. 3.1 Air cleaner assembly and related components – 1.3 litre models (Sec 3)

1	*Retaining screw*	*9*	*Intake elbow*
2	*Washer*	*10*	*Hot air hose*
3	*Top cover*	*11*	*Hot air box*
4	*Paper element*	*12*	*Washer*
5	*Air cleaner body*	*13*	*Bolt*
6	*Sealing washer*	*14*	*Cold air intake*
7	*Thermac unit*		*hose adaptor*
8	*Retaining clip*	*15*	*Cold air intake hose*

4   Lift up the air cleaner and detach the vacuum hose from the connector in the main vacuum line adjacent to the carburettor. Detach the air cleaner hose from the hot air box on the exhaust manifold and remove the air cleaner from the engine.
5   Thoroughly clean the inside of the air cleaner and check the vacuum pipes and thermac switch for condition and security.
6   To test the operation of the air temperature control system, first slacken the retaining clip and remove the cold air intake hose at the elbow.
7   Observe the position of the air temperature control flap which should be set to receive air from the cold air intake. Apply suction to the air cleaner vacuum pipe that was previously disconnected, and ensure that the flap moves to the hot air delivery position. The flap should move to the cold air delivery position when the ambient temperature reaches 85°F (30°C). This can be tested by heating the thermac switch in the air cleaner body with a hair dryer while applying suction to the hose. If the operation of the unit is in doubt it should be renewed.
8   Refitting the air cleaner body and element is the reverse sequence to removal. Ensure that the rubber sealing washer is in position around the lip of the carburettor intake flange on the air cleaner body. Also make sure that the element seats correctly in the body before fitting the cover.

4   Air cleaner and element (1.6 litre models) – removal and refitting

1   To remove the air cleaner element, spring back the retaining clips, lift off the top cover and withdraw the element (photo).
2   If the air cleaner body is to be removed, release the cold air intake hose from its attachments on the front body panel and thermac unit then remove the hose (photo).
3   Undo the three bolts securing the air cleaner mounting bracket to the engine.
4   Detach the air cleaner body from the plenum chamber and hot air ducts, disconnect the vacuum hose from the thermac unit (photo) and remove the air cleaner assembly from the engine.
5   To remove the plenum chamber, disconnect the vacuum hose from the thermac switch at the T-piece (photo).
6   Undo the two screws and one nut securing the plenum chamber and support bracket (photo). Withdraw the bracket then remove the plenum chamber from the carburettor (photo).
7   Thoroughly clean the inside of the air cleaner body and check the vacuum pipes and thermac unit for condition and security.
8   To test the operation of the air temperature control system reconnect the vacuum hose from thermac unit to thermac switch, but leave the other vacuum hose from the thermac switch to the T-piece disconnected.

4.2 Cold air intake hose front body panel attachment

4.4 Vacuum hose attachment at air cleaner thermac unit

4.1 Air cleaner element renewal

4.5 Vacuum hose attachment at vacuum line T-piece

4.6A Plenum chamber retaining screws (A) and nut (B)

4.6B Removing the plenum chamber

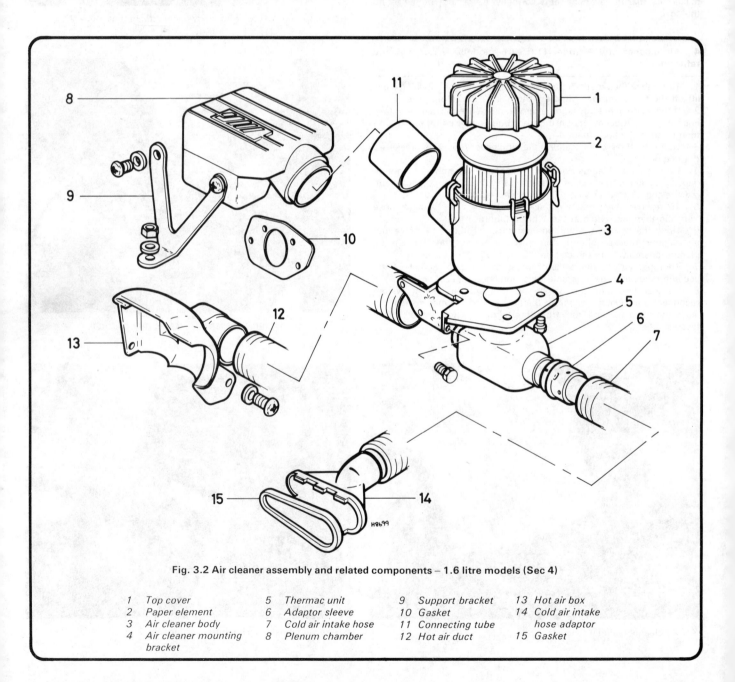

Fig. 3.2 Air cleaner assembly and related components – 1.6 litre models (Sec 4)

1 Top cover	5 Thermac unit	9 Support bracket	13 Hot air box
2 Paper element	6 Adaptor sleeve	10 Gasket	14 Cold air intake
3 Air cleaner body	7 Cold air intake hose	11 Connecting tube	hose adaptor
4 Air cleaner mounting	8 Plenum chamber	12 Hot air duct	15 Gasket
bracket			

9   Observe the position of the air temperature control flap in the thermac unit at the base of the air cleaner body. The flap should be set to receive air from the cold air intake. Apply suction to the disconnected vacuum hose and check that the flap moves to the hot air delivery position. The flap should return to the cold air delivery position when the ambient temperature reaches 85°F (30°C). This can be tested by heating the thermac switch in the plenum chamber with a hair dryer while applying suction to the hose. If the operation of the unit is in doubt, the thermac switch should be renewed.
10 Refitting the plenum chamber and air cleaner is the reverse sequence to removal. Renew the gasket between the plenum chamber and carburettor if the old one shows any sign of deterioration.

5   Fuel pump – removal, testing and refitting

Note: *Refer to the warning note in Section 1 before proceeding.*
1   The mechanical fuel pump is located in the front lower right-hand side of the cylinder block on 1.3 litre engines, and at the rear of the camshaft cover on 1.6 litre units. The pump is of sealed construction and in the event of faulty operation must be renewed as a complete unit.
2   To remove the pump first disconnect the battery negative terminal.
3   Note the location of the fuel inlet pipe and fuel outlet pipe and then remove the pipes from the pump (photo). Plug the pipe ends with a metal rod or old bolt after removal.

5.3 Removing the fuel inlet pipe from the pump nozzle

4   Undo and remove the two bolts and washers and withdraw the pump and insulating block from the engine.
5   To test the pump operation, refit the fuel inlet pipe to the pump inlet and hold a wad of rag near the outlet. Operate the pump lever by hand and if the pump is in a satisfactory condition a strong jet of fuel should be ejected from the pump outlet as the lever is released. If this is not the case, check that fuel will flow from the inlet pipe when it is held below the tank level, if so the pump is faulty.
6   Before refitting the pump clean all traces of old gasket from the pump flange, insulating block and engine face. Use a new gasket on each side of the insulator block and refit the pump, using the reverse sequence to removal. Ensure that the pump lever locates over the top of the camshaft eccentric as the pump is installed. If this proves difficult, turn the engine over until the large offset of the eccentric is facing downward.

6   Fuel tank – removal, servicing and refitting

Note: *Refer to the warning note in Section 1 before proceeding.*
1   Disconnect the battery negative terminal. Remove the fuel tank filler cap.

2   A drain plug is not provided and it will therefore be necessary to syphon, or hand pump, all the fuel from the tank before removal.
3   Having emptied the tank jack up the rear of the car and support it securely on axle stands.
4   Disconnect the electrical leads from the fuel gauge sender unit and release the clips securing the leads to the tank (photo).
5   Using pliers, release the fuel feed hose retaining clip and disconnect the hose from the outlet on the sender unit.
6   Unscrew the retaining clip securing the fuel filler hose to the tank and disconnect the hose.
7   Support the tank on blocks, or with a jack, and undo the two rear and two front retaining bolts and stiffener plates (photo).

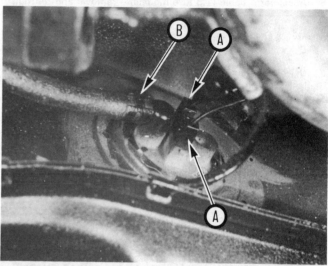

6.4 Electrical leads (A) and fuel feed hose (B) at the fuel gauge sender unit

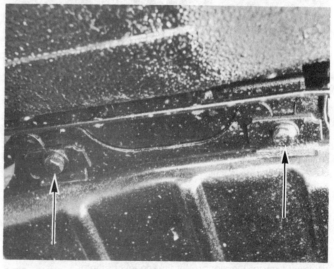

6.7 Fuel tank front retaining bolts and stiffener plates (arrowed)

8   Move the tank to the left to clear the brake pipe and lower it sufficiently to allow the breather hose to be detached. Now lower the tank to the ground and withdraw it from under the car.
9   If the tank is contaminated with sediment or water, remove the sender unit, as described in Section 7, and swill the tank out with clean fuel. If the tank is damaged, or leaks, it should be repaired by a specialist or, alternatively, renewed. **Do not** *under any circumstances solder or weld the tank.*
10 Refitting the tank is the reverse sequence to removal.

7 Fuel gauge sender unit – removal and refitting

Note: *Refer to the warning note in Section 1 before proceeding.*
1   Follow the procedure given in Section 6, paragraphs 1 to 5 inclusive.
2   Engage a screwdriver, flat bar or other suitable tool with the lugs of the locking ring, and turn the ring anti-clockwise to release it.
3   Withdraw the locking ring, seal and sender unit.
4   Refitting is the reverse sequence to removal, but always use a new seal.

8 Accelerator cable (1.3 litre models) – removal and refitting

1   Disconnect the battery negative terminal then remove the air cleaner assembly, as described in Section 3.
2   Slacken the screw to release the inner cable from the connector on the throttle linkage (photo).
3   Release the outer cable from the support bracket and withdraw the inner cable from the connector.
4   Working inside the car, prise the retaining clip from the top of the accelerator pedal and disconnect the inner cable.
5   Release the cable from the engine compartment bulkhead and remove the cable from the car.
6   Refitting is the reverse sequence to removal but, before tightening the throttle linkage connector, adjust the position of the inner cable to provide a small amount of free play. Check that, with the accelerator pedal fully depressed, the throttle linkage is fully open and, with the pedal released, the linkage is in the closed position.

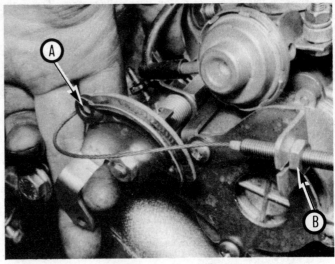

9.1 Carburettor linkage lever slot (A) and cable locknuts (B) – 1.6 litre models

8.2 Accelerator inner cable linkage connector (A) and support bracket fitting (B) – 1.3 litre models

9 Accelerator cable (1.6 litre models) – removal and refitting

1   Open the throttle linkage on the carburettor fully and slip the inner cable end out of the slot on the linkage lever (photo).
2   Slacken the two locknuts securing the cable to the support bracket on the side of the carburettor. Unscrew the nut nearest the end of the cable fully, then slip the cable out of the slot on the bracket.
3   Working inside the car, prise the retaining clip from the top of the accelerator pedal and disconnect the inner cable.
4   Release the accelerator cable from the engine compartment bulkhead and withdraw the complete cable from the car.
5   Refitting is the reverse sequence to removal, but adjust the outer cable lock nuts to provide a small amount of free play of the cable. Check that, with the accelerator pedal fully depressed, the throttle linkage is fully open and, with the pedal released, the linkage is in the closed position.
6   On models equipped with automatic transmission, check the kickdown cable adjustment, as described in Chapter 7.

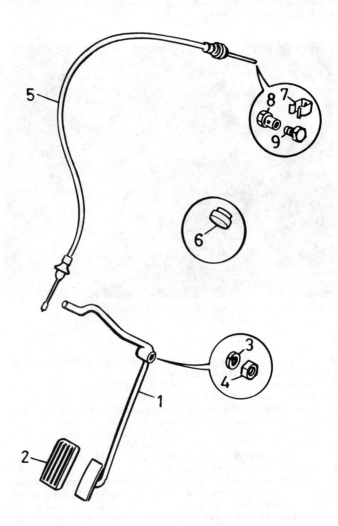

**Fig. 3.3 Accelerator cable and pedal components
(Secs 8, 9 and 10)**

1	Accelerator pedal	6	Pedal stop
2	Pedal pad	7	Cable retaining clip
3	Washer	8	Cable connector
4	Retaining nut	9	Connector screw
5	Accelerator cable		

10 Accelerator pedal – removal and refitting

1 Disconnect the battery negative terminal.
2 Prise the retaining clip from the top of the accelerator pedal and disconnect the accelerator cable from the pedal arm.
3 Unhook the pedal return spring, undo the retaining nut and washer and slide the pedal off the shaft.
4 Refitting is the reverse sequence to removal. After fitting, check the accelerator cable adjustment, as described in Section 8 or 9, as applicable.

11 SU carburettor – description and operation

The SU HIF (Horizontal Integral Float chamber) carburettor is of the variable choke, constant depression type incorporating a sliding piston which automatically controls the mixture of air and fuel supplied to the engine with respect to the throttle valve position and engine speed. In addition the carburettor is equipped with an electronically-operated mixture control device. This alters the mixture strength and engine speed when starting and during slow running, and also controls the operation of a fuel shut-off valve when decelerating or descending a hill.

The carburettor functions as follows. When the engine is started and is allowed to idle, the throttle valve passes a small amount of air. Because the piston is in a low position it offers a larger restriction and the resultant pressure reduction draws fuel from the jet, and atomisation occurs to provide a combustible mixture. Since the inside section of the tapered needle is across the mouth of the jet, a relatively small amount of fuel is passed.

When the throttle valve is opened, the amount of air passing through the carburettor is increased, which causes a greater depression beneath the sliding piston. An internal passageway connects this depression with the suction chamber above the piston, which now rises. The piston offers less of a restriction where the forces of depression, gravity, and spring tension balance out. The tapered needle has now been raised, and more fuel passes from the jet.

Incorporated in the jet adjusting (mixture) screw mechanism is a

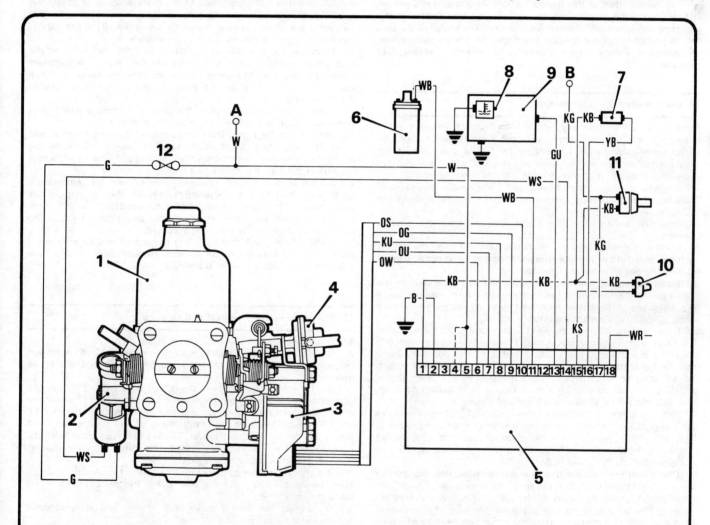

Fig. 3.4 Carburettor and electronic mixture control layout (Sec 11)

1 Carburettor
2 Fuel shut-off valve
3 Mixture control stepping motor
4 Vacuum switch
5 Electronic control unit - ECU

6 Ignition coil
7 Ambient air temperature sensor
8 High engine temperature warning lamp
9 Engine temperature gauge
10 Accelerator pedal switch

11 Coolant temperature thermistor
12 Fuse C5
A From ignition switch
B To ignition ECU
  (1.6 litre models only)

For colour code details refer to the wiring diagrams at the end of this manual bi-metal strip which alters the position of the jet to compensate for varying fuel densities resulting from varying fuel temperatures.

Fuel enrichment for cold starting is by an internal valve which admits more fuel into the airstream passing through the carburettor. This valve is operated by a stepping motor which also controls the engine idling speed. An electronic control unit (ECU) which is a small microprocessor receives inputs from the coolant temperature sensor, ambient air temperature sensor, accelerator pedal switch and ignition coil and adjusts the engine idle speed and mixture accordingly. The ECU also controls the operation of a fuel shut-off valve which comes into operation when decelerating or descending a hill. If, during these conditions, the engine speed is in excess of 1300 rpm, the ambient air temperature and engine temperature are above a predetermined value, and the accelerator pedal switch is closed (pedal released) the valve will be opened and closed at half second intervals. This introduces a partial vacuum to the top of the float chamber thus weakening the mixture. The fuel shut-off circuit is deactivated if the engine speed suddenly drops, ie when declutching. When accelerating, a vacuum-operated switch acts upon the mixture control, allowing more fuel to be drawn through, resulting in the necessary richer mixture.

The overall effect of this type of carburettor is that it will remain in tune during the lengthy service intervals and also under varying operating conditions and temperature changes. The design of the unit and its related systems ensures a fine degree of mixture control over the complete throttle range, coupled with enhanced engine fuel economy.

12 SU carburettor – adjustments

Note: *Before carrying out any carburettor adjustment, ensure that the spark plug gaps, valve clearances and, on 1.3 litre models, the ignition timing are all correctly set. To carry out the following adjustments an accurate tachometer will be required. The use of an exhaust gas analyser (CO meter) is also preferable, although not essential.*

1   Begin by removing the air cleaner on 1.3 litre models, as described in Section 3, or the plenum chamber on 1.6 litre models, as described in Section 4.

2   Unscrew and remove the piston damper from the suction chamber.

3   Undo and remove the three securing screws and lift off the suction chamber, complete with piston and piston spring. After removal avoid rotating the piston in the suction chamber.

4   Invert the suction chamber assembly and drain the oil from the hollow piston rod. Check that the needle guide is flush with the piston face and is secure. Do not be concerned that the needle appears loose in the guide, this is perfectly normal.

5   Observe the position of the jet in relation to the jet guide located in the centre of the carburettor venturi. The jet will probably be slightly below the top face of the jet guide. Turn the mixture adjusting screw until the top of the jet is flush with the top of the jet guide. Now turn the adjusting screw two complete turns clockwise. If the mixture adjusting screw is covered by a small blue or red tamperproof plug, hook this out with a small screwdriver and discard it.

6   Refit the piston and suction chamber assembly, taking care not to turn the piston in the suction chamber any more than is necessary to align the piston groove with its guide. If the piston is turned excessively the spring will be wound up and the assembly will have to be dismantled, as described in Section 14.

7   Check that the piston is free to move in the suction chamber by lifting it and allowing it to drop under its own weight. A definite metallic click should be heard as the piston falls and contacts the bridge in the carburettor body. If this is not the case, dismantle and clean the suction chamber, as described in Section 14. If satisfactory, top up the damper oil with engine oil to the top of the hollow piston rod and refit the damper.

8   Check that the throttle linkage operates smoothly and that there is a small amount of free play in the accelerator cable.

9   Connect a tachometer to the engine in accordance with the manufacturer's instructions, and also a CO meter if this is to be used.

10   Reconnect the vacuum hoses to the air cleaner or plenum chamber and lay the unit alongside the carburettor.

11   Start the engine and run it at a fast idle speed until it reaches its normal operating temperature. Continue to run the engine for a further five minutes before commencing adjustment.

12   Increase the engine speed to 2500 rpm for 30 seconds and repeat this at three minute intervals during the adjustment procedure. This will ensure that any excess fuel is cleared from the inlet manifold.

13   Disconnect the coolant thermistor wiring plug (see Chapter 2 if necessary) and join the two plug terminals together using a suitable length of wire. This will ensure that the mixture control stepping motor is not actuated during adjustment.

14   If the cooling fan is running, wait until it stops then turn the idle speed adjustment screw as necessary until the engine is idling at the specified speed.

15   Switch off the engine.

16   Check the clearance between the fast idle pushrod and fast idle adjustment screw using feeler gauges (position B in Fig. 3.5). Turn the fast idle adjustment screw as necessary to obtain the specified clearance.

17   Check the throttle lever lost motion gap using feeler gauges (position A in Fig. 3.5) and, if necessary, turn the throttle lever adjustment screw to obtain the specified clearance.

18   Start the engine and slowly turn the mixture adjustment screw clockwise (to enrich) or anti-clockwise (to weaken) until the fastest idling speed which is consistent with smooth even running is obtained. If a CO meter is being used, adjust the mixture screw to obtain the specified idling exhaust gas CO content.

19   Reset the idling speed, if necessary, using the idle speed adjustment screw then switch off the engine once more.

20   To adjust the fast idle speed, disconnect the two wires at the ambient air temperature sensor located in the engine compartment behind the left-hand headlamp (photo). Remove the sensor and join the two wires together using a male-to-male connector or suitable length of wire.

21   Remove the wire connecting the coolant thermistor wiring plug terminals together, but leave the plug disconnected.

22   Start the engine again. The mixture control stepping motor should move the fast idle pushrod to the fast idle position. Compare the engine fast idle speed with the specified setting and if necessary adjust by turning the fast idle adjustment screw as required.

23   Switch off the engine, reconnect the ambient air temperature sensor and coolant thermistor wiring plug. **Note**: after carrying out this adjustment, ensure that the specified minimum clearance still exists between pushrod and screw, as described in paragraph 16. Adjust the screw if the clearance is less than specified.

24   Make a final check that the idling speed and mixture are correct after refitting the air cleaner or plenum chamber then switch off the engine and disconnect the instruments.

13 SU carburettor – removal and refitting

1   Disconnect the battery negative terminal.

2   On 1.3 litre models remove the air cleaner assembly, as described in Section 3. On 1.6 litre models remove the plenum chamber, as described in Section 4.

3   On 1.3 litre models slacken the retaining screw and release the accelerator inner cable from the throttle lever connector. Release the outer cable from the support bracket and place the cable to one side. On 1.6 litre models open the throttle by hand and slip the cable end out of the slot on the throttle lever. Slacken the outer cable locknuts, unscrew the inner locknut and release the cable from the support bracket. Place the cable to one side.

4   If automatic transmission is fitted, disconnect the kickdown cable from the throttle lever, slip the cable out of the support bracket and place it to one side.

5   Disconnect the wiring connectors at the fuel shut-off valve solenoid and stepping motor (photos).

6   Detach the ignition vacuum advance hose, crankcase breather hose, fuel inlet hose and float chamber vent hose (photo). Plug the fuel hose after removal.

7   Undo the two nuts, remove the washers and withdraw the carburettors from the manifold studs (photo). Recover the insulating block and gaskets. On 1.3 litre models remove the air cleaner mounting bracket, gaskets and insulating block as required, noting the arrangement of gaskets.

8   Refitting is the reverse sequence to removal. Ensure that all mating faces are perfectly clean and always use new gaskets. After refitting, connect the accelerator cable referring to Section 8 or 9 and, if necessary, adjust the carburettor as described in Section 12.

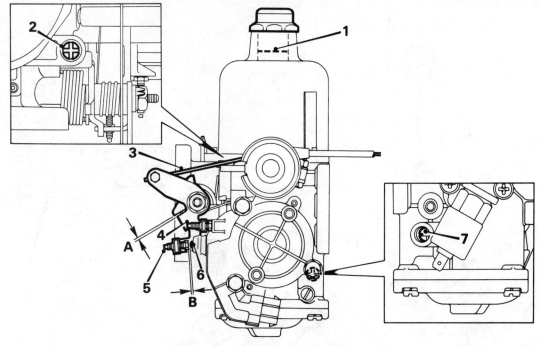

Fig. 3.5 Carburettor adjustment points (Sec 12)

1 Piston damper oil level
2 Idle speed adjustment screw
3 Accelerator cable
4 Throttle lever adjustment screw
5 Fast idle adjustment screw
6 Fast idle pushrod
7 Mixture adjustment screw
A = Lost motion gap
B = Fast idle clearance

12.20 Ambient air temperature sensor location (arrowed) behind left-hand headlamp

13.5A Detach the fuel shut-off valve wiring connector ...

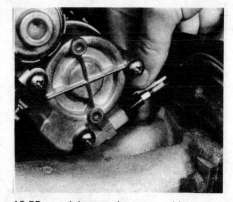

13.5B ... and the stepping motor wiring connector

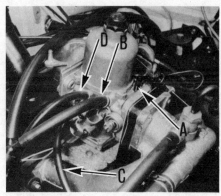

13.6 Ignition vacuum advance hose (A), crankcase breather hose (B), fuel inlet hose (C), and float chamber vent hose (D)

13.7 Removing the carburettor from the manifold – 1.6 litre model

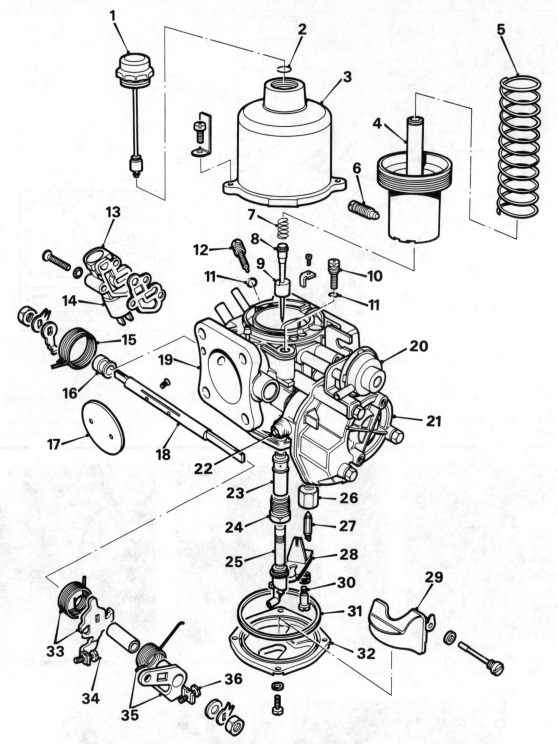

Fig. 3.6 Exploded view of the carburettor (Sec 14)

1 Piston damper
2 Retaining circlip
3 Suction chamber
4 Piston
5 Piston spring
6 Needle guide locking screw
7 Needle bias spring
8 Jet needle
9 Needle guide
10 Idle speed adjustment screw

11 Adjustment screw seal
12 Mixture adjustment screw
13 Fuel shut-off valve and housing
14 Fuel shut-off valve solenoid
15 Throttle return spring
16 Throttle spindle seal
17 Throttle plate
18 Throttle spindle
19 Carburettor body
20 Vacuum switch

21 Mixture control stepping motor
22 Fast idle pushrod
23 Jet bearing
24 Jet bearing nut
25 Jet assembly
26 Float needle seat
27 Float needle
28 Bi-metal jet lever
29 Float
30 Jet lever retaining screw

31 Float chamber cover seal
32 Float chamber cover
33 Throttle lever and return spring
34 Fast idle adjustment screw
35 Lost motion link and return spring
36 Throttle lever adjustment screw

14 SU carburettor – dismantling, overhaul and reassembly

1   Remove the carburettor as described in the previous Section.
2   Clean off the exterior of the carburettor using paraffin or a suitable solvent and wipe dry.
3   Unscrew the three retaining screws and lift off the suction chamber and piston assembly (photo).
4   Unscrew the damper from the suction chamber (photo) and drain the oil from the piston rod.
5   Push the piston up to expose the retaining circlip (photo). Extract the circlip and withdraw the piston and spring assembly (photo).
6   Unscrew the needle guide locking screw and withdraw the needle guide and spring (photo).
7   Mark the relationship of the float chamber cover to the carburettor body. Unscrew the four retaining screws and lift off the cover and O-ring seal (photo).

8   Unscrew the jet adjusting lever retaining screw and withdraw the jet and adjusting lever assembly (photo). Disengage the jet from the lever.
9   Unscrew the float pivot screw (photo) and lift out the float and fuel needle valve (photo). Unscrew the needle valve seat from the base of the float chamber (photo).
10 Unscrew the jet bearing locking nut (photo) and remove the jet bearing (photo).
11 Undo the three retaining screws and remove the fuel shut-off valve and solenoid assembly (photo). Recover the gasket.
12 Dismantling the remaining components is not recommended, as these parts are not available separately. If the throttle levers, linkage or spindle appear worn, or in any way damaged, it will be necessary to renew the complete carburettor. **Do not** remove the mixture control stepping motor or vacuum switch. These components are set to each individual carburettor during manufacture and may not operate correctly if disturbed.

14.3 Removing the carburettor suction chamber and piston assembly

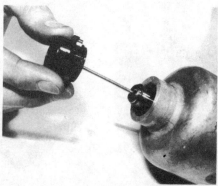
14.4 Unscrew the damper and drain the oil

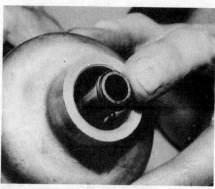
14.5A Extract the retaining circlip ...

14.5B ... to allow removal of the piston and spring

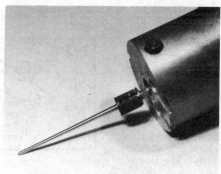
14.6 Slacken the needle guide locking screw and withdraw the needle, guide and spring

14.7 Lift off the float chamber cover after removing the four screws

14.8 Removing the jet and adjusting lever assembly

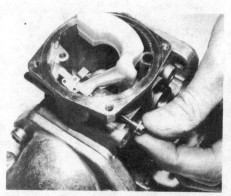
14.9A Unscrew the float pivot screw and remove the float ...

14.9B ... followed by the fuel needle valve ...

14.9C ... then unscrew the needle valve seat

14.10A Unscrew the jet bearing locknut ...

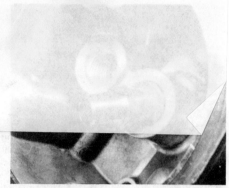
14.10B ... and lift out the jet bearing

14.11 Remove the fuel shut-off valve and solenoid

14.13 Ensure that the filter in the fuel needle valve seat is clean

13 Check the condition of the float needle valve and seat and renew these components if there is any sign of pitting or wear ridges, particularly on the needle. Ensure that the filter in the seat assembly is clean (photo).

14 Examine the carburettor body for cracks and damage, and ensure that the brass fittings and piston guide are secure.

15 Clean the inside of the suction chamber and the outer circumference of the piston with a petrol-moistened rag and allow to air dry. Insert the piston into the suction chamber without the spring. Hold the assembly in a horizontal position and spin the piston. If there is any tendency for the piston to bind, renew the piston and dashpot assembly.

16 Connect a 12 volt supply to the terminals of the fuel shut-off valve solenoid and ensure that the valve closes. If not, renew the solenoid.

17 Check the piston needle and jet bearing for any signs of ovality, or wear ridges.

18 Shake the float and listen for any trapped fuel which may have entered through a tiny crack or fracture.

19 Check the condition of all gaskets, seals and connecting hoses and renew any that show signs of deterioration.

20 Begin reassembly by refitting the jet bearing and retaining nut to the carburettor body.

21 Refit the fuel needle valve and seat, followed by the float and float pivot screw.

22 Allow the float to close the needle valve under its own weight and measure the distance from the centre of the float to the face of the carburettor body, as shown in Fig. 3.7. If the measured dimension is outside the float level height setting given in the Specifications, carefully bend the brass contact pad on the float to achieve the required setting.

23 Engage the jet with the cut-out in the adjusting lever, ensuring that the jet head moves freely. Position the jet in the jet bearing and, at the same time, engage the slot in the adjusting lever with the protruding tip of the mixture adjustment screw. Secure the assembly with the retaining screw.

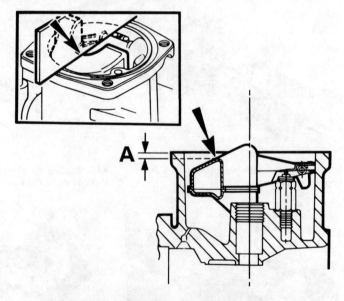
Fig. 3.7 Carburettor float height adjustment (Sec 14)

A = specified float level height
Arrows indicate checking and adjustment points

24 Turn the mixture adjustment screw as necessary to bring the top of the jet flush with the jet bearing upper face when viewed from above. Now turn the adjustment screw two complete turns clockwise to obtain an initial mixture setting.

25 Fit a new O-ring seal to the float chamber cover. Fit the cover with the previously made marks aligned and secure with the four retaining screws.

26 Refit the piston needle, spring and needle guide to the piston (photo), ensuring that the needle guide is flush with the underside of the piston and the triangular etch mark on the guide is between the two transfer holes in the piston (photo). Refit and tighten the locking screw.

27 Temporarily refit the piston and dashpot to the carburettor body without the spring. Engage the piston in its guide and, with the suction chamber in its correct position relative to the retaining screws, mark the piston-to-suction chamber relationship. Remove the suction chamber and piston.

28 Fit the spring to the piston, align the previously made marks and slide the suction chamber over the piston and spring. Avoid turning the piston in the suction chamber, otherwise the spring will be wound up.

29 Push the piston rod up and refit the circlip to the piston rod.

30 Refit the piston and suction chamber, and secure with the three retaining screws tightened evenly. Fill the piston damper with engine oil up to the top of the piston and refit the damper.

31 Refit the fuel shut-off valve using a new gasket, if necessary, and secure the unit with the three screws.

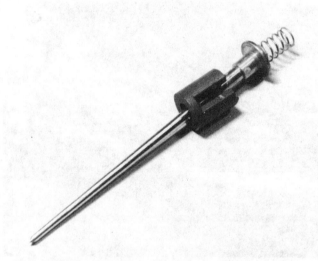

14.26A Refit the needle, spring and guide to the piston ...

14.26B ... with the guide flush with the piston base and the etch mark (arrowed) positioned as shown

15 Inlet and exhaust manifolds (1.3 litre models) – removal and refitting

1 Disconnect the battery negative terminal, and then remove the carburettor, as described in Section 13.

2 Pull the vacuum hose off the banjo union connector in the centre of the inlet manifold (photo). Undo and remove the banjo union bolt and recover the two washers. Place the servo vacuum hose to one side.

3 Remove the cooling system filler cap from the expansion tank. If

15.2 Banjo union components

A Vacuum hose
B Union nut
C Sealing washer
D Banjo union
E Sealing washer

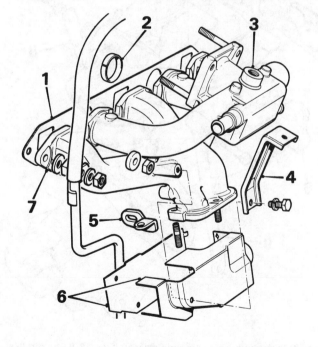

Fig. 3.8 Inlet and exhaust manifold assemblies – 1.3 litre models (Sec 15)

1 Manifold gasket
2 Inlet manifold locating ring
3 Inlet manifold
4 Support bracket
5 Steady bracket
6 Hot air box
7 Exhaust manifold

the engine is hot, unscrew the cap slowly to release the pressure and use a rag as protection against scalding.

4   Place a suitable receptacle beneath the inlet manifold and slacken the two clips securing the water hoses to the manifold. Ease off the hoses and allow the water to drain into the receptacle.

5   Undo and remove the bolt securing the steady bracket to the exhaust manifold and the nuts and large washers securing the inlet manifold to the cylinder head. Withdraw the inlet manifold from the cylinder head and recover the vent hose.

6   To remove the exhaust manifold, remove the front portion of the hot air box to provide access to the exhaust downpipe.

7   Undo and remove the nuts securing the downpipe to the manifold, and the remaining nuts and washers securing the manifold to the cylinder head. Withdraw the exhaust manifold from the cylinder head.

8   With the manifolds removed, recover the gaskets from the cylinder head and exhaust downpipe flange, and remove the inlet manifold locating rings.

9   Refitting the manifolds is the reverse sequence to removal. Use new gaskets and make sure all mating faces are clean. Tighten the manifold-to-cylinder head nuts evenly to the specified torque before fully tightening the downpipe and steady bracket nuts and bolts. Refit the carburettor as described in Section 13, and top up the cooling system, as described in Chapter 2.

16 Inlet and exhaust manifolds (1.6 litre models) – removal and refitting

1   Disconnect the battery negative terminal then remove the air cleaner and plenum chamber, as described in Section 4, and the carburettor, as described in Section 14.

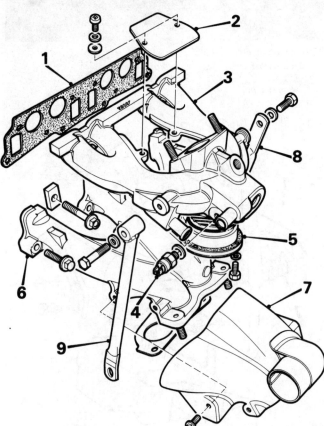

Fig. 3.9 Inlet and exhaust manifold assemblies – 1.6 litre models (Sec 16)

1	Manifold gasket	5	Induction heater
2	Heat shield	6	Exhaust manifold
3	Inlet manifold	7	Hot air box
4	Induction temperature sensor	8	Support strut
		9	Support strut

2   Pull the vacuum hose off the banjo union connector in the centre of the inlet manifold (photo 15.2). Undo and remove the banjo union bolt and recover the two washers. Place the servo vacuum hose to one side.

3   Disconnect the induction heater lead at the wiring connector (photo).

4   Disconnect the two wires at the induction temperature sensor (photo).

5   Remove the cooling system filler cap from the expansion tank. **Note**: If the engine is hot, unscrew the cap slowly to release the pressure and use a rag as protection against scalding.

6   Place a suitable container beneath the inlet manifold and slacken the clips securing the three hoses to the manifold outlets. Ease off the hoses and allow the coolant to drain into the receptacle.

7   Undo the bolts securing the left-hand and right-hand support struts to the manifold (photos).

8   Jack up the front of the car and securely support it on axle stands.

9   Undo the nuts securing the exhaust downpipes to the manifold and ease the flange off the manifold studs.

10 Undo the bolts securing the right-hand engine mounting to the cylinder head. Ease the engine away from the mounting sufficiently to allow removal of the manifold. Wedge the engine in this position.

11 Undo the two screws and remove the heat shield from the inlet manifold (photo).

12 Undo all the bolts, clamp plates and nuts securing the inlet

16.3 Manifold induction heater wiring connector

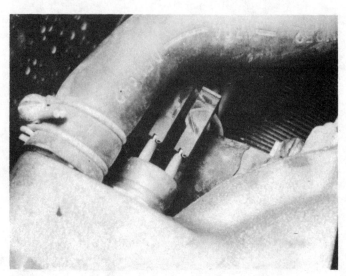

16.4 Wiring connectors at the induction temperature sensor

16.7A Inlet manifold left-hand ...

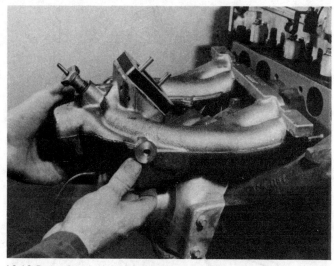

16.12 Removing the inlet manifold ...

16.7B ... and right-hand support strut attachments (arrowed)

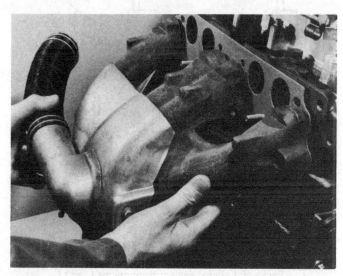

16.13 ... followed by the exhaust manifold

manifold to the cylinder head and remove the manifold (photo). Recover the carburettor vent hose.

13 Undo the remaining nuts and bolts and remove the exhaust manifold (photo).

14 Refitting the manifolds is the reverse sequence to removal. Use new gaskets and make sure all mating faces are clean. Tighten the manifold-to-cylinder head nuts and bolts evenly and to the specified torque before fully tightening the downpipe and steady bracket nuts and bolts. Refit the carburettor, as described in Section 13, the air cleaner and plenum chamber, as described in Section 4, then top up the cooling system, as described in Chapter 2.

17 Exhaust system – checking, removal and refitting

1 The exhaust system should be examined for leaks, damage and security at regular intervals (see Routine Maintenance). To do this, apply the handbrake and allow the engine to idle. Lie down on each side of the car in turn and check the full length of the exhaust system for leaks while an assistant temporarily places a wad of cloth over the end of the tailpipe. If a leak is evident stop the engine and use a proprietary repair kit to seal it. Holts Flexiwrap and Holts Gun Gum exhaust repair systems can be used for effective repairs to exhaust pipes and silencer boxes, including ends and bends (except on models with catalytic converters). Holts Flexiwrap is an MOT approved permanent exhaust repair. If a leak is excessive or damage is evident, renew the section. Check the rubber mountings for deterioration and renew them if necessary.

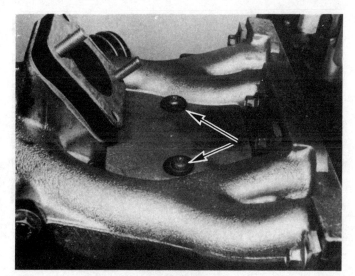

16.11 Heat shield retaining screws (arrowed)

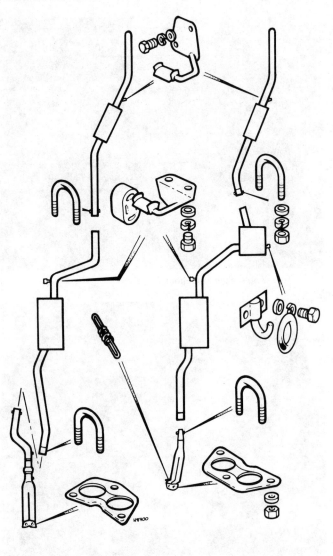

Fig. 3.10 Exhaust system components (Sec 17)

17.3 Exhaust rear section rubber mounting block

17.4A Exhaust downpipe flexible joint – 1.6 litre model

17.4B Front-to-intermediate exhaust section retaining clamp

2   To remove the exhaust system, jack up the front and/or rear of the car and support it securely on axle stands. Alternatively drive the front or rear wheels up on ramps.

3   The system consists of three sections which can be individually removed. If the intermediate section is to be removed it will, however, be necessary to remove the front or rear section first. To remove the rear section of the system, unscrew the retaining nuts and remove the U-shaped retaining clamp. Release the mounting bracket from the rubber mounting block (photo) and twist the section free. If the joint is stubborn, liberally apply penetrating oil and leave it to soak. Tap the joint with a hammer and it should now be possible to twist it free. If necessary, carefully heat the joint with a blowlamp to assist removal, *but shield the fuel tank, fuel lines and underbody adequately from heat.*

4   To remove the front section, first remove the anti-roll bar if working on a 1.6 litre model, as described in Chapter 11. Undo and remove the nuts securing the downpipes to the exhaust manifold (photo) and, where fitted, remove any additional support brackets. Unhook the compression spring from the flexible joint, remove the U-shaped clamp (photo) and release the section.

5   Refitting is the reverse sequence to removal. Position the joints so that there is adequate clearance between all parts of the system and the underbody, and ensure that there is equal load on all mounting blocks. Always use a new gasket on the downpipe-to-manifold joint face.

18 Fault diagnosis – fuel and exhaust systems

Unsatisfactory engine performance, bad starting and excessive fuel consumption are not necessarily the fault of the fuel system or carburettor. In fact they more commonly occur as a result of ignition and timing faults. Before acting on the following, it is necessary to check the ignition system first. Even though a fault may lie in the fuel system, it will be difficult to trace unless the ignition system is correct. The faults below, therefore, assume that, where applicable, this has been attended to first. If during the fault diagnosis procedure it is suspected that the carburettor electronic mixture control or any of its related systems may be at fault, it is recommended that the help of a reputable BL dealer is sought. Accurate testing of the system and its components entails the use of a systematic checking procedure using specialist equipment and this is considered beyond the scope of the average home mechanic.

Symptom	Reason(s)
Difficult starting when cold	Faulty electronic mixture control system or related component Carburettor piston sticking Fuel tank empty or pump defective Incorrect float chamber fuel level
Difficult starting when hot	Faulty electronic mixture control system or related component Air cleaner choked Carburettor piston sticking Float chamber flooding or incorrect fuel level Fuel tank empty or pump defective Carburettor idle mixture adjustment incorrect
Excessive fuel consumption	Leakage from tank, pipes, pump or carburettor Air cleaner choked Carburettor idle mixture adjustment incorrect Carburettor float chamber flooding Faulty fuel shut-off solenoid or control system Carburettor worn Excessive engine wear or other internal fault Tyres underinflated Brakes binding
Fuel starvation	Fuel level flow Leak on suction side of pump Fuel pump faulty Float chamber fuel level incorrect Fuel tank breather restricted Fuel tank inlet or carburettor inlet filter blocked
Poor performance, hesitation or erratic running	Carburettor idle mixture adjustment incorrect Faulty carburettor vacuum switch Carburettor piston damper oil level low Leaking manifold gasket Fuel starvation Carburettor worn Excessive engine wear or other internal fault

Chapter 4 Ignition system

For modifications, and information applicable to later models, see Supplement at end of manual

Contents

Specifications

Part A: 1.3 litre models

System type .... Electronic breakerless, inductive type

Ignition coil
Type .... Ducellier 520029A
Current consumption – engine idling .... 2.3 to 2.7 amps
Primary resistance at 20°C (68°F) .... 0.82 ohms ± 5%

Distributor
Type .... Lucas 59 DM4
Direction of rotation .... Anti-clockwise
Reluctor air gap .... 0.008 to 0.014 in (0.20 to 0.35 mm)
Ignition amplifier .... Lucas AB14 or Ducellier 54403505
Firing order 1-3-4-2
Location of No 1 cylinder .... Crankshaft pulley end
Lubricant type/specification .... Multigrade engine oil, viscosity SAE 10W/40 (Duckhams QXR, Hypergrade, or 10W/40 Motor Oil)

Spark plugs
Type .... Champion RN9YCC or RN9YC
Electrode gap .... 0.032 in (0.8 mm)

HT leads .... Champion LS-02 boxed set

Ignition timing
Stroboscopic at 1500 rpm with vacuum pipe disconnected .... 12° BTDC
Timing marks .... Crankshaft pulley groove and 4° pointers on timing cover scale

Torque wrench settings	lbf ft	Nm
Spark plugs	18	24
Distributor clamp bolt	16	22

Part B: 1.6 litre models

System type	Microprocessor-controlled programmed electronic ignition

Ignition coil

Type	Lucas 45328
Current consumption – engine idling	2.3 to 2.7 amps
Primary resistance at 20°C (68°F)	0.82 ohms ± 5%

Programmed ignition system

Electronic control unit type	Lucas AB17-84185
Distributor cap type	Lucas 544-03944
Rotor arm type	Lucas 544-04286
Rotor arm rotation	Anti-clockwise
Knock sensor type	Lucas or Lamerholm VP50/1-M12
Crankshaft sensor type	Lucas 547-42886
Firing order	1-3-4-2
Location of No 1 cylinder	Crankshaft pulley end

Spark plugs

Type	Champion RC9YCC or RC9YC
Electrode gap	0.032 in (0.8 mm)

HT leads	Champion LS-05 boxed set

Ignition timing*

With vacuum pipe disconnected	10° to 16° BTDC @ 1000 rpm
With vacuum pipe connected	23° to 27° BTDC @ 1000 rpm

* Non-adjustable; for reference purposes only – see text

Torque wrench settings	lbf ft	Nm
Spark plugs	13	18
Knock sensor	9	12

PART A: 1.3 LITRE MODELS

1 General description

A breakerless inductive type electronic ignition system is used; consisting of the battery, coil, distributor, ignition amplifier, spark plugs and HT leads. The distributor is located on the rear facing side of the engine and is driven by a driveshaft in mesh with the camshaft.

In order that the engine can run correctly, it is necessary for an electrical spark to ignite the fuel/air mixture in the combustion chamber at exactly the right moment in relation to engine speed and load. The ignition system is based on feeding low tension voltage from the battery to the coil, where it is converted to high tension voltage. The high tension voltage is powerful enough to jump the spark plug gap in the cylinders many times a second under high compression, providing that the system is in good condition and that all adjustments are correct.

The ignition system is divided into two circuits, the low tension circuit and the high tension circuit. The low tension circuit consists of the battery, lead to the ignition switch, lead from the ignition switch to the low tension coil windings, leads from the coil windings to the ignition amplifier and leads from the ignition amplifier to the pick-up coil assembly in the distributor. The high tension circuit consists of the high tension coil windings, the heavy lead from the coil to the distributor cap, the rotor arm, spark plug leads and spark plugs.

The system functions in the following manner. Low tension voltage is changed in the coil into high tension voltage by the action of the ignition amplifier in conjunction with the pick-up coil assembly. As each of the reluctor teeth passes through the magnetic field of the pick-up coil in the distributor an electrical signal is sent to the ignition amplifier which triggers the coil in the same way as the opening of the contact breaker points in a conventional system. High tension voltage is then fed via the carbon brush in the centre of the distributor cap to the distributor rotor arm. The voltage passes across to the appropriate metal segment in the cap and via the spark plug lead to the spark plug where it finally jumps the spark plug gap to earth. The ignition is advanced and retarded automatically, to ensure that the spark occurs at just the right instant for the particular load at the prevailing engine speed.

The ignition advance is controlled both mechanically and by a vacuum-operated system. The mechanical governor mechanism consists of two weights, which move out from the distributor shaft as the engine speed rises due to centrifugal force. As they move outwards they rotate the cam relative to the distributor shaft, and so advance the spark. The weights are held in position by two light springs and it is the tension of the springs which is largely responsible for correct spark advancement.

The vacuum control consists of a diaphragm, one side of which is connected via a small bore tube to the carburettor, and the other side to the distributor baseplate. Depression in the inlet manifold and carburettor, which varies with engine speed and throttle opening, causes the diaphragm to move, so moving the baseplate, and advancing or retarding the spark. A fine degree of control is achieved by a spring in the vacuum assembly.

Due to the nature of the electronic ignition system, no current is consumed by the ignition coil with the ignition switched on and the engine stationary. A ballast resistor is not required in the starting circuit.

Warning: The voltages produced by the electronic ignition system are considerably higher than those produced by conventional systems. Extreme care must be used when working on the system with the ignition switched on. Persons with surgically-implanted cardiac pacemaker devices should keep well clear of the ignition circuits, components and test equipment.

2 Maintenance and inspection

1 At the intervals specified in Routine Maintenance at the beginning of this manual remove the distributor cap and thoroughly clean it inside and out with a dry lint-free rag. Examine the four HT lead segments inside the cap. If the segments appear badly burnt or pitted, renew the cap. Make sure that the carbon brush in the centre of the cap is free to move and that it protrudes by approximately 0.1 in (3 mm) from its holder.

2 With the distributor cap removed, lift off the rotor arm and the plastic anti-flash shield. Carefully apply two drops of engine oil to the felt pad in the centre of the cam spindle. Also lubricate the centrifugal advance mechanism by applying two drops of oil through the square hole in the baseplate. Wipe away any excess oil and refit the anti-flash shield, rotor arm and distributor cap.

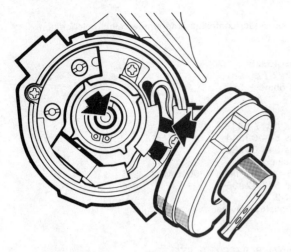

Fig. 4.1 Distributor lubrication points (Sec 2)

3   Renew the spark plugs at the specified interval. Using a stroboscopic timing light check and, if necessary, reset the ignition timing, as described in Section 5.

3   Distributor – removal and refitting

1   Disconnect the battery negative terminal.
2   Pull off the HT lead and remove No 1 spark plug (nearest the crankshaft pulley).
3   Place a finger over the plug hole and turn the engine in the normal direction of rotation (clockwise from the crankshaft pulley end) until pressure is felt in No 1 cylinder. This indicates that the piston is commencing its compression stroke. The engine can be turned with a socket and bar on the crankshaft pulley bolt after removing the access cover from the inner wheel arch.
4   Continue turning the engine until the notch in the crankshaft pulley is aligned with the TDC pointer on the timing scale. This is the last pointer on the scale, nearest the front of the car.
5   Make a reference mark on the side of the distributor body adjacent to the No 1 cylinder spark plug lead position in the cap. Spring back the clips, or remove the two screws, and lift off the cap. Check that the rotor arm is pointing toward the reference mark.
6   Make a further mark on the cylinder block in line with the mark on the distributor body.
7   Detach the vacuum advance and disconnect the ignition amplifier wiring harness at the connector.
8   Unscrew the distributor clamp retaining bolt, lift away the clamp and withdraw the distributor from the engine.
9   To refit the distributor, first check that the engine is still at the TDC position with No 1 cylinder on compression. If the engine has been turned while the distributor was removed, return it to the correct position, as previously described.
10 With the vacuum unit pointing toward No 1 spark plug slide the distributor into the cylinder block and turn the rotor arm slightly until the offset slot on the distributor drive dog positively engages with the driveshaft.
11 Turn the distributor body until the rotor arm is pointing toward the No 1 spark plug lead segment in the distributor cap, or if the original distributor is being refitted, align the previously made reference marks. Hold the distributor in this position and refit the clamp and retaining bolt.
12 Reconnect the vacuum advance pipe and the wiring harness connector. Refit the distributor cap, No 1 spark plug and the HT lead.
13 Reconnect the battery and, if removed, refit the access panel to the inner wheel arch.
14 The ignition timing should now be adjusted, as described in Section 5.

4   Distributor – dismantling and reassembly

1   Remove the distributor, as described in the previous Section.
2   Lift off the rotor arm, followed by the anti-flash shield (photo).
3   Using circlip pliers, extract the retaining circlip (photo) and lift off the washer and O-ring (photo).

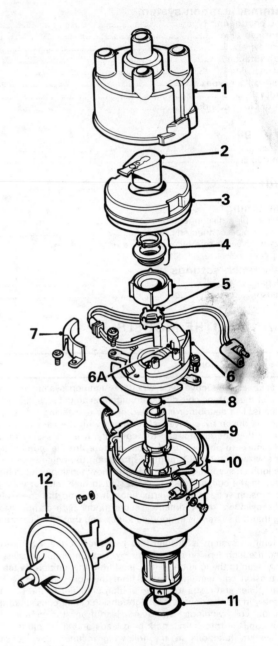

Fig. 4.2 Exploded view of the distributor (Sec 4)

1	Distributor cap	6A Pick-up limb
2	Rotor arm	7 Wiring guide
3	Anti-flash shield	8 Felt pad
4	O-ring, washer and circlip	9 Distributor shaft
5	Reluctor and coupling ring	10 Distributor body
6	Pick-up coil and baseplate assembly	11 O-ring
		12 Vacuum unit

4.2 Lift off the rotor arm and anti-flash shield

4.3A Extract the circlip ...

4.3B ... and lift off the washer and O-ring

4 Withdraw the reluctor and coupling ring (photos) using a screwdriver very carefully to ease them off the shaft if they are initially tight.

5 Undo and remove the screws securing the vacuum unit to the distributor body. Disengage the vacuum unit operating link from the peg on the underside of the baseplate using a twisting movement and withdraw the unit (photo).

6 Release the wiring harness rubber grommet and remove the two baseplate securing screws. Lift the baseplate out of its location in the distributor body (photo).

7 This is the limit of dismantling, as the parts located below the baseplate, the distributor shaft and distributor body can only be renewed as an assembly.

8 With the distributor dismantled, renew any parts that show signs of wear, or damage, and any that are known to be faulty. Pay close attention to the centrifugal advance mechanism (photo), checking for loose or broken springs, wear in the bob weight pivots and play in the distributor shaft.

9 Begin reassembly by lubricating the distributor shaft, bob weight pivots, vacuum link and baseplate sliding surfaces with engine oil.

10 Place the baseplate assembly in position, with the peg on the underside adjacent to the vacuum unit aperture. Refit and tighten the two securing screws.

11 Refit the coupling ring to the underside of the reluctor and slide this assembly over the distributor shaft. Align the broad lug of the coupling ring with the broad slot in the shaft and push the ring and reluctor fully into place.

12 Position the O-ring and washer over the shaft and secure them with the circlip.

13 Insert the vacuum unit operating link into its aperture and manipulate the unit and baseplate until the link can be engaged with the peg. Refit and tighten the retaining screws.

14 Refit the harness leads and grommet to the slot in the distributor body.

15 Position the reluctor so that one of the teeth is adjacent to the

4.4A Withdraw the reluctor ...

4.4B ... followed by the coupling ring

4.5 Disengage the vacuum unit operating link from the baseplate peg and withdraw the unit

4.6 Remove the screws and lift out the baseplate

4.8 Check the components below the baseplate for wear

4.15 Using feeler gauges to measure the reluctor air gap

limb on the pick-up assembly. Using feeler gauges, preferably of plastic or brass, measure the air gap between the reluctor tooth and pick-up assembly (photo). If the measured dimension is outside the tolerance given in the Specifications, slacken the adjusting nuts on the pick-up assembly and reposition the unit as necessary.

16 Refit the anti-flash shield and rotor arm, then refit the distributor, as described in Section 3.

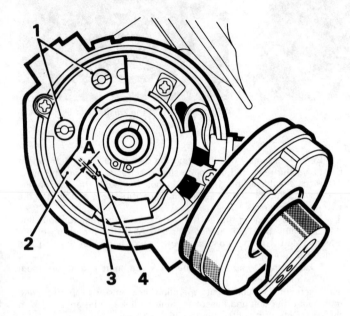

Fig. 4.3 Reluctor air gap adjustment (Sec 4)

Dimension A = specified air gap   3 Pick-up limb
1 Adjusting nuts   4 Reluctor tooth
2 Pick-up coil assembly

---

5 Ignition timing – adjustment

Note: *With electronic ignition systems the only suitable method which may be used to accurately time the ignition is with a stroboscopic timing light. However, for initial setting up purposes (ie after major overhaul, or if the timing has been otherwise completely lost) a basic initial static setting may be used to get the engine started. Once the engine is running, the timing should be accurately set using the timing light. A further method, employing the light emitting diode (LED) sensor bracket, and timing disc on the crankshaft pulley may be used, but the equipment for use with this system is not normally available to the home mechanic.*

1 In order that the engine can run efficiently, it is necessary for a spark to occur at the spark plug and ignite the fuel/air mixture at the instant just before the piston on the compression stroke reaches the top of its travel. The precise instant at which the spark occurs is determined by the ignition timing, and this is quoted in degrees before top-dead-centre (BTDC).

2 If the timing is being checked as a maintenance or servicing procedure, refer to paragraph 11. If the distributor has been dismantled or renewed, or if its position on the engine has been altered, obtain an initial static setting as follows.

Static setting
3 Pull off the HT lead and remove No 1 spark plug (nearest the crankshaft pulley).
4 Place a finger over the plug hole and turn the engine in the normal direction of rotation (clockwise from the crankshaft pulley end) until pressure is felt in No 1 cylinder. This indicates that the piston is commencing its compression stroke. The engine can be turned with a socket and bar on the crankshaft pulley bolt after removing the access cover from the inner wheel arch.
5 Continue turning the engine until the notch in the crankshaft

pulley is aligned with the TDC pointer on the timing scale (photo 5.12). This is the last pointer on the scale, nearest the front of the car.
6 Remove the distributor cap and check that the rotor arm is pointing toward the No 1 spark plug HT lead segment in the cap.
7 Lift off the rotor arm and anti-flash shield and observe the position of the reluctor in relation to the pick-up coil. One of the teeth on the reluctor should be aligned with, or very near to, the small pip, or limb, of the pick-up coil.
8 Slacken the distributor clamp retaining bolt and turn the distributor body until the reluctor tooth and pick-up limb are directly in line (photo).
9 Tighten the distributor clamp, or pinch-bolt, refit the anti-flash shield, rotor arm and distributor cap. Refit No 1 spark plug and HT lead.
10 It should now be possible to start and run the engine enabling the timing to be accurately checked with a timing light as follows.

Stroboscopic setting
11 Disconnect the vacuum advance pipe at the distributor and plug its end.
12 The timing marks are located on a scale just above the crankshaft pulley, with a corresponding V-notch in the rim of the pulley (photo). The pointer peak at the far right of the scale (ie nearest the front of the

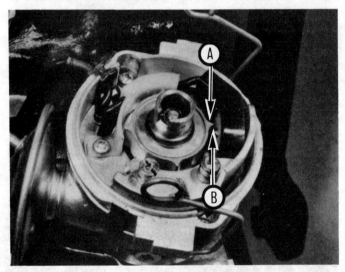

5.8 Reluctor tooth (A) aligned with pick-up limb (B)

5.12 Crankshaft pulley timing notch and timing scale
Pulley notch shown positioned at TDC

car) is the TDC position and is marked with a '0'. The pointer peaks to the left of the TDC are in increments of 4° BTDC. If a timing light with a strong light source is being used the marks will be clearly visible from above; however, for greater clarity, the access panel under the inner wheel arch may be removed – enabling the timing light to be held very close to the marks.

13 Refer to the Specifications for the timing setting applicable to the engine being worked on, and then highlight the appropriate mark and pulley V-notch with white chalk or paint.

14 Connect a timing light to the engine in accordance with the manufacturer's instructions (usually between No 1 spark and its HT lead). If the car is not equipped with a tachometer, connect a suitable unit to the engine in accordance with the manufacturer's instruction.

15 Start the engine and run it at the speed specified for ignition timing.

16 Point the timing light at the timing marks and they should appear to be stationary with the crank pulley notch in alignment with the appropriate pointer.

17 If adjustment is necessary (ie the pulley notch does not line up with the appropriate point), loosen the distributor clamp retaining bolt, or pinch-bolt, and turn the distributor body clockwise to advance the timing and anti-clockwise to retard it. Tighten the securing bolt when the setting is correct.

18 Gradually increase the engine speed while still pointing the timing light at the marks. The pulley notch should appear to advance further, indicating that the centrifugal advance mechanism is operating correctly. If the timing marks remain stationary when the engine speed is increased, or if the movement is erratic or jerky, then the distributor should be dismantled for inspection, as described in Section 4.

19 Reconnect the vacuum pipe to the distributor and check that the advance alters when the pipe is connected. If not, the vacuum unit on the distributor may be faulty.

20 Switch off the engine and disconnect the timing light and tachometer. Where applicable, refit the access panel to the inner wheel arch.

6 Ignition coil – description and testing

1 The coil is bolted to the centre of the engine compartment bulkhead and it should be periodically wiped over to prevent high tension (HT) voltage loss through arcing.

2 To ensure correct HT polarity at the spark plugs, the LT coil leads must always be connected correctly. The LT leads from the ignition amplifier should be connected as follows. White/black leads to the coil **negative** terminal. White leads to the coil **positive** terminal. Incorrect connections can cause bad starting, misfiring and short spark plug life.

3 Apart from the tests of the low tension circuit contained in the ignition system test procedure (Section 9), accurate checking of the coil output requires special equipment and for the home mechanic the easiest test is by substitution of a new unit.

4 If a new coil is to be fitted, ensure that it is of the correct type and suitable for use on electronic ignition systems. Failure to do so could cause irreparable damage to the ignition amplifier or distributor pick-up assembly.

5 To remove the coil, disconnect the HT and LT wires from the terminals. Undo and remove the two retaining bolts and lift away the coil complete with bracket. Refitting is the reverse sequence to removal.

7 Ignition amplifier – general

1 The ignition amplifier is mounted on the engine compartment bulkhead, just below the coil. The amplifier controls the function of the ignition coil in response to signals received from the pick-up coil in the distributor.

2 The unit may be tested using the procedure described in Section 9. If the amplifier is found to be faulty it must be renewed as a complete unit, repairs to the components or circuitry are not possible due to its sealed construction.

3 To remove the unit, disconnect the battery negative terminal and then disconnect the wiring plug from the end of the amplifier. Remove the two small retaining screws and withdraw the unit from the car. Refitting is the reverse sequence to removal.

8 Spark plugs and HT leads – general

1 The correct functioning of the spark plugs is vital for the correct running and efficiency of the engine. It is essential that the plugs fitted are appropriate for the engine, and the suitable type is specified at the beginning of this chapter. If this type is used and the engine is in good condition, the spark plugs should not need attention between scheduled replacement intervals. Spark plug cleaning is rarely necessary and should not be attempted unless specialised equipment is available as damage can easily be caused to the firing ends.

2 To remove the plugs, first mark the HT leads to ensure correct refitment, and then pull them off the plugs. Using a spark plug spanner, or suitable deep socket and extension bar, unscrew the plugs and remove them from the engine.

3 The condition of the spark plugs will also tell much about the overall condition of the engine.

4 If the insulator nose of the spark plug is clean and white, with no deposits, this is indicative of a weak mixture, or too hot a plug. (A hot plug transfers heat away from the electrode slowly – a cold plug transfers it away quickly.)

5 If the tip and insulator nose are covered with hard black-looking deposits, then this is indicative that the mixture is too rich. Should the plug be black and oily, then it is likely that the engine is fairly worn, as well as the mixture being too rich.

6 If the insulator nose is covered with light tan to greyish brown deposits, then the mixture is correct and it is likely that the engine is in good condition.

7 The spark plug gap is of considerable importance, as, if it is too large or too small, the size of the spark and its efficiency will be seriously impaired. The spark plug gap should be set to the figure given in the Specifications at the beginning of this Chapter.

8 To set it, measure the gap with a feeler gauge, and then bend open, or close, the *outer* plug electrode until the correct gap is achieved. The centre electrode should *never* be bent as this may crack the insulation and cause plug failure, if nothing worse.

9 When fitting new plugs screw them in by hand initially and then fully tighten to the specified torque. If a torque wrench is not available, tighten the plugs until initial resistance is felt as the sealing washer contacts its seat and then tighten by a further eighth of a turn. Refit the HT leads in the correct order, ensuring that they are a tight fit over the plug ends. Periodically wipe the leads clean to reduce the risk of HT leakage by arcing.

9 Fault diagnosis – electronic ignition system

There are two main symptoms indicating ignition faults. Either the engine will not start or fire, or the engine is difficult to start and misfires. If it is a regular misfire, ie the engine is only running on two or three cylinders, the fault is almost sure to be in the high tension circuit. If the misfiring is intermittent, the fault could be in either the high or low tension circuits. If the car stops suddenly, or will not start at all, it is likely that the fault is in the low tension circuit. Loss of power and overheating, apart from faulty carburation settings, are normally due to faults in the distributor or incorrect ignition timing.

The first part of this Section deals with the diagnosis of faults in the high tension circuit. If these tests prove negative or indicate a possible fault in the electronic ignition or low tension circuit, a separate test procedure should be followed. This is contained in the second part of this Section and entails the use of a 0 to 12 volt voltmeter and an ohmmeter.

PART 1

Engine fails to start

1 If the engine fails to start and the car was running normally when it was last used, first check there is fuel in the petrol tank. If the engine turns over normally on the starter motor and the battery is evidently well charged, then the fault may be either in the high or low tension circuits. First check the HT circuit. If the battery is known to be fully charged, the ignition light comes on and the starter motor fails to turn the engine, check the tightness of the leads on the battery terminals and the security of the earth lead to its connection to the body. It is quite common for the leads to have worked loose, even if they look

and feel secure. If one of the battery terminal posts gets very hot when trying to work the starter motor, this is a sure indication of a faulty connection to that terminal.

2 One of the most common reasons for bad starting is wet or damp spark plug leads and distributor. Remove the distributor cap. If condensation is visible internally a moisture dispersant such as Holts Wet Start can be very effective. Refit the cap. To prevent the problem recurring, Holts Damp Start can be used to provide a sealing coat, so excluding any further moisture from the ignition system. In extreme difficulty, Holts Cold Start will help to start a car when only a very poor spark occurs.

3 If the engine still fails to start, check that current is reaching the plugs, by disconnecting each plug lead in turn at the spark plug end, and holding the end of the cable about 0.2 in (5 mm) away from the cylinder block. Spin the engine on the starter motor.

4 Sparking between the end of the cable and the block should be fairly strong with a regular blue spark. (Hold the lead with rubber to avoid electric shocks). If current is reaching the plugs, then remove and regap them. The engine should now start.

5 If there is no spark at the plug leads, take off the HT lead from the centre of the distributor cap and hold it to the block as before. Spin the engine on the starter once more. A rapid succession of blue sparks between the end of the lead and the block indicates that the coil is in order and that the distributor cap is cracked, the rotor arm faulty or the carbon brush in the top of the distributor cap is not making good contact with the rocker arm.

6 If there are no sparks from the end of the lead from the coil, check the connection at the coil end of the lead. If it is in order carry out the checks contained in the electronic ignition test procedure (Part 2).

Engine misfires

7 If the engine misfires regularly, run it at a fast idling speed. Pull off each of the plug caps in turn and listen to the note of the engine. Hold the plug cap in a dry cloth or with a rubber glove as additional protection against a shock from the HT supply.

8 No difference in engine running will be noticed when the lead from the defective circuit is removed. Removing the lead from one of the good cylinders will accentuate the misfire.

9 Remove the plug lead from the end of the defective plug and hold it about 0.2 in (5 mm) away from the block. Restart the engine. If the sparking is fairly strong and regular, the fault must lie in the spark plug.

10 The plug may be loose, the insulation may be cracked, or the points may have burnt away giving too wide a gap for the spark to jump. Worse still, one of the points may have broken off. Either renew the plug, or clean it, reset the gap, and then test it.

11 If there is no spark at the end of the plug lead, or if it is weak and intermittent, check the ignition lead from the distributor to the plug. If the insulation is cracked or perished, renew the lead. Check the connections at the distributor cap.

12 If there is still no spark, examine the distributor cap carefully for tracking. This can be recognised by a very thin black line running between two or more electrodes, or between an electrode and some other part of the distributor. These lines are paths which now conduct electricity across the cap, thus letting it run to earth. The only answer in this case is a new distributor cap.

13 Apart from the ignition timing being incorrect, other causes of misfiring have already been dealt with under the section dealing with the failure of the engine to start. To recap, these are that:

(a) The coil may be faulty giving an intermittent misfire
(b) There may be a damaged wire or loose connection in the low tension circuit
(c) There may be a fault in the electronic ignition system
(d) There may be a mechanical fault in the distributor

14 If the ignition timing is too far retarded it should be noted that the engine will tend to overheat, and there will be a quite noticeable drop in power. If the engine is overheating and the power is down, and the ignition timing is correct, then the carburettor should be checked, as it is likely that this is where the fault lies.

PART 2

Electronic ignition system test procedure

Test	Remedy
1 Is the reluctor air gap set to the specified dimension?	Yes: Proceed to Test 2 No: Adjust the gap, as described in Section 4
2 Is the battery voltage greater than 11.5 volts?	Yes: Proceed to Test 3 No: Recharge the battery
3 Is the voltage at the coil '+' terminal more than 1 volt below battery voltage?	Yes: Faulty wiring or connector between ignition switch and coil. Faulty ignition switch No: Proceed to Test 4
4 Is the voltage at the coil − terminal more than 1 volt	Yes: Disconnect the wiring connector between distributor and ignition amplifier and proceed to Test 7 No: Disconnect the ignition amplifier lead at the coil − terminal and proceed to Test 5
5 Is the voltage at the coil − terminal now more than 1 volt below battery voltage?	Yes: Proceed to Test 6 No: Renew the ignition coil
6 Is the voltage at the ignition amplifier earth more than 0.1 volts?	Yes: Clean and/or repair the earth connection No: Renew the ignition amplifier
7 Is the pick-up coil resistance measured at the wiring connector terminals between 2.2 k ohms and 4.8 k ohms?	Yes: Reconnect the wiring connector between distributor and ignition amplifier and proceed to Test 8 No: Renew the pick-up coil assembly in the distributor
8 Does the voltage at the coil − terminal drop when the starter motor is operated?	Yes: Check and adjust the ignition timing. If the fault still exists the problem may lie with the engine internal components No: Renew the ignition amplifier

PART B: 1.6 LITRE MODELS

10 General description

The programmed electronic ignition system operates on an advanced principle whereby the main functions of the distributor are replaced by an electronic control unit (ECU).

The mechanical operation of the contact breaker points in a conventional distributor is simulated electronically by the reluctor disc on the periphery of the clutch pressure plate and by the crankshaft sensor whose inductive head runs between the reluctor disc teeth. 34 teeth are used on the reluctor disc, spaced at 10° intervals with two spaces 180° apart which correspond to TDC for Nos 1 and 4 pistons and Nos 2 and 3 pistons respectively. As the crankshaft rotates, the reluctor disc teeth pass over the crankshaft sensor which transmits a pulse to the ECU every time a tooth passes over it. The ECU recognises the absence of a pulse every 180° and consequently establishes the TDC position. Each subsequent pulse then represents 10° of crankshaft rotation. This, and the time interval between pulses, allows the ECU to accurately determine engine position and speed.

A small bore pipe connecting the inlet manifold to a pressure transducer within the ECU, supplies the unit with information on engine load. From this constantly changing data the ECU selects a particular advance setting from a range of ignition characteristics stored in its memory. This basic setting can be further advanced or

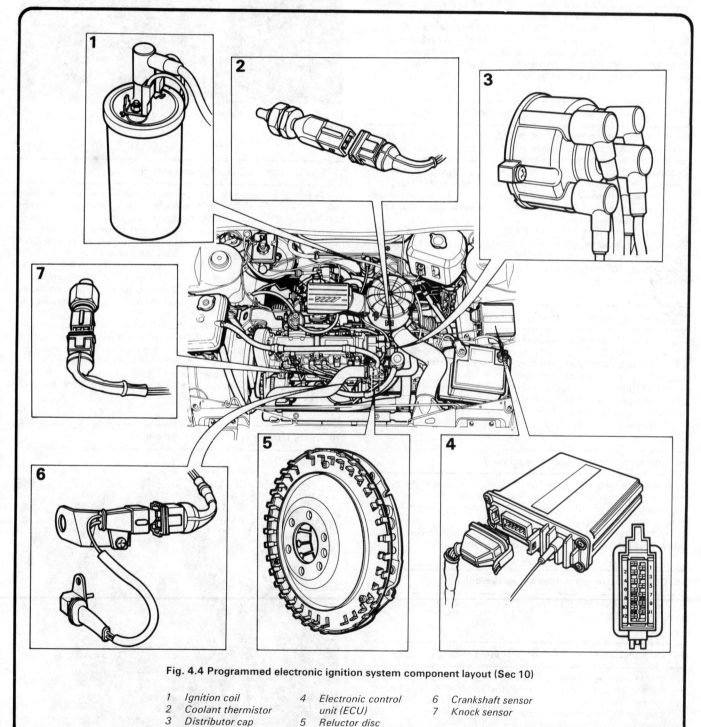

Fig. 4.4 Programmed electronic ignition system component layout (Sec 10)

1 Ignition coil	4 Electronic control	6 Crankshaft sensor
2 Coolant thermistor	unit (ECU)	7 Knock sensor
3 Distributor cap	5 Reluctor disc	

retarded according to information sent to the ECU from the coolant temperature thermistor and knock sensor.

With the firing point established, the ECU triggers the ignition coil which delivers HT voltage to the spark plugs in the conventional manner. The cycle is then repeated, many times a second for each cylinder in turn.

11 Maintenance and inspection

1   The only components of the system which require periodic maintenance are the distributor cap, HT leads and spark plugs. The spark plugs and HT leads should be treated in the same way as for 1.3 litre models, as described in Section 8. Attend to the distributor cap and rotor arm as described in Section 12.
2   On this system, dwell angle and ignition timing are a function of the electronic control unit and there is no provision for adjustment. It is possible to check the ignition advance using a stroboscopic timing light, but this should only be necessary as part of a fault finding procedure. Further details will be found in Section 19.

12 Distributor cap and rotor arm – removal and refitting

1   Undo the two screws and lift the cap off the camshaft carriers. Thoroughly clean the cap inside and out with a dry lint-free rag. Examine the four HT lead segments inside the cap. If the segments appear badly burned or pitted, renew the cap. Make sure that the carbon brush in the centre of the cap is free to move and stands proud of its holders. If renewal of the cap is necessary, mark the position of the HT leads then pull them off. Transfer the leads to a new cap, fitting them in the same position.
2   To remove the rotor arm, undo the retaining screw using a suitable Allen key and withdraw the rotor arm from the end of the camshaft. If necessary remove the rotor arm shield.
3   Refitting the shield, rotor arm and distributor cap is the reverse sequence to removal.

13 Crankshaft sensor – removal and refitting

1   Disconnect the battery negative terminal.
2   Disconnect the wiring multi-plug and undo the wiring plug screw. Undo the two bolts and withdraw the unit from the gearbox or automatic transmission adaptor plate.
3   To refit the sensor, ensure that the correct spacer is fitted to the sensor then position the unit on the adaptor plate and secure with the two retaining bolts and one screw.
4   Reconnect the wiring multi-plug and the battery negative terminal.

14 Knock sensor – removal and refitting

1   The knock sensor is located on the front facing side of the cylinder block in the centre. To remove the unit, disconnect the battery negative terminal and the wiring multi-plug then unscrew the sensor from its location.
2   Refitting is the reverse sequence to removal, but ensure that the sensor and cylinder block mating faces are clean.

15 Coolant thermistor – removal and refitting

1   Removal and refitting procedures for this component are contained in Chapter 2.

16 Electronic control unit – removal and refitting

1   Disconnect the battery negative terminal.
2   Release the catch and disconnect the wiring multi-plug from the front of the unit (photo).
3   Disconnect the ignition vacuum supply hose.
4   Undo the retaining screw, slip the unit out of the retaining tags and remove it from the engine compartment.
5   Refitting is the reverse sequence to removal.

16.2 Electronic control unit wiring multi-plug (A) and vacuum connection (B)

17 Ignition coil – description and testing

1   The coil is bolted to the centre of the engine compartment bulkhead (photo) and it should be periodically wiped over to prevent high tension (HT) voltage loss through arcing.
2   To ensure correct HT polarity at the spark plugs, the LT coil leads must always be connected correctly ie white lead to the coil **positive** terminal and white/black lead to the coil **negative** terminal. Incorrect connections can cause bad starting, misfiring and short spark plug life.
3   Apart from the tests of the low tension circuit contained in the fault diagnosis test procedure (Section 19), accurate checking of the coil output requires special equipment and for the home mechanic the easiest test is by substitution of a new unit.
4   If a new coil is to be fitted, ensure that it is of the correct type, specifically for use on the programmed electronic ignition system. Failure to do so could cause irreparable damage to the electronic control unit.
5   To remove the coil, disconnect the HT and LT wiring, undo the two retaining bolts and lift away the coil. Refitting is the reverse sequence to removal.

18 Spark plugs and HT leads – general

Refer to Section 8.

17.1 Ignition coil location on bulkhead

19 Fault diagnosis – programmed electronic ignition system

Problems associated with the programmed electronic ignition system can usually be grouped into one of two areas, those caused by the more conventional HT side of the system such as spark plugs, HT leads, rotor arm and distributor cap, and those caused by the LT circuitry including the electronic control unit and its related components.

It is recommended that the checks described in Section 9 under the headings 'Engine fails to start' or 'Engine misfires' should be carried out first, according to the symptoms. If the fault still exists the following step-by-step test procedure should be used. For these tests a good quality 0 to 12 volt voltmeter and an ohmmeter will be required.

Engine fails to start

Test	Remedy
1 Connect a voltmeter across pins 9 (+) and 12 – of the electronic control unit (ECU) wiring connector. Does the voltmeter indicate battery voltage 10 seconds after switching on the ignition?	Yes: Proceed to Test 2 No: Check the wiring between the ignition switch and pin 9, and between pin 12 and earth. Rectify as required
2 Connect a voltmeter across pins 10 (+) and 12 – of the ECU wiring connector. Does the voltmeter indicate battery voltage 10 seconds after switching on the ignition?	Yes: Proceed to test 3 No: Check the wiring between the ignition switch and coil (+) terminal and between pin 10 and the coil – terminal. Rectify as required
3 Connect an ohmmeter across the coil terminals. Is the coil primary winding resistance between 0.73 and 0.83 ohms?	Yes: Proceed to test 4 No: Renew the coil
4 Connect a voltmeter between the battery (+) terminal and the coil – terminal. Does the reading on the voltmeter increase when the engine is cranking?	Yes: Engine should start. If not check ignition HT components, fuel system and engine internal components No: Proceed to test 5
5 Switch ignition off and connect an ohmmeter across terminals 4 and 6 of the ECU. Does the ohmmeter register 1.5 k ohms approximately?	Yes: Probable ECU fault No: Check crankshaft sensor wiring and connections. If satisfactory, sensor is suspect

Engine misfires and performance is unsatisfactory

Test	Remedy
1 Set engine at TDC with No 1 cylinder on compression. Highlight mark on crankshaft pulley with white chalk. Make a corresponding mark on timing belt cover. Connect a stroboscopic timing light, disconnect vacuum pipe at manifold and start engine. Does the pulley mark advance as engine speed is increased?	Yes: Proceed to test 2 No: Probable ECU fault
2 With the engine operating as in test 1, apply suction to the end of the vacuum pipe. Does the pulley mark advance as vacuum is increased?	Yes: Ignition system is satisfactory, fault lies elsewhere No: Check for leaks in vacuum pipe and connections. If satisfactory, ECU is faulty

Chapter 5 Clutch

For modifications, and information applicable to later models, see Supplement at end of manual

Contents

Specifications

Type .... Cable-operated single dry plate, with diaphragm spring

Clutch disc diameter
1.3 litre models .... 7.48 in (190.0 mm)
1.6 litre models .... 7.87 in (200.0 mm)

Clutch pedal free travel (all models) .... 0.47 to 1.10 in (12.0 to 28.0 mm)

Torque wrench settings	lbf ft	Nm
Pressure plate to crankshaft	55	75
Flywheel to pressure plate:		
1.3 litre models	15	20
1.6 litre models	11	15
Gearbox end cover (four-speed gearbox)	11	15

1 General description

Unlike the clutch on most engines, the clutch pressure plate is bolted to the crankshaft flange. The flywheel, which is dish shaped, is bolted to the pressure plate with the friction disc being held between them. This is, in effect, the reverse of the more conventional arrangement where the flywheel is bolted to the crankshaft flange and the clutch pressure plate bolted to the flywheel.

The release mechanism consists of a metal disc, or release plate, which is clamped in the centre of the pressure plate by a retaining ring. In the centre of the release plate is a boss into which the clutch pushrod is fitted. The pushrod passes through the centre of the gearbox input shaft and is actuated by a release bearing located in the gearbox end housing. A single finger lever presses on this bearing when the shaft to which it is splined is turned by operation of the cable from the clutch pedal. In effect the clutch lever pushes the clutch pushrod, which in turn pushes the centre of the release plate inwards towards the crankshaft. The outer edge of the release plate presses on the pressure plate fingers forcing them back towards the engine and removing the pressure plate friction face from the friction disc, thus disconnecting the drive. When the clutch pedal is released the pressure plate reasserts itself, clamping the friction disc firmly against the flywheel and restoring the drive.

Clutch adjustment is carried out automatically by a ratchet and spring type self-adjusting mechanism incorporated in the clutch cable.

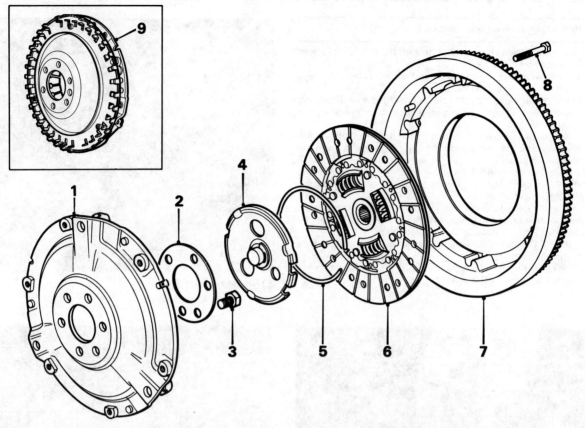

Fig. 5.1 Exploded view of the clutch assembly (Sec 1)

1   Pressure plate (1.3 litre models)
2   Locking plate
3   Pressure plate retaining bolts
4   Release plate
5   Retaining ring
6   Friction disc
7   Flywheel
8   Flywheel retaining bolt
9   Pressure plate (1.6 litre models)

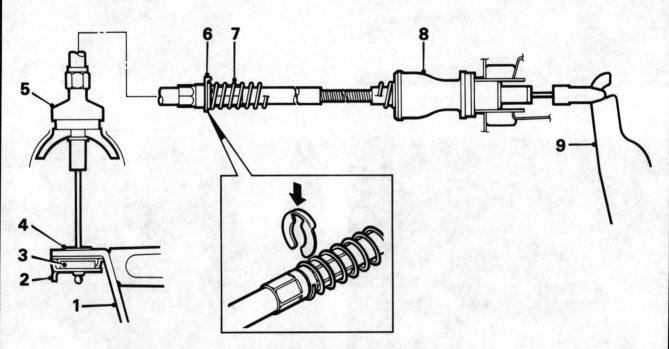

Fig. 5.2 Clutch cable and self-adjusting mechanism (Sec 2)

1   Clutch operating lever
2   Cable retaining clip
3   Cable seating plate
4   Rubber pad
5   Outer cable - gearbox end
6   Spring retaining clip
7   Self-adjusting spring
8   Outer cable - bulkhead end
9   Clutch pedal

2   Clutch cable – removal and refitting

1   Working in the engine compartment, release the clutch cable from its retaining clips and cable ties.

2   Using pliers, withdraw the retaining clip from the cable at the end of the self-adjusting spring (photo).

3   Release the inner cable from the clutch operating lever by sliding out the retaining clip and cable seating plate located on the underside of the lever (photos).

4   Withdraw the inner cable end from the operating lever rubber pad, release the rubber retainer and withdraw the cable assembly from the gearbox bracket (photos).

5   From inside the car, unhook the cable end from the clutch pedal and withdraw the cable into the engine compartment. Remove the cable assembly from the car.

6   To refit the cable, feed the hooked end through the engine compartment bulkhead and connect it to the pedal. Ensure that the outer cable is located correctly in the bulkhead tube.

7   Route the cable through the engine compartment, locating it in its retaining clips and cable ties.

8   Feed the cable through the transmission bracket until the guide sleeve is seated squarely in the bracket.

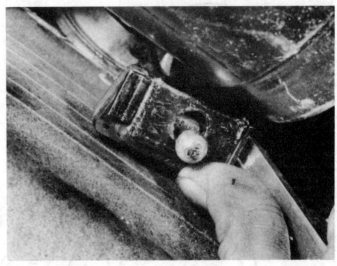

2.3B ... and seating plate

2.2 Remove the cable self-adjusting spring retaining clip

2.4A Withdraw the inner cable from the operating lever rubber pad ...

2.3A Slide out the cable retaining clip ...

2.4B ... release the rubber retainer and remove the cable from the gearbox

9   Feed the inner cable through the rubber pad of the operating lever and slide on the cable seating plate and retaining clip.
10 Refit the retaining clip to the cable at the end of the self-adjusting spring. Press down on the clutch operating lever and at the same time pull up on the cable to operate the self-adjusting mechanism.

3 Clutch pedal – removal and refitting

The clutch pedal and brake pedal are hinged on a common pivot and reference should be made to Chapter 9, Section 21 for the removal and refitting procedures.

4 Clutch assembly – removal and refitting

1   Refer to Chapter 6 and remove the gearbox.
2   In a diagonal sequence, half a turn at a time, slacken the six bolts securing the flywheel to the pressure plate. Use a screwdriver or stout bar engaged with the ring gear teeth and a suitable bolt to prevent the crankshaft turning (photos).
3   When all the bolts are slack, take them out and lift off the flywheel and the clutch friction disc. It may be necessary to carefully ease the flywheel off using a screwdriver, due to the tight fit of the locating dowels.
4   Note the fitted position of the clutch release plate retaining ring as a guide to reassembly, and then prise the ring out using a screwdriver. Lift off the release plate.
5   It is not necessary to remove the pressure plate unless it is visibly worn or is to be renewed for other reasons. If the plate is to be removed, undo the retaining bolts in a diagonal sequence, lift off the locking plate and withdraw the pressure plate. To prevent the crankshaft turning as the bolts are undone, lock it using a spanner or socket on the pulley bolt if a 1.3 litre model is being worked on. On 1.6 litre models use a screwdriver or stout bar located in one of the U-shaped cut-outs in the pressure plate and in contact with a bolt engaged in the starter top mounting bolt hole (photo). **Do not,** under any circumstances, lock the crankshaft using the reluctor ring teeth on the pressure plate periphery. Note that after removal of the pressure plate retaining bolts, new bolts must be obtained before reassembly. The bolts are of the encapsulated type containing a thread locking and sealing compound the properties of which are impaired after removal.
6   With the clutch assembly removed from the engine, refer to Section 5 and carry out a careful inspection of the components.
7   To refit the clutch, place the pressure plate and locking plate in position on the end of the crankshaft and secure with new encapsulated bolts, tightened to the specified torque (photos).
8   Position the release plate over the pressure plate diaphragm fingers (photo) and refit the retaining ring. Ensure that the retaining ring is correctly located, as noted during removal, and as shown in the accompanying illustrations.
9   Hold the clutch friction disc against the flywheel (photo), with the greater projecting boss incorporating the torsion springs facing away from the engine. Refit the flywheel and the six retaining bolts. Tighten the bolts finger tight so that the friction disc is gripped, but can still be moved.

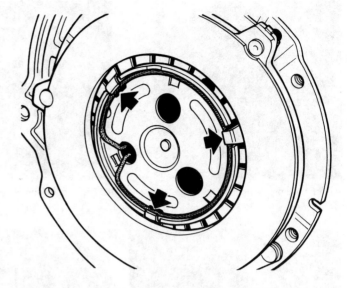

Fig. 5.3 Fitted position of release plate retaining ring (Sec 4)

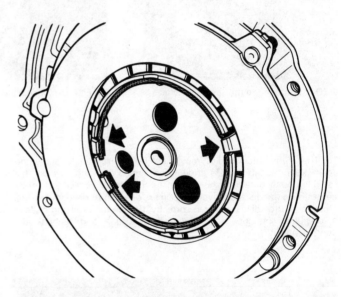

Fig. 5.4 Fitted position of alternative type release plate retaining ring 8Sec 4)

4.2A Undo the flywheel-to-pressure plate retaining bolts (arrowed) ...

4.2B ... while locking the ring gear to prevent rotation

4.5 Lock the pressure plate using a screwdriver or flat bar engaged with the U-shaped cut-out while undoing the bolts

4.7A Fit the pressure plate ...

4.7B ... and locking plate ...

4.7C ... followed by new encapsulated bolts ...

4.7D ... tightened to the specified torque

4.8 Fit the release plate ...

4.9 ... and then the friction disc and flywheel

10 The clutch friction disc must now be centralised to allow the gearbox input shaft to pass through the friction disc hub splines. This can be done by obtaining the manufacturer's special service tool, or by centralising the disc using vernier calipers or a pair of dividers (photo). Once the clutch disc is centred correctly, tighten the retaining bolts in a diagonal sequence to the specified torque and recheck the centralisation.

4.10 Using vernier calipers to check the clutch friction disc centralisation

5 Clutch assembly – inspection

1 With the clutch assembly removed, clean off all traces of asbestos dust using a dry cloth. This is best done outside or in a well ventilated area; *asbestos dust is harmful, and must not be inhaled.*

2 Examine the linings of the friction disc for wear and loose rivets, and the disc for rim distortion, cracks, broken torsion springs and worn splines. The surface of the friction linings may be highly glazed, but, as long as the friction material pattern can be clearly seen, this is satisfactory. If there is any sign of oil contamination, indicated by a continuous, or patchy, shiny black discolouration, the disc must be renewed and the source of the contamination traced and rectified. This will be either a leaking crankshaft oil seal or gearbox input shaft oil seal – or both. Renewal procedures are given in Chapter 1 and Chapter 6 respectively. The disc must also be renewed if the lining thickness has worn down to, or just above, the level of the rivet heads.

3 Check the machined faces of the flywheel and pressure plate. If either is grooved, or heavily scored, renewal is necessary. The pressure plate must also be renewed if any cracks are apparent, or if the diaphragm spring is damaged or its pressure suspect.

4 With the gearbox removed it is advisable to check the condition of the release bearing, as described in the following Sections.

6 Clutch release bearing (four-speed gearbox) – removal, inspection and refitting

1 If the gearbox has not been removed, apply the handbrake, prise off the wheel trim and slacken the left-hand front roadwheel nuts. Jack up the front left-hand side of the car and support it securely on axle stands. Remove the roadwheel.

2   From under the wheel arch, undo and remove the retaining screws and lift away the access panel from the inner wing.
3   Refer to Section 2 and disconnect the clutch cable from the gearbox operating lever and bracket.
4   Place a suitable receptacle beneath the gearbox end cover as there is likely to be some oil spillage when the cover is removed.
5   Undo and remove the four retaining bolts and lift off the end cover. Recover the gasket.
6   Move the operating lever sufficiently to allow the release bearing and sleeve to be withdrawn, and then remove the bearing from the sleeve.
7   Check the release bearing for smoothness of operation and renew it if there is any roughness or harshness as the bearing is spun.
8   Refitting is the reverse sequence to removal, bearing in mind the following points.

(a) Use a new end cover gasket, lightly smeared with jointing compound
(b) Tighten the end cover retaining bolts to the specified torque in a progressive diagonal sequence
(c) Reconnect the clutch cable, as described in Section 2
(d) Check the gearbox oil level and top up if necessary

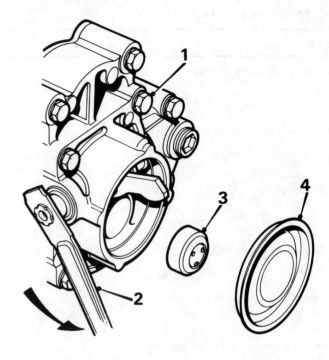

Fig. 5.6 Release bearing components – five-speed gearbox (Sec 6)

1   Gearbox
2   Operating lever
3   Release bearing
4   End cap

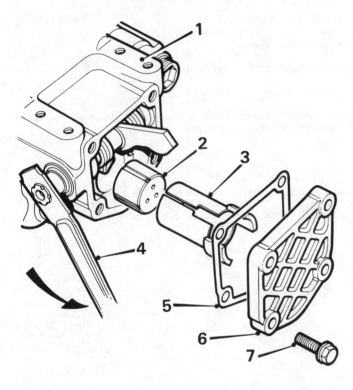

Fig. 5.5 Release bearing components – four-speed gearbox (Sec 6)

1   Gearbox
2   Release bearing
3   Sleeve
4   Operating lever
5   Gasket
6   End cover
7   End cover bolt

7   Clutch release bearing (five-speed gearbox) – removal, inspection and refitting

1   If the gearbox has not been removed, apply the handbrake, prise off the wheel trim and slacken the left-hand front roadwheel nuts. Jack up the front left-hand side of the car and support it securely on axle stands. Remove the roadwheel.
2   From under the wheel arch, undo and remove the retaining screws and lift away the access panel from the inner wing.
3   Refer to Section 2 and disconnect the clutch cable from the gearbox operating lever and bracket.
4   Place a suitable receptacle beneath the gearbox end cap as there is likely to be some oil spillage when the cover is removed.
5   Using a suitable sharp instrument, pierce the end cap and lever it out of its location in the gearbox housing. Discard the end cap, as a new one will be needed when refitting.
6   Remove the operating lever stop clip and then move the lever sufficiently to allow the release bearing to be withdrawn.
7   Check the release bearing for smoothness of operation and renew it if there is any roughness or harshness as the bearing is spun.
8   Refitting is the reverse sequence to removal, bearing in mind the following points.

(a) Use a block of wood or suitable drift to drive in a new end cap and ensure that the cap is kept square as it is fitted
(b) Reconnect the clutch cable, as described in Section 2
(c) Check the gearbox oil level and top up if necessary

8   Fault diagnosis – clutch

Symptom	Reason(s)
Judder when taking up drive	Loose or worn engine/gearbox mountings Friction disc linings contaminated with oil, or worn clutch cable sticking or defective Friction disc hub sticking on input shaft splines
Clutch fails to disengage (clutch spin)*	Clutch cable sticking or defective Excessive free play in the cable or release mechanism Friction disc linings contaminated with oil Friction disc hub sticking on input shaft splines
Clutch slips	Clutch cable sticking or defective Clutch release mechanism sticking or partially seized Faulty pressure plate or diaphragm assembly Friction disc linings worn or contaminated with oil
Noise when depressing clutch pedal	Worn release bearing Defective release mechanism Faulty pressure plate or diaphragm assembly
Noise when releasing clutch pedal	Faulty pressure plate assembly Broken friction disc torsion springs Gearbox internal wear
Self-adjusting mechanism not working	Adjuster body not fully home in bulkhead Incorrect free play in clutch pedal Cable end fitting at pedal not in contact with trigger tube of adjusting mechanism when clutch engaged

*This condition may also be due to the friction disc being rusted to the flywheel or pressure plate. It is possible to free it by applying the handbrake, engaging top gear, depressing the clutch and operating the starter motor. If really badly corroded, then the engine will not turn over, but in the majority of cases the friction disc will free. Once the engine starts, rev it up and slip the clutch several times to clear the rust deposits.

Chapter 6 Manual gearbox

For modifications, and information applicable to later models, see Supplement at end of manual

Contents

Specifications

Type ...... Four or five forward speeds (all synchromesh) and reverse. Final drive integral with main gearbox

Gearbox ratios

1.3 litre models:
1st	3.45 : 1
2nd	1.94 : 1
3rd	1.29 : 1
4th	0.91 : 1
Reverse	3.17 : 1

1.6 litre models:
1st	3.45 : 1
2nd	1.94 : 1
3rd	1.29 : 1
4th	0.91 : 1
5th	0.71 : 1
Reverse	3.17 : 1

Final drive ratios

1.3 litre models	4.25 : 1
1.6 litre models	3.94 : 1

Gearbox overhaul data

3rd gear axial movement	0 to 0.008 in (0 to 0.20 mm)
Axial movement adjustment	Selective circlips

Minimum baulk ring-to-gear clearance:

1st and 2nd gear:
New ring	0.043 to 0.067 in (1.1 to 1.7 mm)
Wear limit	0.020 in (0.5 mm)

3rd gear:
New ring	0.045 to 0.069 in (1.15 to 1.75 mm)
Wear limit	0.020 in (0.5 mm)

4th and 5th gear:
New ring	0.051 to 0.075 in (1.3 to 1.9 mm)
Wear limit	0.020 in (0.5 mm)

Lubricant

Type/specification .... Hypoid gear oil, viscosity SAE 80EP (Duckhams Hypoid 80)
Capacity:
    Four-speed gearbox .... 2.75 pints (1.6 litres)
    Five-speed gearbox .... 3.5 pints (2.0 litres)

Torque wrench settings

	lbf ft	Nm
Gearbox to adaptor plate:		
M7 bolts	4	5
M10 bolts	38	52
M12 bolts	67	91
Main casing to gear carrier housing	18	24
End cover bolts:		
Four-speed gearbox	11	15
Five-speed gearbox	18	24
Selector shaft locking screw	15	20
Engine mounting bracket to end cover or casing	30	41
Mainshaft bearing retaining nuts or screws	11	15
Pinion shaft bearing retaining plate bolts	30	41
Shift fork set retaining bolts	18	24
Reverse idler shaft retaining bolt:		
Four-speed gearbox	15	20
Five-speed gearbox	22	30
5th gear synchro-hub retaining screw	110	149
Oil filler/level plug and drain plug	18	24
Reverse lamp switch	22	30

1 General description

The gearbox is of Volkswagen manufacture and is equipped with either four forward and one reverse gear or five forward and one reverse gear, according to model. Synchromesh gear engagement is used on all forward gears.

The final drive (differential) unit is integral with the main gearbox and is located between the main casing and the bearing housing. The gearbox and differential both share the same lubricating oil.

Gearshift is by means of a floor-mounted lever connected by a remote control housing and shift rod to the gearbox selector shaft and relay lever.

If gearbox overhaul is necessary, due consideration should be given to the costs involved since it is often more economical to obtain a service exchange or good secondhand gearbox rather than fit new parts to the existing unit.

2 Maintenance and inspection

1 At the intervals specified in Routine Maintenance at the beginning of this manual, inspect the gearbox joint faces and oil seals for any sign of damage, deterioration or oil leakage.
2 At the same service interval check and, if necessary, top up the gearbox oil. The filler plug is located on the end of the gearbox casing and the oil level should be maintained up to the level of the filler plug orifice. Removal of the filler plug entails the use of a large Allen key. A suitable alternative, however, is to lock two nuts together on the

thread of a suitable bolt, insert the hexagonal head of the bolt into the filler plug and then unscrew the plug using a spanner on the innermost nut.
3 Although not considered necessary by the manufacturers, the diligent home mechanic may wish to renew the gearbox oil, in the interests of extended gearbox life, every 24 000 miles (40 000 km) or two years.
4 It is also advisable to check for excess free play or wear in the gear linkage joints and levers, and check the gear lever adjustment, as described in Section 24. If any repair or adjustment work is to be carried out on the gear linkage, lubricate all sliding joints and contact surfaces (with the exception of the nylon balljoints) using Unipart Solid Lubricating Paste.

3 Gearbox – removal and refitting

1 Disconnect the battery negative terminal.
2 Slide back the rubber cap and disconnect the two reversing lamp switch wires from the switch terminals.
3 Release the clutch cable adjustment by removing the retaining clip at the end of the cable self-adjusting spring (photo).
4 Slide out the clip securing the clutch cable to the gearbox operating lever then withdraw the seating plate. Disengage the rubber retainer from the guide sleeve and pull the cable and guide sleeve out of the operating lever and gearbox bracket (photo).
5 Undo and remove the bolt securing the speedometer cable to the gearbox casing. Carefully withdraw the cable and pinion assembly and place them aside (photo).

3.3 Remove the cable self-adjusting spring retaining clip

3.4 Remove the cable and guide sleeve from the gearbox bracket

3.5 Withdraw the speedometer cable and pinion assembly

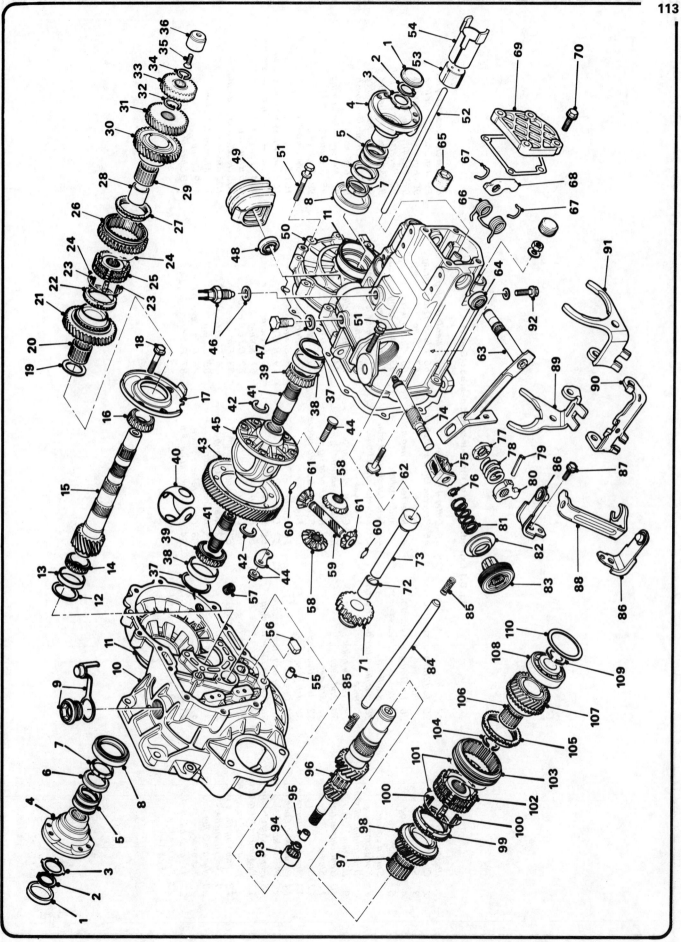

Fig. 6.1 Exploded view of the four-speed gearbox (Sec 1)

1 Plastic cap
2 Circlip
3 Spring washer
4 Drive coupling
5 Spring collar
6 Spring
7 Konusring
8 Differential oil seal
9 Blanking plug
10 Gear carrier housing
11 Sleeve
12 Shim
13 Bearing outer race
14 Pinion shaft bearing
15 Pinion shaft
16 Pinion shaft bearing
17 Bearing outer race and retaining plate assembly
18 Retaining plate bolt
19 Thrust washer
20 Needle roller bearing
21 1st gear
22 1st gear baulk ring
23 Synchro key

24 Spring
25 1st/2nd gear synchroniser hub
26 1st/2nd gear synchroniser sleeve and reverse gear
27 Baulk ring
28 Needle roller bearing inner race
29 Needle roller bearing
30 2nd gear
31 3rd gear
32 Selective circlip
33 4th gear
34 Circlip
35 Bearing retaining screw
36 Needle roller bearing
37 Shim
38 Bearing outer race
39 Taper roller bearing
40 Thrust washer
41 Differential shaft
42 Circlip
43 Final drivegear
44 Bolt, lockplate and nut

45 Differential cage
46 Reverse lamp switch and washer
47 Selector shaft peg bolt and washer
48 Selector shaft oil seal
49 Selector shaft boot
50 Main casing
51 Casing retaining bolt
52 Clutch pushrod
53 Release bearing
54 Release bearing sleeve
55 Dowel
56 Magnet
57 Drain plug
58 Differential gear
59 Pinion pin
60 Circlip
61 Differential pinion
62 Mainshaft bearing retaining clamps
63 Clutch release shaft
64 Release shaft oil seal

65 Filler/level plug
66 Clutch return spring
67 Circlip
68 Clutch lever
69 Main casing end cover
70 End cover bolt
71 Reverse idler gear
72 Bush
73 Reverse idler shaft
74 Selector shaft
75 Yoke
76 Circlip
77 Detent plate
78 Spring
79 Roll pin
80 Selector finger
81 Spring
82 Oil deflector
83 End cover
84 Selector rod
85 Spring
86 Relay lever pillar
87 Pillar retaining bolt

88 Reverse gear relay lever
89 1st/2nd selector fork
90 Reverse selector fork
91 3rd/4th selector fork
92 Reverse idler shaft bolt
93 Needle roller bearing
94 Clutch pushrod oil seal
95 Clutch pushrod bush
96 Mainshaft
97 Needle roller bearing
98 3rd gear
99 Baulk ring
100 Synchro key
101 Spring
102 3rd/4th synchroniser hub
103 3rd/4th synchroniser sleeve
104 Circlip
105 Baulk ring
106 Needle roller bearing
107 4th gear
108 Mainshaft bearing
109 Circlip
110 Shim (if fitted)

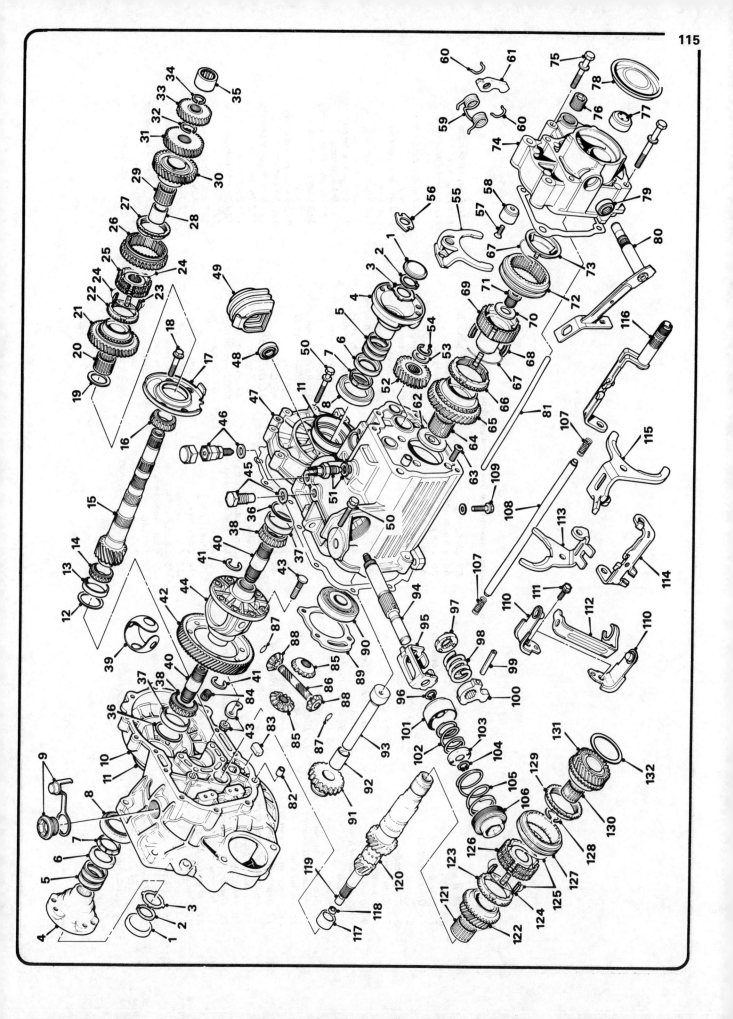

Fig. 6.2 Exploded view of the five-speed gearbox (Sec 1)

1 Plastic cap
2 Circlip
3 Spring washer
4 Drive coupling
5 Spring
6 Spring collar
7 Konusring
8 Differential oil seal
9 Blanking plug
10 Gear carrier housing
11 Sleeve
12 Shim
13 Bearing outer race
14 Pinion shaft bearing
15 Pinion shaft
16 Pinion shaft bearing
17 Bearing outer race and retaining plate assembly
18 Retaining plate bolt
19 Thrust washer
20 Needle roller bearing
21 1st gear
22 1st gear baulk ring
23 Synchro key
24 Spring
25 1st/2nd gear synchroniser hub
26 1st/2nd gear synchroniser sleeve and reverse gear
27 Baulk ring
28 Needle roller bearing inner race
29 Needle roller bearing
30 2nd gear
31 3rd gear
32 Selective circlip
33 4th gear
34 Circlip
35 Needle roller bearing
36 Shim
37 Bearing outer race
38 Taper roller bearing
39 Thrust washer
40 Differential shaft
41 Circlip
42 Final drivegear
43 Bolt, lockplate and nut
44 Differential cage
45 Selector shaft peg bolt and washer
46 5th gear detent plunger and washer
47 Main casing
48 Selector shaft oil seal
49 Selector shaft boot
50 Casing retaining bolt
51 Reverse lamp switch and washer
52 5th gear
53 Thrust washer

54 Circlip
55 5th gear selector fork
56 Lock plate
57 Bearing retainer screw
58 Needle roller bearing
59 Clutch return spring
60 Circlip
61 Clutch lever
62 Thrust washer
63 Bearing retaining plate screw
64 Needle roller bearing
65 5th gear
66 Baulk ring
67 Spring
68 Synchro key
69 5th gear synchroniser hub
70 Washer
71 Synchroniser hub screw
72 5th gear synchroniser sleeve
73 Stop plate
74 End cover
75 End cover bolt
76 Filler/level plug
77 Clutch release bearing
78 End cap
79 Release shaft oil seal
80 Clutch release shaft
81 Clutch pushrod
82 Dowel
83 Magnet
84 Drain plug
85 Differential gear
86 Pinion pin
87 Circlip
88 Differential pinion
89 Mainshaft bearing retaining plate
90 Mainshaft bearing
91 Reverse idler gear
92 Bush
93 Reverse idler shaft
94 Selector shaft
95 Yoke
96 Circlip
97 Detent plate
98 Spring
99 Roll pin
100 Selector finger
101 Sleeve
102 Spring
103 Spring plate
104 Circlip
105 Spring
106 End cover
107 Spring
108 Selector rod
109 Reverse idler shaft bolt

110 Relay lever pillar
111 Pillar bolt
112 Reverse gear relay lever
113 1st/2nd selector fork
114 Reverse selector fork
115 3rd/4th selector fork
116 5th gear selector link and tube
117 Needle roller bearing
118 Clutch pushrod oil seal
119 Clutch pushrod bush
120 Mainshaft
121 Needle roller bearing
122 3rd gear
123 Baulk ring
124 Synchro key
125 Spring
126 3rd/4th synchroniser hub
127 3rd/4th synchroniser sleeve
128 Circlip
129 Baulk ring
130 Needle roller bearing
131 4th gear
132 Shim (if fitted)

3.6 Undo the bolt and release the cable harness clip

3.15 Remove the left-hand engine mounting through-bolt and flat washers (1.3 litre model shown)

3.7 Remove the gearchange rod clip (A) and the selector rod balljoint (B) – shown removed – to allow removal of the linkage

6   Disconnect the battery earth cable and cable harness retaining clip from the top of the gearbox (photo).
7   Extract the clip securing the gearchange rod to the selector shaft lever and slide the rod out of the lever bush (photo). Prise off the rear selector rod nylon balljoint from the relay lever using a screwdriver and move the rod to one side.
8   Make a note of the wiring harness connections at the starter motor solenoid and disconnect them.
9   Undo the starter motor retaining bolts, withdraw the starter and, where fitted, the front snubber and its bracket.
10   Prise off the left-hand front wheel trim and slacken the wheel nuts. Jack up the car, support it securely on axle stands and remove the roadwheel.
11   Undo and remove the retaining screws and lift off the access panel from under the wheel arch.
12   From underneath the front of the car, mark the drive flange to inner constant velocity joint flange relationship using paint or a file.
13   Lift off the protective covers and then undo and remove the bolts securing the constant velocity joints to the drive flanges, using an Allen key. Tie the driveshafts out of the way using string or wire.
14   Using a suitable jack and interposed block of wood, support the engine and gearbox assembly under the engine sump.
15   Undo and remove the left-hand engine mounting through-bolt

(photo). On 1.6 litre models undo the bolts securing the left-hand mounting to the gearbox and remove the mounting. Undo the nuts and bolts securing the rear engine mounting to the crossmember and gearbox and remove the mounting.
16   Position a second jack beneath the gearbox and remove all the bolts securing the gearbox to the engine adaptor plate. Make a note of the different lengths of bolts and their locations and also, on 1.6 litre models, the arrangement of nuts and washers at the inlet manifold support strut.
17   With all the bolts removed, make a final check that everything attached to the gearbox has been disconnected.
18   With the help of an assistant, lower the jacks until sufficient clearance exists to enable the gearbox to be drawn off the side of the engine. Keep the gearbox supported on the jack, as it is quite heavy, and, after releasing the gearbox from the adaptor plate dowels, lower the unit slowly and carefully to the ground.
19   Refitting the gearbox is the reverse sequence to removal, bearing in mind the following points.

(a) *Tighten all retaining and mounting bolts to the specified torque*
(b) *Refill the gearbox with the specified lubricant to the level of the filler plug orifice*
(c) *Align the marks on the drive flanges and inner constant velocity joints before refitting the retaining bolts*
(d) *Lubricate the selector linkage rod and gearchange rod with a lithium-based grease before refitting*
(e) *On models equipped with a front snubber, slacken the snubber cup retaining bolts, position the cup centrally around the snubber then tighten the bolts*
(f) *With the clutch cable connected, refit the spring retaining clip then press down on the operating lever and pull up on the outer cable to operate the self-adjusting mechanism*

4   Gearbox overhaul – general

The overhaul of the gearbox requires the use of a number of special tools, and also certain critical adjustments, if the job is to be done successfully. For this reason it is not recommended that a complete overhaul be attempted by the home mechanic unless he has access to the tools required and feels reasonably confident after studying the procedure. However, the gearbox can at least be dismantled into its major assemblies without too much difficulty, and the following Sections describe this, and the overhaul procedures.

Repair or overhaul of the differential unit is not considered to be within the scope of the home mechanic, since special jigs and fixtures are required for this work.

Before starting any repair work on the gearbox, drain the oil and thoroughly clean the exterior of the casings, using paraffin or a suitable solvent. Dry the unit with a lint-free rag. Make sure that an uncluttered working area is available with some small containers and trays handy to store the various parts. Label everything as it is removed.

Before starting reassembly all the components must be spotlessly clean and should be lubricated with the recommended grade of gear oil during assembly.

5 Gearbox (four-speed) – separating the housings

1   Remove the clutch pushrod. Undo the four bolts securing the gearbox end cover plate and remove the cover plate. This will give access to the clutch release mechanism. Lift out the clutch release bearing and sleeve.

2   There are two circlips, one on each side of the clutch release lever.

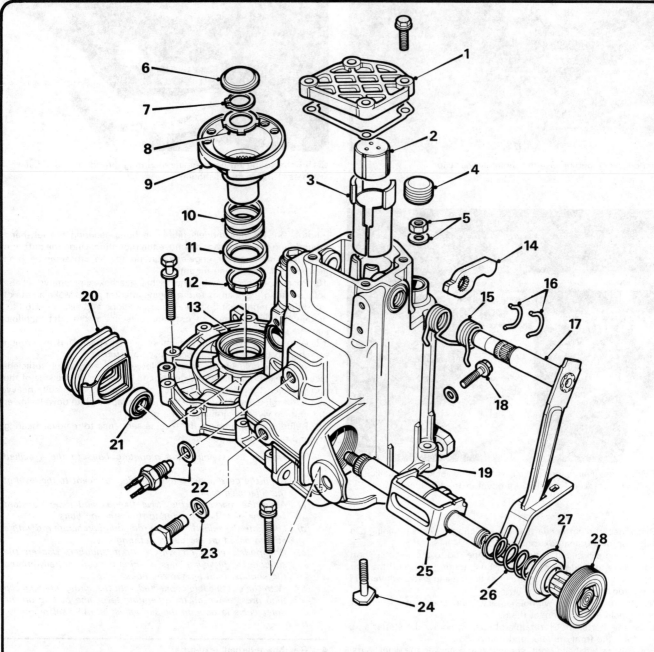

Fig. 6.3 Main casing components – four-speed gearbox (Sec 5)

1 Main casing end cover	8 Spring washer	16 Circlip	23 Selector shaft peg bolt
2 Clutch release bearing	9 Drive coupling	17 Clutch operating shaft	24 Mainshaft bearing retaining
3 Release bearing sleeve	10 Spring	18 Reverse idler shaft bolt	clamp
4 Rubber plug	11 Spring collar	19 Main casing	25 Selector shaft assembly
5 Bearing retaining nut	12 Konusring	20 Selector shaft rubber boot	26 Spring
and washer	13 Oil seal sleeve	21 Selector shaft seal	27 Oil deflector
6 Plastic cap	14 Clutch lever	22 Reverse lamp switch	28 End cap
7 Circlip	15 Return spring		

Remove these and slide the operating shaft out of the main casing, collecting the return spring and release lever as the shaft is withdrawn. Note that there is a master spline on the shaft and that the release lever will fit on the shaft in one way only.

3   Prise out the plastic cap from the centre of the left-hand side drive flange and remove the circlip and spring washer. Withdraw the flange using a suitable puller (photos) then lift out the spring, spring collar and Konusring. There is no need to remove the opposite driveshaft flange at this stage.

4   Remove the selector shaft peg bolt and the lockbolt for the reverse gear shaft (photo). Remove the reversing lamp switch (photo).

5   On the side of the main casing below the clutch withdrawal shaft is the cover for the selector shaft. Using a plug spanner, remove this cover and lift off the oil deflector, then remove the detent spring (photo). Note that the gearbox must be in neutral for this operation.

6   Withdraw the selector shaft from the main casing (photo).

7   On the end of the main casing (where the clutch withdrawal

mechanism is located) are two plastic caps. Prise these out and undo the nuts underneath them (photo). There is a third nut inside the casing from which the clutch withdrawal mechanism was removed; this must also be removed (photo). If these nuts are not removed, the mainshaft bearing cannot be pulled out of the casing and the casing will fracture if pressure is applied to draw it off.

8   Undo and remove the 14 bolts securing the two casings together. Twelve of these bolts are M8 x 50 and two are M8 x 36, note where the shorter bolts are fitted.

9   The casings are now ready for separation. Photos 15.16a and 15.16b show a suitable tool being used to draw the casing off the mainshaft on a five-speed gearbox. The arrangement is the same on the four-speed unit. Secure the tool in the holes for the end cover with two 7 mm bolts and then screw the centre screw down on the top of the mainshaft until it just touches. Fasten a bar or piece of angle iron across the clutch housing, in such a manner as to support the other end of the mainshaft, and then continue to screw in the centre screw

5.3A Prise out the drive flange plastic cap ...

5.3B ... extract the circlip and spring washer ...

5.3C ... then remove the flange with a puller

5.4A Remove the reverse gear shaft lockbolt ...

5.4B ... and the reversing lamp switch

5.5 Unscrew the selector shaft cover, oil deflector and spring

5.6 Withdraw the selector shaft

5.7A Remove the plastic caps over the mainshaft bearing clamp retaining nuts

5.7B A third nut is located inside the casing

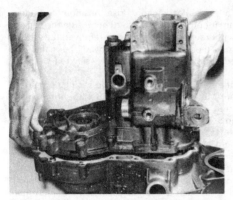

5.9 Lift off the main casing after releasing the bearing

6.2A Extract the selector rod circlips (upper circlip arrowed) ...

6.2B ... and lift away the selector fork assembly

of the tool until the casing is pulled away, leaving the mainshaft bearing complete on the mainshaft. Lift away the main casing (photo). On top of the bearing there may be one or more shims, collect them and label them to ensure that they can be identified at reassembly. The needle bearing for the pinion shaft will remain in the main casing; it can be removed, if necessary, using a suitable extractor.

10 Recover the three clamping screws which retain the mainshaft bearing – they will drop into the gearbox as the casing is being removed. Remove the magnet from the gear carrier housing.

6 Gearbox (four-speed) mainshaft, pinion shaft and differential – removal

1 The mainshaft assembly can be removed quite easily, but the pinion shaft assembly must be partially dismantled before the pinion shaft and the differential unit can be removed.

2 Extract the two circlips (where fitted), securing the selector rod to the selector fork assembly. Withdraw the selector rod from the forks and gear carrier housing then lift away the selector fork assembly (photos). Recover the selector rod spring from the gear carrier housing.

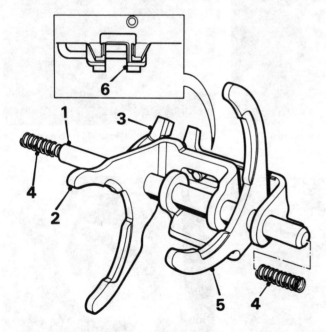

Fig. 6.4 Selector fork assembly – four-speed gearbox (Sec 6)

1 Selector rod
2 1st/2nd selector fork
3 Reverse selector fork
4 Spring
5 3rd/4th selector fork
6 Alignment of selector recesses

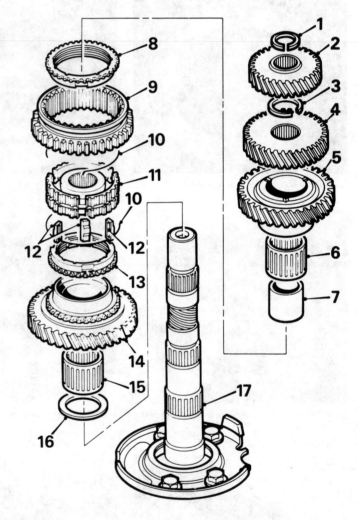

Fig. 6.5 Pinion shaft components – four-speed gearbox (Sec 6)

1 Circlip
2 4th gear
3 Selective circlip
4 3rd gear
5 2nd gear
6 Needle roller bearing
7 Needle roller bearing
   inner race
8 Baulk ring
9 1st/2nd synchro sleeve
10 Retaining spring
11 1st/2nd synchro-hub
12 Synchro key
13 1st gear baulk ring
14 1st gear
15 Needle roller bearing
16 Thrust washer
17 Pinion shaft

6.4 Extract the 4th gear retaining circlip

6.5A Extract the 3rd gear retaining circlip ...

6.5B ... and remove 3rd gear from the pinion shaft

6.6A Lift off 2nd gear ...

6.6B ... followed by its needle roller bearing

6.9A Lift off 1st gear needle roller bearing ...

6.9B ... and the thrust washer

6.10A Unscrew the retaining bolts ...

6.10B ... and withdraw the pinion bearing retaining plate

3   Undo the relay lever pillar retaining bolts, remove the two pillars and the reverse gear relay lever.

4   Remove the circlip retaining 4th gear on the pinion shaft (photo) then lift the mainshaft out of its bearing in the gear carrier housing and at the same time remove 4th gear from the pinion shaft. The mainshaft needle bearing and oil seal will remain in the gear carrier housing.

5   Remove the circlip retaining 3rd gear on the pinion shaft (photo). This circlip is used to adjust the axial movement of 3rd gear and must be refitted in the same position, so label it for identification at reassembly. Remove 3rd gear (photo).

6   Remove 2nd gear and then the needle bearing from over its inner sleeve (photos).

7   To remove the rest of the gears a long-legged puller will be required. Before pulling off the synchroniser unit and 1st gear, remove the reverse gear by tapping the reverse gear shaft out of its seating, then lift the shaft and gear away.

8   Remove the plastic stop button from the end of the pinion shaft and fit the puller under 1st gear. Note that the pinion shaft bearing retainer has two notches to accommodate the puller legs. Pull the gear and synchro-hub off the shaft. Tape the synchro unit together to prevent it coming apart.

9   Remove the needle bearing and thrust washer (photos). Note that the flat side of the washer is towards 1st gear.

10 Remove the four nuts or bolts securing the pinion bearing retainer and lift off the retainer (photos). Note that the retainer incorporates the reverse gear stop. The pinion shaft is seated in a taper roller bearing, and can now be removed from the gear carrier housing.

11 Remove the second drive flange, as described in Section 5, paragraph 3, and then lift the differential unit out of the gear carrier housing (photo). We do not recommend trying to overhaul the differential unit; if it is in any way suspect seek advice from your BL dealer.

6.11 Lift out the differential

7.2A Fit a new differential drive flange oil seal to the gear carrier housing ...

7.2B ... and drive it fully into place

7 Gearbox (four-speed) gear carrier housing – overhaul

1   Clean the housing using paraffin, or a suitable solvent.
2   Prise or drift out the oil seals and fit new seals using a socket or tube and hammer to drive them in (photos). Fill the space between the seal lips with a multi-purpose grease before fitting.
3   The mainshaft needle roller bearing may be removed, if necessary, using a suitable extractor. Do not remove the bearing unless it is defective as it is likely to be damaged during removal (photo).
4   If the outer races of the differential bearings are in need of renewal, then this, and the renewal of the corresponding inner races and bearings on the differential, should be left to a BL dealer as a complicated setting up procedure is involved.
5   The fit of the starter motor armature should be tried in the starter bush. If undue wear is apparent, renew the bush.

7.3 Mainshaft needle roller bearing in the gear carrier housing

8 Gearbox (four-speed) main casing – overhaul

1   The three oil seals in the main casing should be renewed by prising them out, noting their fitted direction, and fitting new seals using a block of wood, large socket or tube to drive them in. Fill the lips of the seals with multi-purpose grease before fitting.
2   The needle roller bearing for the pinion shaft is retained by a screw and can be withdrawn for renewal after removal of the screw.

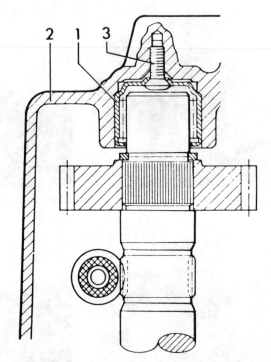

Fig. 6.6 Sectional view of the pinion shaft needle roller bearing – four-speed gearbox (Sec 8)

1   Needle roller bearing   3   Self-tapping screw
2   Main casing

3   If the differential bearing outer race requires renewal, this should be left to a BL dealer, as the bearings also require accurate setting up using jigs and fixtures.

9 Gearbox (four-speed) pinion shaft bearings – renewal

1   The large and small bearings accurately locate the pinion shaft gear with the crownwheel of the differential. If either bearing is defective then both must be renewed. In the removal process the bearings are destroyed. New ones have to be shrunk on and the shim under the smaller bearing changed for one of the correct size.
2   This operation is quite complicated and requires special equipment for preloading of the shaft and measurement of the torque required to rotate the new bearings. In addition the shim at the top of the mainshaft and the axial play at the circlip of the 3rd gear on the pinion shaft will be affected. This will mean selection of a new shim and circlip. There are six different thicknesses of circlip. Therefore it is recommended that if these bearings require renewal, the work should be left to your BL agent.

10 Gearbox (four-speed) mainshaft – dismantling and reassembly

1   Remove the ball-bearing retaining circlip and then, with the legs of a two-legged puller positioned under 4th gear, pull the bearing and gear off the mainshaft.

2   Withdraw the 4th gear needle bearing and baulk ring (photo).

3   Remove the circlip, then support 3rd gear and press the mainshaft through the 3rd/4th synchro-hub. Tape the synchro unit together to prevent its coming apart.

4   Remove the needle bearing to complete the dismantling of the shaft (photos).

5   If the gears on either shaft are to be renewed then the mating gear on the other shaft must be renewed as well. They are supplied in pairs only.

6   The inspection of the synchro units is dealt with in Section 11.

7   When reassembling the mainshaft, lightly oil all the parts.

8   Fit the 3rd gear needle bearing, 3rd gear and the 3rd gear baulk ring. Press on the 3rd/4th gear synchro-hub and fit the retaining circlip (photos). When pressing on the synchro-hub and sleeve, turn the rings so that the keys and grooves line up. The chamfer on the inner splines of the hub must face 3rd gear.

9   The mainshaft ball-bearing should now be pressed into the main casing (photo). Ensure that the same shim(s) removed at dismantling are refitted between the bearing and the casing. The bearing is fitted with the closed side of the ball-bearing cage towards 4th gear. Insert the clamping screws and tighten the clamping screw nuts to the specified torque (photo).

Note: *The endplay will have to be adjusted if either of the bearings, the thrust washer or mainshaft has been renewed, so the help of a BL agent with the necessary special tools and gauges will be required.*

10.2 Withdraw the 4th gear needle roller bearing

10.4A Open the cage and slide off the 3rd gear needle roller bearing

10.4B The mainshaft completely dismantled

10.8A Fit the 3rd gear ...

10.8B ... the 3rd gear baulk ring ...

10.8C ... the 3rd/4th synchro unit ...

10.8D ... and the retaining circlip to the mainshaft

10.9A Fit the mainshaft ball-bearing into the casing ...

10.9B ... and secure with the clamps and retaining nuts

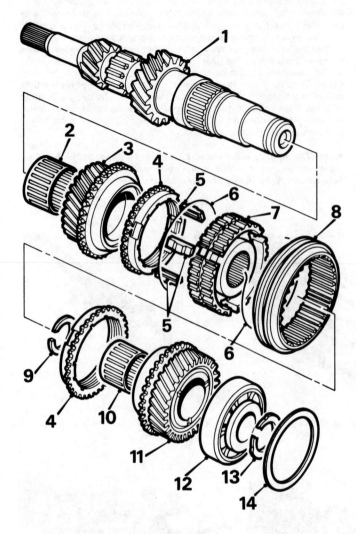

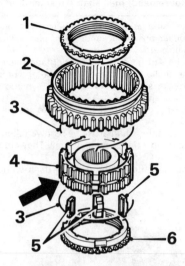

Fig. 6.8 Exploded view of the 1st/2nd synchroniser unit – four and five-speed gearboxes (Sec 11)

1 2nd gear baulk ring	5 Synchro key
2 Synchro sleeve	6 1st gear baulk ring
3 Retaining spring	Arrow indicates identification
4 Synchro-hub	grooves in synchro-hub

---

11 Gearbox (four-speed) synchroniser units – inspection

1   The synchroniser unit hubs and sleeves are supplied as a matched set and must not be interchanged. Before dismantling the units, mark the sleeve and hub in relation to each other.

2   When renewing the synchro baulk rings it is advisable to fit new sliding keys and retaining springs.

3   When examining the units for wear, bear in mind the following:

   (a) With the keys removed, the hub and sleeve should slide easily with minimum backlash or axial rock
   (b) With the baulk rings in position the clearance from the face of the baulk ring to the face of the gear, measured with feeler gauges (photo), must not be less than the dimension given in the Specifications
   (c) Check for excess movement between the selector forks and their grooves in the sleeve. If in doubt compare the clearance with that of new components

4   To reassemble the synchroniser units lay them out on the bench with the identifying marks made during removal uppermost (photo).

5   Slide the synchro sleeve over the hub, with the slots in the sleeve

Fig. 6.7 Mainshaft components – four-speed gearbox (Sec 10)

1 Mainshaft	8 Synchro sleeve
2 Needle roller bearing	9 Circlip
3 3rd gear	10 Needle roller bearing
4 Baulk ring	11 4th gear
5 Synchro key	12 Mainshaft bearing
6 Retaining spring	13 Circlip
7 Synchro-hub	14 Shim (where fitted)

11.3 Checking baulk ring wear using feeler gauges

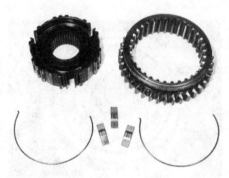

11.4 Components of the synchroniser laid out prior to assembly

11.5A The slots in the synchro sleeve must align with the cutaways in the hub

11.5B Identifying grooves (arrowed) in synchro-hub

11.6A Fit the keys to the grooves in the hub and sleeve ...

11.6B ... and secure with the retaining springs with their hooked ends (arrowed) engaged with the keys

splines (photo) aligned with the cutaways in the hub. Note also that the selector fork groove on the 1st/2nd synchro unit must be fitted away from the side of the hub having the identifying grooves (photo).

6   Fit the keys into the hub and sleeve grooves (photo) and then place one of the retaining springs in position with its hooked end engaged with a key (photo).

7   Turn the synchro unit over and fit the other retaining spring, in the opposite direction to the first, and with its end engaged in a different key.

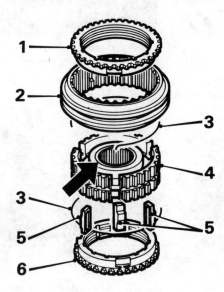

Fig. 6.9 Exploded view of the 3rd/4th synchroniser unit – four- and five-speed gearboxes (Sec 11)

1	4th gear baulk ring	4	Synchro-hub
2	Synchro sleeve	5	Synchro key
3	Retaining spring	6	Baulk ring

Arrow indicates identification groove on hub boss

12 Gearbox (four-speed) differential unit – inspection

1   As described earlier in this Chapter, dismantling and overhaul of the differential are considered beyond the scope of the home mechanic. If the differential assembly is obviously in need of renewal it will be necessary to obtain a complete unit or entrust the work to a dealer, as critical adjustments are required.

2   If the problem with the differential is purely one of noise, and if the noise is more of a whine than a rumble, the unit may continue working

for some considerable time without getting any worse. Again, the help or advice of a dealer is recommended for deciding the course of action to be taken.

13 Gearbox (four-speed) differential, pinion shaft and mainshaft reassembly

1   Refit the differential unit into the gear carrier housing. Refit the Konusring, tapered side towards the differential, followed by the spring collar, plain face towards the differential, and the spring. Refit the drive flange to the differential shaft and, using a long bolt and suitable tube, draw the flange fully onto the differential shaft. **Do not** attempt to drift the flange into place as the differential pinion pin may be damaged. Refit the spring washer, retaining circlip and a new plastic cap.

2   Check that the mainshaft ball-bearing is correctly fitted in the main casing, plastic cage towards the casing, and that the bearing clamp bolt nuts are tight.

3   Fit the pinion shaft complete with its taper bearings into the gear carrier housing, so that the pinion gear meshes with the crownwheel (photo).

4   Fit the bearing retaining plate and the four securing bolts (photo). Fit the 1st gear thrust washer with its flat side up (towards 1st gear). Fit the needle roller cage.

5   Slide 1st gear over the needle bearing and fit the 1st gear baulk ring (photos). The synchro-hub will slide on if heated to 248°F (120°C) – it can then be tapped into position. Make sure that the cutouts are in line with the synchro keys in the 1st/2nd synchro unit to avoid damage to the baulk ring on reassembly. The shift fork groove in the operating sleeve should be nearer 2nd gear and the groove on the hub nearer 1st gear (photos). Fit the 2nd gear synchro baulk ring.

6   The inner race for the 2nd gear needle bearing must be fitted next and pressed down as far as it will go.

7   Fit the reverse idler gear and shaft with the shaft aligned as shown in Fig. 6.10. Use a plastic hammer to drive the shaft into the casing.

8   Fit the 2nd gear needle bearing on the pinion shaft and the 2nd gear with the shoulder downwards.

9   Warm the 3rd gear and press it down over the splines with the collar thrust face towards the 2nd gear.

10 Fit the 3rd gear retaining circlip and, using feeler gauges, measure the play between the gear and circlip (3rd gear axial movement). If the play is in excess of the specified amount, an oversize circlip must be fitted. Circlips are available in a range of thicknesses from 0.098 in (2.5 mm) to 0.118 in (3.0 mm), in 0.004 in (0.1 mm) increments.

11 At this stage the mainshaft must be fitted in position on the gear carrier housing. Slide it into the needle bearing in the casing and locate the selector forks in their appropriate synchroniser sleeve grooves. Locate the selector rod spring in the gear carrier housing, slide the selector rod into place and, where fitted, secure with the retaining circlips (photos).

12 Fit the reverse gear relay lever fork and pillars, ensuring that the relay lever locates correctly (photos). Adjust the position of the pillars so that the relay lever is free to turn, but without any endfloat. Tighten the pillar retaining bolts.

13.3 Fit the pinion shaft to the gear carrier housing

13.4 Tighten the pinion bearing retaining plate bolts to the specified torque

13.5A Fit the 1st gear ...

13.5B ... the 1st gear baulk ring

13.5C ... and the 1st/2nd synchro unit ...

13.5D ... then drive the synchro unit onto the shaft using a suitable tube and hammer

13.11A Fit the assembled mainshaft to the gear carrier housing ...

13.11B ... locate the selector forks in their grooves ...

13.11C ... then slide the selector rod into place

13.12A Fit the reverse gear relay lever fork and pillars

13.12B ... ensuring that the relay lever locates correctly

13 Fit the 4th gear and its retaining circlip on the pinion shaft. Finally insert the stop button, where fitted, for the pinion shaft needle bearing in the end of the pinion shaft. Fit the magnet in its location in the gear carrier housing.

14 The gear carrier housing and shafts are now ready for the assembly of the main casing.

14 Gearbox (four-speed) – reassembling the housings

1 Check that the reverse gear shaft is in the correct position (see Fig. 6.10) and set the geartrain in neutral. Fit a new gasket on the gear carrier housing flange.

2 Lower the main casing over the gears, checking that the pinion shaft is aligned with the pinion shaft needle bearing in the casing. Drive the mainshaft into its bearing, using a suitable mandrel on the inner race. A piece of suitable diameter steel tube can be used. Ensure that the mainshaft is supported on a block of wood when driving the mainshaft into the bearing.

3 Insert the 14 bolts which secure the two housings together and tighten them to the specified torque.

4 Fit the circlip over the end of the mainshaft, working through the release bearing hole (photo). Insert the clutch pushrod into the mainshaft (photo). Ensure that the circlip is properly seated, then fit the clutch release bearing and sleeve assembly.

5 Fit the clutch release shaft and lever. Ensure that the spring is hooked over the lever in the centre and that the angled ends rest against the casing. The shaft can be inserted in the lever in one position only. Fit the two circlips, one each side of the lever.

6 Fit the clutch release sleeve and bearing.

7 Position a new gasket on the end of the casing and fit the end cover plate and four securing screws. Tighten the screws to the specified torque.

8 Lubricate the selector shaft and insert it into the casing. When it is in position, fit the spring(s) and screw in the shaft cover with a plug box spanner, tightening it to the specified torque.

9 Fit the selector shaft peg bolt, reverse gear shaft lockbolt and the reversing lamp switch.

10 Refit the remaining drive flange to the differential shaft using the procedure described in Section 13, paragraph 1.

15 Gearbox (five-speed) – separating the housings

1 Remove the clutch pushrod from the centre of the mainshaft.

2 Undo and remove the retaining bolts and withdraw the main casing end cover. Remove the clutch release bearing and the gasket.

3 Move the selector shaft lever to the neutral position.

4 Unscrew the selector shaft peg bolt, the 5th gear detent plunger and the reversing lamp switch.

5 Undo and remove the nut securing the selector shaft lever to the shaft. Withdraw the lever and the rubber boot.

6 Using a suitable box spanner, such as a spark plug spanner, undo and remove the selector shaft end cap and lift out the spring. Carefully slide out the selector shaft assembly.

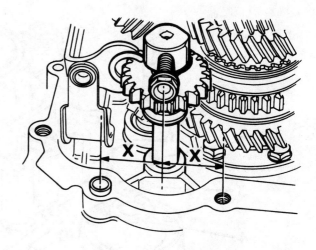

Fig. 6.10 Correct positioning of reverse idler shaft – four- and five-speed gearboxes (Sec 14)

Dimensions X must be equal

7 Undo and remove the reverse idler shaft lockbolt.

8 Using a screwdriver inserted through the selector shaft opening, engage 5th and reverse gears simultaneously by moving the front and rear selector forks towards the clutch housing (photo).

9 Using a Torx driver socket bit, unscrew the 5th gear synchro-hub retaining screw. **Note:** *this screw is extremely tight and is also retained with a thread locking compound. It will be necessary to engage the help of an assistant to hold the gearbox as the screw is undone.* Remove the washer from under the screw head.

10 Carefully prise the 5th gear selector fork lockplate upwards to release it from the selector tube.

11 Using a pair of right-angle circlip pliers, or a similar tool engaged in the slots of the selector tube, turn the tube anti-clockwise to unscrew it from the selector fork.

12 When the fork is released, lift off the synchro hub assembly complete with 5th gear, thrust washer, needle roller bearing and selector fork. *Do not pull the selector rod out of the selector tube, otherwise the selector fork assembly will fall apart inside the gearbox.*

13 Extract the circlip and withdraw the thrust washer and 5th gear from the pinion shaft. Lever the gear up carefully, using two screwdrivers if it is tight.

14 Prise the plastic cap from the centre of the left-hand side drive flange, extract the circlip and dished washer then withdraw the flange using a two-legged puller. Lift out the spring, spring collar and Konusring.

15 Using a Torx driver bit and ratchet or bar, undo and remove the four bolts securing the mainshaft bearing retainer to the gearbox housing.

14.4A Fit the mainshaft circlip working through the release bearing hole ...

14.4B ... then insert the clutch pushrod

15.8 Engage two gears simultaneously by moving the selector forks with a screwdriver

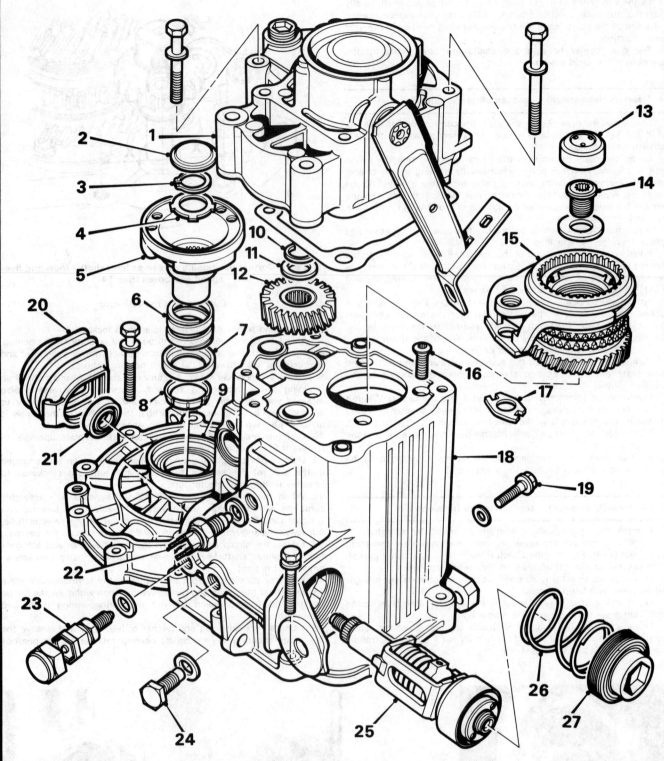

Fig. 6.11 Main casing components – five-speed gearbox (Sec 15)

1	Main casing end cover	9	Oil seal sleeve	15	5th gear selector fork and synchroniser assembly	21	Selector shaft seal
2	Plastic cap	10	Circlip			22	Reverse lamp switch
3	Circlip	11	Thrust washer	16	Mainshaft bearing retaining screw	23	5th gear detent plunger
4	Spring washer	12	5th gear			24	Selector shaft peg bolt
5	Driveshaft coupling	13	Clutch release bearing	17	Lockplate	25	Selector shaft assembly
6	Spring	14	5th gear synchro-hub retaining screw	18	Main casing	26	Spring
7	Spring collar			19	Reverse idler shaft bolt	27	End cap
8	Konusring			20	Selector shaft rubber boot		

15.16A Suitable apparatus for removing the main casing – in position

15.16B By tightening the centre screw the casing is drawn off the bearing

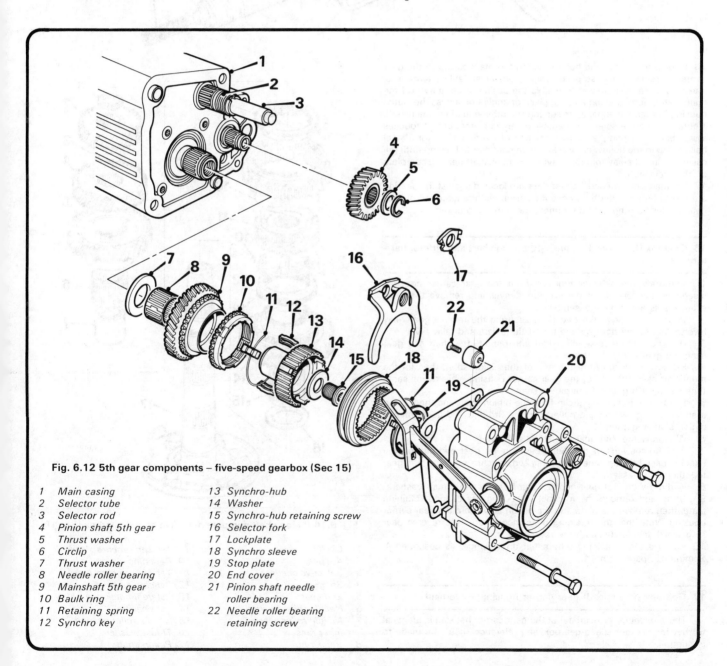

Fig. 6.12 5th gear components – five-speed gearbox (Sec 15)

1	Main casing	13	Synchro-hub
2	Selector tube	14	Washer
3	Selector rod	15	Synchro-hub retaining screw
4	Pinion shaft 5th gear	16	Selector fork
5	Thrust washer	17	Lockplate
6	Circlip	18	Synchro sleeve
7	Thrust washer	19	Stop plate
8	Needle roller bearing	20	End cover
9	Mainshaft 5th gear	21	Pinion shaft needle
10	Baulk ring		roller bearing
11	Retaining spring	22	Needle roller bearing
12	Synchro key		retaining screw

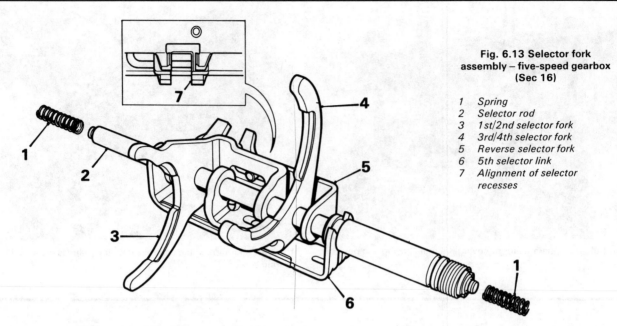

Fig. 6.13 Selector fork
assembly – five-speed gearbox
(Sec 16)

1   Spring
2   Selector rod
3   1st/2nd selector fork
4   3rd/4th selector fork
5   Reverse selector fork
6   5th selector link
7   Alignment of selector
    recesses

16 Undo and remove the bolts securing the main casing to the gear carrier housing. Using strips of angle iron, or suitable alternatives, make up a removal bracket to enable the casing to be drawn off the mainshaft (photo). Fasten a bar or piece of angle iron across the clutch housing in such a manner as to support the other end of the mainshaft. Tighten the centre screw of the casing separator tool until it touches a suitable thrust pad placed over the end of the mainshaft (or a steel ball placed in the hollow centre) and then continue tightening until the casing is pulled away (photo), leaving the mainshaft bearing complete on the mainshaft.
17 Lift the casing off and recover the shim located against the bearing outer race (where fitted). Remove the gasket and the magnet from the gear carrier housing, and disconnect the removal brackets.

16 Gearbox (five-speed) mainshaft, pinion shaft and differential – removal

1   Withdraw the selector fork rod from the gear carrier housing, disengage the forks from the synchro sleeves and remove the forks sideways as a complete assembly.
2   Undo and remove the two bolts securing the reverse gear relay lever pillars to the housing and lift off the pillars and relay lever.
3   Withdraw the reverse idler shaft and remove the shaft and gear from the gear carrier housing.
4   Extract the circlip from the end of the pinion shaft, then lift the mainshaft assembly out of the gear carrier housing while at the same time sliding 4th gear off the pinion shaft.
5   Extract the remaining circlip from the pinion shaft and then, using a suitable long two-legged puller, pull the 3rd and 2nd gears off the pinion shaft together.
6   Withdraw the 2nd gear needle roller bearing.
7   Position the legs of the puller beneath 1st gear and pull off the needle roller bearing inner race, 1st/2nd synchro unit and 1st gear together as an assembly.
8   Withdraw the 1st gear needle roller bearing and thrust washer.
9   Undo and remove the four bolts securing the bearing retaining plate, then remove the plate and the pinion shaft from the gear carrier housing. Note that the retaining plate incorporates the reverse gear stop which locates beneath the reverse gear.
10 Remove the remaining differential drive flange, as described in Section 15, paragraph 14.

17 Gearbox (five-speed) gear carrier housing – overhaul

1   The procedure for overhaul of the gear carrier housing is identical to that for the four-speed gearbox, and reference should be made to Section 7.

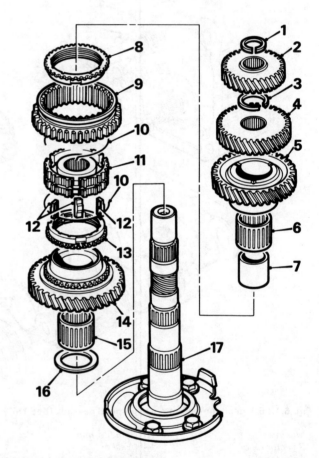

Fig. 6.14 Pinion shaft components – five-speed gearbox (Sec 16)

1   Circlip
2   4th gear
3   Selective circlip
4   3rd gear
5   2nd gear
6   Needle roller bearing
7   Needle roller bearing
    inner race
8   Baulk ring
9   1st/2nd synchro sleeve
10   Retaining spring
11   1st/2nd synchro-hub
12   Synchro key
13   1st gear baulk ring
14   1st gear
15   Needle roller bearing
16   Thrust washer
17   Pinion shaft

17.7A Refit the mainshaft bearing to the main casing ...

17.7B ... fit the bearing retaining plate ...

17.7C ... then secure the plate with the Torx screws

2   The three oil seals in the main casing should be removed by prising them out, noting their fitted direction, and fitting new seals using a block of wood, large socket or tube to drive them in. Fill the lips of the seals with multi-purpose grease before fitting.
3   If required the needle roller bearing for the pinion shaft can be removed using a suitable extractor. Do not remove the bearing unless renewal is necessary as it will probably be damaged during removal.
4   If the differential outer race requires renewal, this should be left to a BL dealer, as the bearings must also be renewed – requiring accurate setting up using jigs and fixtures.
5   If the mainshaft has been dismantled, or if the mainshaft bearing has been removed, it should now be fitted to the main casing.
6   Position the bearing on the casing with the plastic side facing outward. If there were any shims recovered during dismantling, these should be fitted between the bearing and casing.
7   Drive the bearing into position and then refit the bearing retaining plate (photos). Secure the plate with the four Torx screws (photo).
8   To remove the clutch release mechanism from the end cover, pierce the end cap using a sharp tool and prise it out. A new cap must be obtained when refitting.
9   Extract the two circlips retaining the release lever in position on the shaft.
10 Withdraw the release arm and shaft and remove the lever and spring.
11 Inspect the components for signs of wear and renew as necessary. Renew the shaft oil seal.
12 Refit the components of the release mechanism to the end cover using the reverse sequence to removal. Tap a new end cap into place using a hammer and block of wood.

18 Gearbox (five-speed) pinion shaft bearings – renewal

The procedure is identical to that described for the four-speed gearbox, and reference should be made to Section 9.

19 Gearbox (five-speed) mainshaft – dismantling and reassembly

1   Extract the circlip securing the ball-bearing to the mainshaft.
2   Using a press or a hydraulic puller with its legs positioned beneath 4th gear, draw the gear and bearing off the mainshaft.
3   Remove the 4th gear needle roller bearing and extract the circlip securing the 3rd/4th gear synchro-hub to the shaft.
4   Again using a press or hydraulic puller remove 3rd gear and the 3rd/4th synchro-hub from the mainshaft as an assembly.
5   Withdraw the 3rd gear needle roller bearing.
6   Begin reassembly by sliding the 3rd gear needle roller bearing and 3rd gear onto the mainshaft (photos). Place the baulk ring onto 3rd gear (photo).
7   Heat the 3rd/4th synchro-hub in an oven to approximately 248°F (120°C). Place the hub on the mainshaft with the identifying groove in the hub boss towards 4th gear, and drive the synchro-hub assembly into position using a hammer and suitable tube. Make sure that the slots in the baulk ring engage with the keys in the synchro-hub assembly as it is fitted.

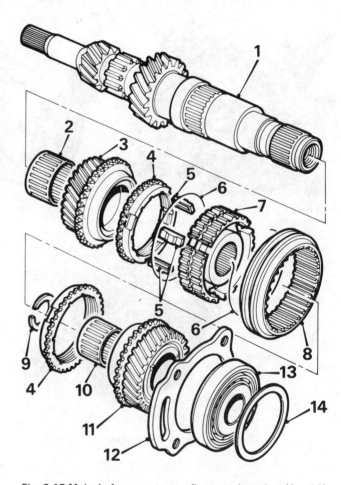

Fig. 6.15 Mainshaft components – five-speed gearbox (Sec 19)

1	*Mainshaft*	*8*	*Synchro sleeve*
2	*Needle roller bearing*	*9*	*Circlip*
3	*3rd gear*	*10*	*Needle roller bearing*
4	*Baulk ring*	*11*	*4th gear*
5	*Synchro key*	*12*	*Bearing retaining plate*
6	*Retaining spring*	*13*	*Mainshaft bearing*
7	*Synchro-hub*	*14*	*Shim (where fitted)*

8   Secure the synchro-hub in position using the circlip (photo).
9   Place the baulk ring against the synchro-hub assembly with the slots aligned with the keys (photo).
10 Slide the needle roller bearing onto the mainshaft and refit 4th gear (photos).

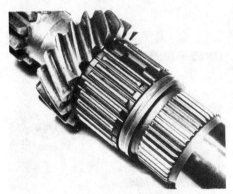

19.6A Slide the 3rd gear needle roller bearing ...

19.6B ... and 3rd gear onto the mainshaft

19.6C Place the baulk ring on the gear

19.8 Refit the 3rd/4th synchro unit and secure with the circlip

19.9 Fit the baulk ring to the synchro unit

19.10A Slide the needle roller bearing onto the mainshaft ...

19.10B ... and fit 4th gear

22.1 Lay the differential in the gear carrier housing

20 Gearbox (five-speed) synchroniser units – inspection

1   The procedure is the same as for the four-speed gearbox, and reference should be made to Section 11.
2   Note that when reassembling the 5th gear synchroniser, the chamfer on the sleeve splines must face toward the hub boss.

21 Gearbox (five-speed) differential unit – inspection

The procedure is the same as for the four-speed gearbox, and reference should be made to Section 12.

22 Gearbox (five-speed) differential, pinion shaft and mainshaft – reassembly

1   Carefully lay the differential unit in the gear carrier housing (photo).
2   Refit the Konusring, tapered side towards the differential, followed by the spring collar, plain face towards the differential, and the spring. Refit the drive flange to the differential shaft and, using a long bolt and suitable tube, draw the flange fully onto the differential shaft. **Do not** attempt to drive the flange into place or the differential pinion pin may be damaged.
3   Refit the spring washer, retaining circlip and a new plastic cap.

4 With the bearings in position on the pinion shaft (photo), place the shaft in the gear carrier housing and mesh it with the differential gear (photo).

5 Fit the retaining plate and tighten the bolts to the specified torque (photos).

6 Place the thrust washer over the pinion shaft with the shoulder on its internal diameter towards the pinion (photo).

7 Slide on the 1st gear needle roller bearing followed by 1st gear and the baulk ring (photo).

8 Heat the assembled 1st/2nd synchro unit in an oven to approximately 248°F (120°C) and then, with the identifying groove in the hub circumference towards 1st gear, drive the unit onto the pinion shaft using a hammer and a suitable tube or press. As the unit is fitted align the slots in the 1st gear baulk ring so that they engage with the keys of the synchro unit.

9 Heat the 2nd gear needle roller bearing inner race and slide it over the pinion shaft and position it in contact with the shoulder on the shaft (photo).

10 Position the needle roller bearing over the inner race (photo) and then refit the baulk ring and 2nd gear (photos).

11 Place the 3rd gear on the pinion shaft with its boss facing 2nd gear (photo), and drive or press it onto the shaft (photo).

12 Fit the circlip to the pinion shaft groove about 3rd gear and then, using feeler gauges, measure the clearance between the gear and circlip (3rd gear axial movement) (photo). If the clearance is in excess of the specified amount, an oversize circlip must be fitted. Circlips are available in a range of thicknesses from 0.098 in (2.5 mm) to 0.118 in (3.0 mm) in 0.004 in (0.1 imm) increments.

13 At this stage the mainshaft must be fitted in position on the gear carrier housing (photo). Engage its end into the needle roller bearing and bring the gears into mesh.

14 Place 4th gear, boss upwards, on the pinion shaft (photo) and secure with the retaining circlip (photo).

15 Ensure that the reverse gear stop is in place on the pinion shaft bearing retaining plate and then refit the reverse idler gear and shaft (photo).

22.4A With the pinion bearing in position ...

22.4B ... place the pinion shaft in the gear carrier housing

22.5A Refit the bearing retaining plate ...

22.5B ... and tighten the bolts to the specified torque

22.6 Fit the thrust washer to the pinion shaft

22.7A Slide on the 1st gear needle roller bearing ...

22.7B ... followed by 1st gear ...

22.7C ... and the baulk ring

22.9 Fit the 2nd gear needle roller bearing inner race

22.10A Place the needle roller bearing over the inner race ...

22.10B ... fit the baulk ring ...

22.10C ... and slide on 2nd gear

22.11A Place 3rd gear on the pinion shaft ...

22.11B ... then drive it into place using a tube and hammer

22.12 Fit the circlip then measure the 3rd gear axial movement

22.13 Position the assembled mainshaft on the gear carrier housing

22.14A Fit 4th gear to the pinion shaft ...

22.14B ... and secure with the retaining circlip

22.15 Fit the reverse idler gear and shaft

22.16 Place the magnet in its location

22.17 Refit the selector forks and rod to the housing

16 Locate the magnet in its location in the gear carrier housing (photo).
17 Engage the selector forks with their respective gears and position the forks and selector rod in the housing (photo).
18 Engage the forked end of the reverse gear relay lever with the reverse idler gear, and secure the relay lever pillars with the two retaining bolts (photo) ensuring that the relay lever is free to turn, but without any endfloat.

23 Gearbox (five-speed) – reassembling the housings

1 Check that the reverse idler shaft is positioned correctly, as shown in Fig. 6.10 and set the geartrain by moving the selector forks to neutral.
2 Fit a new gasket on the gear carrier flange and then carefully lower the main casing over the gears, shafts and selector rod (photo).
3 Support the clutch housing end of the mainshaft with an angle iron bracket and thrust screw, as described during removal, or on a block of wood. Using a tube against the mainshaft bearing inner race in the main casing, drive the bearing and the casing fully into position (photo).
4 Refit and tighten the reverse idler shaft lockbolt (photo), and then secure the gear carrier housing and main casing together – with the retaining bolts tightened to the specified torque.
5 Heat the 5th gear to 248°F (120°C) and fit it into the pinion shaft with the groove near its centre facing away from the main casing (photo).
6 Secure 5th gear with the thrust washer and circlip (photo).
7 Place the needle roller bearing and thrust washer in the mainshaft 5th gear, and place a new lockplate on the selector fork. With the fork engaged with the synchroniser fit this assembly over the mainshaft and engage the selector fork and lockplate over the selector rod (photo).
8 Hold the selector rod and, using right-angled circlip pliers or another suitable tool, turn the selector tube clockwise to engage the

22.18 Refit the reverse gear relay lever and pillars

fork (photo). Screw the tube into the fork until the upper edge of the tube protrudes above the fork face by 0.197 in (5.0 mm) (photo).
9 Engage 5th and reverse gears simultaneously, as described in the removal procedure.
10 Apply a thin smear of thread locking compound to the threads of a new 5th gear synchroniser retaining screw and refit the screw and washer (photo). Tighten the screw to the specified torque.
11 Move the selector forks back to the neutral position and refit the

23.2 Fit the main casing to the gear carrier housing

23.3 Drive the bearing onto the mainshaft

23.4 Refit the reverse idler shaft lockbolt

23.5 Fit 5th gear to the pinion shaft

23.6 Secure 5th gear with the thrust washer and circlip

23.7 Fit the 5th gear synchronizer and selector fork assembly to the mainshaft

23.8A Turn the selector tube clockwise ...

23.8B ... until the tube protrudes above the fork by the specified amount

23.10 Fit the 5th gear synchronizer retaining screw and washer

selector shaft assembly, spring and end cover (photos). Apply a thin smear of thread locking compound to the threads of the end cover and tighten it using a spark plug box spanner or a suitable bolt with two nuts locked together on the bolt thread (photo).

12 Refit the selector shaft peg bolt and the reversing lamp switch (photos).

13 Refit the 5th gear detent plunger (photo) and adjust it as follows if the casing, selector shaft or detent plunger have been renewed, or if too little or too much force is needed to overcome the 5th gear detent.

(a)   Select neutral and remove the plunger plastic cap
(b)   Slacken the locknut and screw in the plunger until the centre pin starts to lift
(c)   Slacken the plunger ⅓ of a turn and tighten the locknut

14 Make sure that the selector tube protrusion is still as described in paragraph 8, and then engage 5th gear. Check that, with 5th gear engaged and the selector fork held away from the gearbox to take up any free play, the synchro sleeve overlaps the hub teeth by at least 0.040 in (1.0 mm).

15 Using two pairs of pliers, one at each end of the lockplate and clamping the lockplate and selector fork bridge, secure the lockplate to the splines of the selector tube.

16 Place the clutch release bearing in its housing (photo) and, with a new gasket in place, refit the end cover to the main casing (photo).

17 Refit and tighten the end cover retaining bolts, noting that the two short bolts are fitted either side of the oil level/filler plug.

18 Refit the remaining differential drive flange, as described in Section 22, paragraph 2.

19 Refit the selector shaft rubber boot (photo), position the selector shaft lever on the shaft and secure with the retaining nut (photo).

20 Attach the linkage support bracket to the main casing (photo).

21 Finally slide the clutch operating pushrod into the mainshaft (photo).

23.11A Refit the selector shaft assembly ...

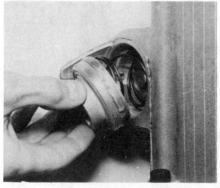
23.11B ... and the spring end cover

23.11C Tighten the end cover securely

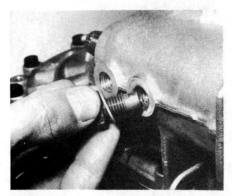
23.12A Refit the selector shaft peg bolt ...

23.12B ... and reversing lamp switch

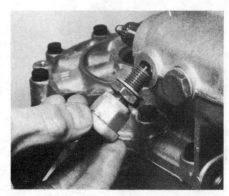
23.13 Refit the 5th gear detent plunger

23.16A Fit the clutch release bearing to the end cover ...

23.16B ... and fit the end cover to the main casing

23.19A Refit the selector shaft rubber boot ...

23.19B ... and selector shaft lever

23.20 Attach the linkage support bracket to the main casing

23.21 Slide the clutch operating pushrod into the mainshaft

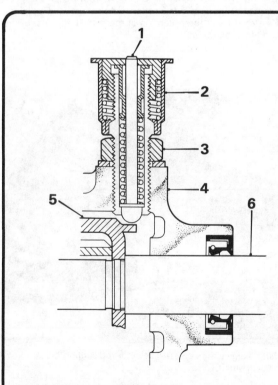

Fig. 6.16 Sectional view of the 5th gear detent plungers – five-speed gearbox (Sec 23)

1	Centre pin	4	Selector shaft housing
2	Plunger	5	Selector shaft yoke
3	Locknut	6	Selector shaft

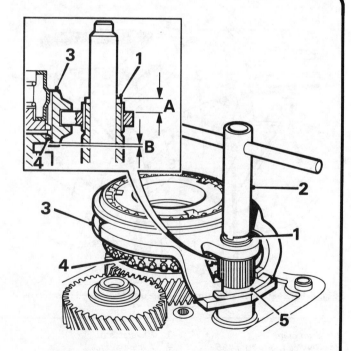

Fig. 6.17 5th gear adjustment – five-speed gearbox (Sec 23)

1	Selector tube	A	Selector tube protrusion =
2	Tool for turning selector tube		0.197 in (5.0 mm)
3	Synchro sleeve	B	Synchro sleeve overlap =
4	Baulk ring		0.040 in (1.0 mm)
5	Lockplate		

24 Gear lever – adjustment

Note: *It will be necessary to obtain BL service tool 18G 1455 to enable the following adjustment to be carried out.*

1   Jack up the front of the car and support it securely on axle stands.
2   From under the car remove the plastic cap from the gear lever rubber boot. Ensure that the gearbox is in neutral.
3   Slacken the clamp bolt (arrowed in Fig. 6.18) which secures the selector rod to the linkage lever, and make sure that the linkage slides

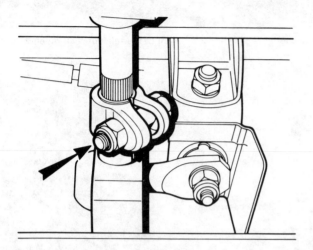

Fig. 6.18 Selector rod to linkage clamp bolt location – four- and five-speed gearboxes (Sec 24)

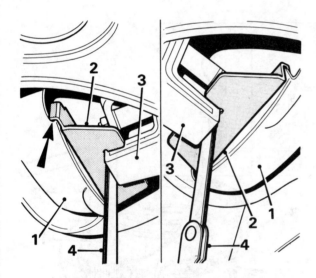

Fig. 6.19 Gear lever adjustment – four- and five-speed gearboxes (Sec 24)

1   *Stop plate*   3   *Reverse stop*
2   *Service tool 18G 1455*   4   *Feeler gauges*

freely on the rod. The angular setting of the linkage cannot be altered, as the splines on the selector rod will have cut unique grooves in the bore of the lever. A new lever is the only solution.
4   Refer to Fig. 6.19 and fit tool 18G 1455 over the lip of the stop plate and with the gear linkage reverse stop engaged with the cutaway portion of the tool.
5   Pull the tool rearward so that the lug on the tool leading edge is in firm contact with the edge of the stop plate.
6   Insert an 0.080 in (2.0 mm) thick feeler gauge between the end of the reverse stop and the tool cutaway. Move the reverse stop until the feeler gauge is a tight sliding fit.
7   Have an assistant hold the selector shaft lever in the vertical position and tighten the selector rod clamp bolt.
8   Remove the feeler gauge and tool, refit the plastic cap to the rubber boot and lower the car to the ground.

25 Gear lever and linkage – removal and refitting

1   Jack up the front of the car and support it securely on stands.
2   From under the car, remove the plastic cap from the gear linkage rubber boot. Undo and remove the nut, washer and through-bolt securing the gear lever to the selector rod.
3   From inside the car, remove the centre console, as described in Chapter 12, undo and remove the two nuts and washers securing the gear lever to the floor and lift out the lever assembly.
4   To remove the selector rod and remote control linkage, extract the retaining clip and slide the gearchange rod out of the bush on the gearbox selector lever.
5   Disconnect the rear selector rod from the relay lever on the gearbox housing by prising off the balljoint using a screwdriver.
6   Undo and remove the two bolts securing the selector rod support bracket to the vehicle floor, and the nuts and bolts securing the linkage support bracket to the steering gear (photo).
7   Withdraw the linkage from under the car.
8   Refitting is the reverse sequence to removal, but adjust the gear lever, as described in the previous Section. Lubricate all sliding joints and contact surfaces (with the exception of the nylon balljoints) using Unipart Solid Lubricating Paste.

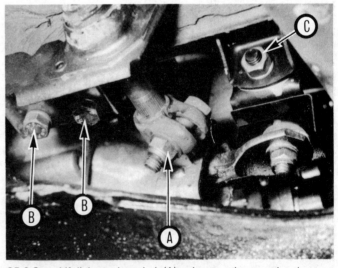

25.6 Gearshift linkage clamp bolt (A), selector rod support bracket retaining bolts (B) and (C)

Fault diagnosis appears overleaf

26 Fault diagnosis – manual gearbox

Symptom	Reason(s)
Gearbox noisy in neutral	Mainshaft (input shaft) bearings worn
Gearbox noisy only when moving (in all gears)	Pinion shaft (output shaft) bearings worn Differential bearings worn Countershaft bearings worn
Gearbox noisy in only one gear	Worn, damaged or chipped gear teeth Worn bearings
Jumps out of gear	Worn synchro-hub or baulk rings Gearshift mechanism out of adjustment Worn selector forks Worn detent mechanisms
Ineffective synchromesh	Worn baulk rings or synchro-hubs
Difficulty in engaging gears	Clutch fault Gearshift mechanism out of adjustment or broken Worn synchromesh

Chapter 7 Automatic transmission

Contents

Specifications

Type .... Hydraulically-controlled epicyclic geartrain with three element torque converter

Gear ratios
1st 2.71:1
2nd 1.50:1
3rd 1.00:1
Reverse 2.43:1
Final drive ratio 3.409:1

Lubricant
Type/specification:
Automatic transmission .... Dexron IID type ATF (Duckhams Uni-Matic or D-Matic)
Final drive .... Hypoid gear oil, viscosity SAE 90EP (Duckhams Hypoid 90S)
Capacity:
Automatic transmission:
Drain and refill .... 5.25 pints (3.0 litres)
Total (dry unit including torque converter) .... 10.5 pints (6.0 litres)
Final drive .... 1.25 pints (0.71 litre)

Torque wrench settings

	lbf ft	Nm
Torque converter to driveplate	22	30
Oil strainer to valve block	2	3
Oil pan to transmission	15	20
Transmission to engine and adaptor plate:		
M10 bolts	33	45
M12 bolts	66	89
Dipstick tube lower mounting bolt	7	9
Dipstick tube upper mounting bolt	18	24

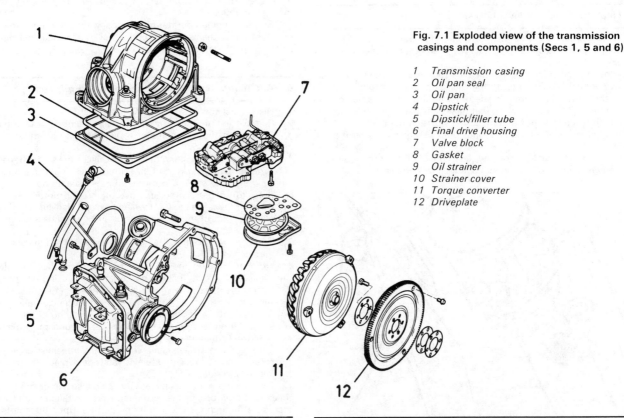

Fig. 7.1 Exploded view of the transmission casings and components (Secs 1, 5 and 6)

1 *Transmission casing*
2 *Oil pan seal*
3 *Oil pan*
4 *Dipstick*
5 *Dipstick/filler tube*
6 *Final drive housing*
7 *Valve block*
8 *Gasket*
9 *Oil strainer*
10 *Strainer cover*
11 *Torque converter*
12 *Driveplate*

1 General description

A three-speed automatic transmission is available as an option on Montego 1.6 HL models. The transmission consists of a torque converter, an epicyclic geartrain and hydraulically-operated clutches and brakes.

The torque converter provides a fluid coupling between engine and transmission which acts as an automatic clutch and also provides a degree of torque multiplication when accelerating.

The epicyclic geartrain provides either of the three forward or one reverse gear ratios according to which of its component parts are held stationary or allowed to turn. The components of the geartrain are held or released by brakes and clutches which are activated by hydraulic valves. An oil pump within the transmission provides the necessary hydraulic pressure to operate the brakes and clutches.

Driver control of the transmission is by a six position selector lever which allows fully automatic operation with a hold facility on the first and second gear ratios.

Due to the complexity of the automatic transmission any repair or overhaul work must be left to a BL dealer or automatic transmission specialist with necessary equipment for fault diagnosis and repair. The contents of the following Sections are therefore confined to supplying general information and any service information and instructions that can be used by the owner.

2 Maintenance and inspection

1   At the intervals specified in Routine Maintenance, carefully inspect the transmission joint faces and oil seals for any signs of damage, deterioration or oil leakage.
2   At the same service intervals check the transmission fluid level and the final drivegear oil level, as described in Sections 3 and 4.
3   At less frequent intervals (see Routine Maintenance) drain the transmission fluid, clean the oil strainer then refill with fresh fluid, using the procedure described in Section 5.
4   Carry out a thorough road test ensuring that all gear changes occur smoothly and, when under kickdown acceleration, at the speeds given in Section 7. With the vehicle at rest, check the operation of the parking pawl when P is selected.

3 Automatic transmission fluid level checking

1   The automatic transmission fluid level should be checked when the engine is at normal operating temperature, preferably after a short journey.
2   With the car standing on level ground and with the engine running, apply the handbrake and slowly move the selector lever through all gear positions.
3   Return the selector lever to N and with the engine still idling, withdraw the dipstick from the filler tube and wipe it on paper or a non-fluffy cloth.
4   Reinsert the dipstick, withdraw it immediately and observe the fluid level. This should be between the upper and lower 'O' marks on the dipstick.
5   If topping-up is necessary, switch off the engine and add the required quantity of the specified fluid through the dipstick tube. Use a funnel with a fine mesh screen to avoid spillage and to ensure that any foreign matter is trapped. Take care not to overfill the transmission; noting that the difference between the upper and lower 'O' marks on the dipstick is 0.75 pint (0.4 litre).
6   After topping-up, re-check the level again, as described above, refit the dipstick and switch off the engine.

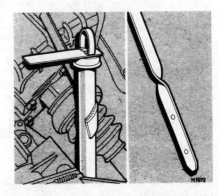

Fig. 7.2 Transmission fluid dipstick/filler tube details (Sec 3)

4   Final drive gear oil level checking

1   With the car raised to provide working clearance, but in a level position, wipe clean the area around the filler/level plug then unscrew the plug.

2   Allow any oil lodged behind the plug to trickle out and then check the level which should be up to the filler plug orifice. If topping-up is necessary, inject the correct grade of oil until it just runs out of the orifice then refit the plug. Note that gear oil is used for lubrication of the final drivegears – **not** automatic transmission fluid.

3   After checking the level and topping-up as required, lower the car to the ground.

4   Draining and refilling of the final drive oil is not a service requirement and no provision is made for this purpose.

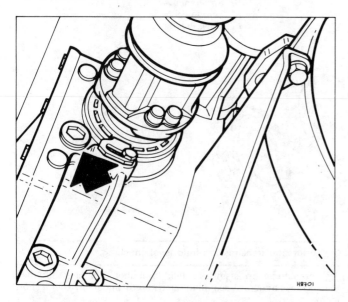

Fig. 7.3 Final drivegear oil level/filler plug location (Sec 4)

5   Automatic transmission fluid – draining and refilling

Note: *To avoid the risk of scalding, only drain the fluid when cold or after the vehicle has been standing for some time.*

1   Jack up the front of the car and support it on stands.

2   Wipe around the oil pan-to-transmission case joint to avoid the risk of dirt or grit entry.

3   Place a suitable container beneath the transmission oil pan and slacken the oil pan retaining bolts. Carefully separate the oil pan-to-transmission joint if it is stuck, and allow the fluid to drain.

4   When most of the fluid has drained, remove the bolts and lift off the oil pan. Recover the seal.

5   Undo the bolts securing the oil strainer to the valve block and withdraw the strainer, strainer cover and gasket.

6   Thoroughly clean the strainer in paraffin, or a suitable solvent, and dry with compressed air.

7   Using a new gasket if necessary, refit the strainer assembly and secure with the retaining bolts tightened to the specified torque.

8   Ensure that the mating faces of the oil pan and transmission casing are clean and refit the pan, using a new seal if necessary. Tighten the retaining bolts to the specified torque.

9   Lower the car to the ground and fill the transmission with the specified type and quantity of transmission fluid, through the dipstick/filler tube. Use a funnel with a fine mesh screen to avoid spillage and to ensure that any foreign matter is trapped.

10 With the car standing on level ground, apply the handbrake, select P and start the engine.

11 Move the selector lever through all gear positions, pausing at each position.

12 With the engine idling, select N, withdraw the dipstick and wipe it on paper or a lint-free cloth.

13 Reinsert the dipstick fully then withdraw it again and observe the fluid level. Top up if necessary until the level reaches the lower 'O' mark on the dipstick.

14 Run the engine until normal operating temperature is reached or, preferably, drive the car for a short journey. Carry out a final level check and top up, as described in Section 3.

6   Automatic transmission – removal and refitting

1   Disconnect the battery negative terminal.

2   Remove the air cleaner assembly, as described in Chapter 3.

3   Disconnect the speedometer cable from its transmission attachment.

4   Undo the bolts securing the dipstick/filler tube to the transmission and remove the dipstick and tube. Note the O-ring at the base of the tube and also the cable clip on the upper retaining bolt.

5   Refer to Chapter 10 and remove the starter motor.

6   Prise off the left-hand front wheel trim and slacken the wheel nuts. Jack up the front of the car, support it on axle stands and remove the roadwheel.

7   Undo and remove the retaining screws and lift off the access panel from under the wheel arch.

8   From underneath the front of the car, mark the drive flange to inner constant velocity joint flange relationship using paint or a file.

9   Lift off the protective covers and then undo and remove the bolts securing the constant velocity joints to the drive flanges, using an Allen key. Tie the driveshafts out of the way using string or wire.

10 Undo the nut securing the selector cable trunnion to the selector lever. Slip the trunnion out of the lever.

11 Release the kickdown cable end from the transmission lever, undo the cable support bracket retaining bolts and place the cables and bracket to one side.

12 Turn the crankshaft as necessary using a socket or spanner on the pulley bolt until one of the torque converter retaining bolts becomes accessible through the starter motor aperture. Undo the bolt then turn the crankshaft and remove the remaining two bolts in the same way.

13 Place a jack beneath the engine sump with a block of wood between the jack head and sump.

14 Slacken, but do not remove, the upper front transmission-to-engine retaining bolt then remove all the other remaining bolts securing the transmission to the engine and adaptor plate.

15 Undo the bolts securing the left-hand mounting bracket to the body. Lower the engine slightly, undo the mounting-to-transmission bolts, remove the earth cable and the mounting assembly.

16 Remove the engine front snubber bracket and the rear mounting-to-crossmember assembly.

17 Place a second jack beneath the transmission with interposed block of wood and just take the weight of the unit.

18 Remove the remaining transmission-to-engine bolt then lower both jacks until the transmission is clear of the body side-member.

19 Withdraw the transmission and torque converter from the engine and remove the assembly from under the car.

20 Refitting the transmission is the reverse sequence to removal, bearing in mind the following points:

(a) *Tighten all retaining and mounting bolts to the specified torque where applicable*

(b) *Top up or refill the transmission and final drive with the specified lubricants as described earlier in this Chapter*

(a) *Align the marks on the drive flanges and inner constant velocity joints before refitting the retaining bolts*

(d) *Adjust the front snubber cup position so that it is central around the snubber rubber*

(e) *Check and, if necessary, adjust the kick-down cable and selector cable, as described in Sections 7 and 8*

7   Kickdown cable – adjustment

1   Fully depress the accelerator pedal and check that there is full movement at the carburettor linkage. If the kickdown cable is preventing full movement, slacken the locknut and back off the cable adjuster slightly (Fig. 7.4).

2   Drive the car until normal operating temperature is reached and then find a quiet, straight stretch of road.

3   Accelerate the car from rest with the selector lever in D and the

accelerator pedal fully depressed in the kickdown position. Note the road speeds at which the 1st/2nd and 2nd/3rd gear changes occur, these should be:

1st to 2nd .... 38 to 41 mph (61 to 66 km/h)
2nd to 3rd .... 67 to 69 mph (108 to 112 km/h)

If the gearchange speeds are too low, decrease the cable tension. If the gearchange speeds are too high, increase the cable tension.
4   To adjust the cable tension, switch off the engine and slacken the kickdown cable locknut. Turn the cable adjuster to increase or decrease the cable tension as required then screw the locknut back up to the bracket. Hold the locknut and tighten the adjuster by hand. Do not hold the adjuster and tighten the locknut.
5   Repeat the procedure described in paragraphs 3 and 4, making small adjustments each time until the gear changes occur at the specified speeds.

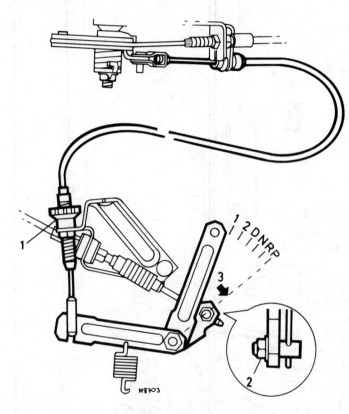

Fig. 7.4 Kickdown cable and selector cable adjustment points (Secs 7 and 8)

1   Kickdown cable adjuster and locknut
2   Selector cable trunnion
3   Move selector lever towards left-hand roadwheel to adjust

8   Selector cable – adjustment

1   Move the selector lever to the P position.
2   Slacken the selector cable trunnion at the transmission selector lever then move the lever towards the left-hand roadwheel as far as it will go (Fig. 7.4).
3   Hold the lever in this position and tighten the trunnion.
4   Check that the trunnion pivots are without any free play, that the selector lever moves through all gear positions and that the starter only operates with the lever in the P or N positions.

9   Selector cable – removal and refitting

1   Refer to Chapter 12 and remove the centre console.
2   Move the selector lever to the P position.

3   Extract the retaining spring clip and disconnect the cable end from the selector lever. Undo the outer cable locknut and release the cable from the selector lever housing.
4   Attach a drawstring to the disconnected cable end.
5   Working in the engine compartment, slacken the cable-to-transmission selector lever trunnion and the locknut securing the cable to the support bracket. Remove the cable from the trunnion and bracket.
6   Release the grommet from the engine compartment bulkhead, pull the cable through and into the engine compartment then untie the drawstring. Remove the cable from the car.
7   Attach the drawstring to the new cable and insert the cable into the bulkhead as far as possible.
8   Pull the carpet up from behind the left-hand side of the heater ducts. Pull the drawstring and guide the cable below the heater ducts and through the selector lever housing. Remove the drawstring.
9   Lubricate both ends of the inner cable with multi-purpose grease and connect the cable using the reverse of the removal sequence.
10   Refit the centre console then adjust the cable, as described in Section 8.

10   Starter inhibitor/reversing lamp switch – removal and refitting

1   Refer to Chapter 12 and remove the centre console.
2   Place the selector lever in N then undo the retaining Allen screw and remove the selector lever handle.
3   Remove the selector panel and light screen. Withdraw the bulb holders and remove the guide plate.
4   Disconnect the switch wiring connectors, undo the two screws and lift out the switch.
5   Before refitting the switch, check the condition and operation of the switch contact on the selector lever and renew the contact if worn. Also check that the selector cable trunnion on the transmission selector lever pivots freely but without free play. Free play in the trunnion or selector cable will affect the operation of the starter inhibitor/reversing lamp switch. If free play is evident, renew the nylon bush in the selector lever trunnion.
6   Fit the switch and secure with the two screws, finger tight only at this stage.
7   With the selector lever at N align the mark on the upper face of the switch with the mark on the contact bracket, then tighten the retaining screws.
8   Connect the wiring, refit the selector handle and check that the starter only operates in P or N and the reversing lamps operate in R. Realign the switch if necessary.
9   Remove the selector handle, refit the guide plate, bulb holders, light screen and selector panel.
10   Refit the selector handle, followed by the centre console.

11   Selector lever assembly – removal and refitting

1   Refer to Chapter 12 and remove the centre console.
2   Place the selector lever in the P position.
3   Extract the spring clip securing the selector cable to the lever and disconnect the cable end. Undo the outer cable retaining locknut and withdraw the selector cable from the selector lever assembly.
4   Disconnect the electrical wiring at the starter inhibitor/reversing lamp switch.
5   Undo the retaining bolts and remove the selector lever assembly from the car.
6   Refitting is the reverse sequence to removal. Check the selector cable adjustment, as described in Section 8 after refitting.

12   Fault diagnosis – automatic transmission

In the event of a fault occurring on the transmission, it is first necessary to determine whether it is of a mechanical or hydraulic nature and to do this the transmission must be in the car. Special test equipment is necessary for this purpose, together with a systematic test procedure, and the work should be entrusted to a suitably equipped BL dealer or automatic transmission specialist.

Do not remove the transmission from the car for repair or overhaul until professional fault diagnosis has been carried out.

144

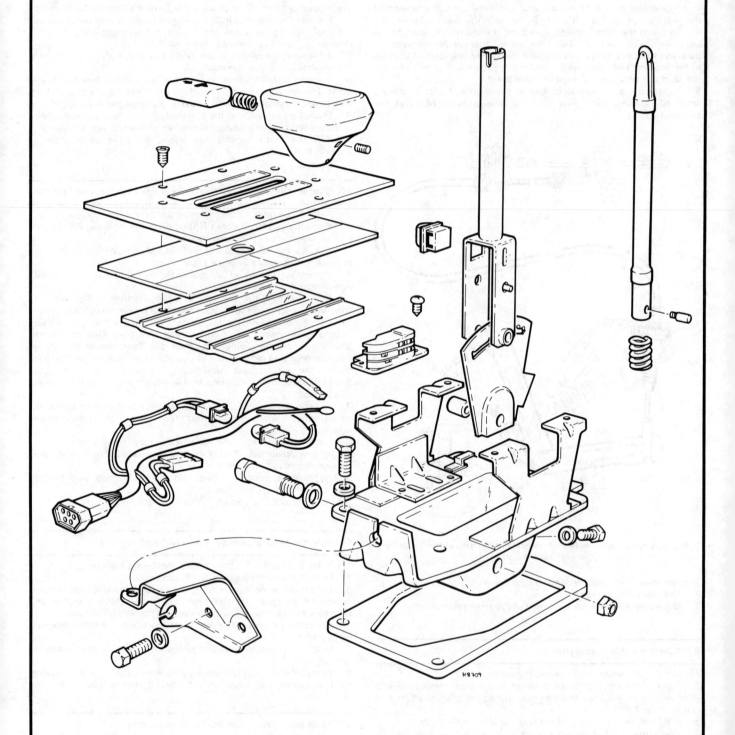

Fig. 7.5 Exploded view of the selector lever assembly (Sec 11)

Chapter 8 Driveshafts

For modifications, and information applicable to later models, see Supplement at end of manual

Contents

Specifications

Type
.... Unequal length, solid (left-hand), tubular (right-hand), splined to inner and outer constant velocity joints

Lubrication
Assembly only – see text
Type:
    Outer constant velocity joint .... Molycote grease VN2461/C, or equivalent
    Inner constant velocity joint .... Mobil 525 grease, or equivalent
    Hub bearing water shield .... Duckhams LBM 10, or equivalent
Quantity:
    Outer constant velocity joint:
        1.3 litre models .... 78 cc
        1.6 litre models .... 90 cc
    Inner constant velocity joint:
        All models .... 180 cc

Torque wrench settings

	lbf ft	Nm
Driveshaft nut*	150	203
Inner joint to drive flange	33	45
Suspension strut to swivel hub nuts	66	89
Roadwheel nuts	53	72

Refer to Chapter 13, Section 10

1 General description

Drive is transmitted from the differential to the front wheels by means of two unequal length driveshafts. The right-hand driveshaft is a hollow construction and is of a larger diameter than the solid left-hand shaft. This is necessary to balance the torsional stiffness of both driveshafts so that an equal torque is applied to the front wheels under acceleration.

Both driveshafts are fitted with constant velocity joints at each end. The outer joints are of the Rzeppa ball and cage type and are splined to accept the driveshaft and wheel hub drive flange. The inner joints are of the sliding tripod type allowing lateral movement of the driveshaft during suspension travel. The inner joints are splined to the driveshafts and bolted to the differential drive flanges.

2 Maintenance and inspection

1 At regular intervals (see Routine Maintenance) carry out a thorough inspection of the driveshafts and joints as follows.

2 Jack up the front of the car and support it securely on axle stands.

3 Slowly rotate the roadwheel and inspect the condition of the outer joint rubber boots. Check for signs of cracking, splits or deterioration of the rubber which may allow the grease to escape and lead to water and grit entry into the joint. Also check the security and condition of the retaining clips. Repeat these checks on the inner constant velocity joints. If any damage or deterioration is found, the joints should be attended to, as described in Sections 4 or 5.

4 Continue rotating the roadwheel and check for any distortion or damage to the driveshafts. Check for any free play in the joints by holding the driveshaft firmly and attempting to rotate the wheel. Repeat this check whilst holding the differential drive flange. Any noticeable movement indicates wear in the joints, wear in the driveshaft splines or loose joint retaining bolts or hub nut.

5 Check the tightness of the inner joint retaining bolts using an Allen key, after removing the protective caps.

6 Lower the car to the ground, prise off the wheel trim, extract the split pin and check the tightness of the hub retaining nut. Fit a new split pin after aligning the holes and slots in the joint and hub nut. Refit the wheel trim.

7 Road test the car and listen for a metallic clicking from the front, as the car is driven slowly in a circle on full lock. If a clicking noise is

heard this indicates wear in the outer constant velocity joint caused by excessive clearance between the balls in the joint and the recesses in which they operate. Remove and inspect the joint, as described in Section 5.

8   If vibration, consistent with road speed, is felt through the car when accelerating, there is a possibility of wear in the inner constant velocity joint. Remove and inspect the joint, as described in Section 4.

3   Driveshaft – removal and refitting

Note: *First refer to Chapter 13, Section 10 for details of modification*
1   While the vehicle is standing on its wheels, firmly engage the handbrake and put the transmission in gear or in the P position if automatic transmission is fitted.
2   Prise off the wheel trim and extract the driveshaft nut retaining split pin. Using a suitable socket and bar, slacken the nut, but do not remove it at this stage.
3   Slacken the wheel nuts, jack up the front of the car and support it on axle stands. Remove the roadwheel and return the transmission to neutral. Remove the driveshaft nut and washer.
4   From underneath the car, make an alignment mark between the inner constant velocity joint flange and the differential drive flange, as an aid to reassembly.
5   Remove the protective caps over the inner joint retaining bolts and using an Allen key of the appropriate size, unscrew and remove the bolts (photos).
6   The procedure now varies slightly depending on whether the left-hand or right-hand driveshaft is being removed.

Left-hand driveshaft
7   On vehicles with manual transmission ease the inner constant velocity joint away from the differential drive flange and lower the inner end of the shaft. Withdraw the outer constant velocity joint from the wheel hub, lower the driveshaft to the ground and remove it from under the car. Withdraw the bearing water shield from the outer joint.
8   On vehicles with automatic transmission undo and remove the nuts and washers, then withdraw the two bolts securing the

suspension strut to the upper part of the swivel hub. Separate the swivel hub from the strut, ease the inner constant velocity joint away from the differential drive flange and withdraw the outer constant velocity joint from the wheel hub. Remove the driveshaft from under the car and slide the bearing water shield off the outer joint.

Right-hand driveshaft
9   On vehicles with manual transmission, select reverse gear. Check that the linkage has moved sufficiently to allow clearance for the inner joint to be raised and the outer CV joint withdrawn from the wheel hub. If the clearance is insufficient, extract the retaining clip and flat washer from the gearchange linkage pivot point directly above the inner constant velocity joint, then release the linkage from the pivot location and place it to one side.
10 On all vehicles undo and remove the retaining screws and lift off the access cover from the inner wheel arch (photo).
11 Ease the inner constant velocity joint away from the differential drive flange, raise the inner end of the shaft and allow the joint to rest on the flange.
12 Withdraw the outer constant velocity joint from the wheel hub (photo) and remove the driveshaft from under the wheel arch. Withdraw the bearing water shield from the outer joint.

Refitting
13 Refitting both driveshafts is the reverse sequence to removal, bearing in mind the following points:

   (a) *Fill the bearing water shield with the specified grease and position the shield on the outer joint flange before fitting the joint to the hub (photo)*
   (b) *If the original inner constant velocity joint is being refitted, ensure that the marks on the joint and differential drive flange made during removal are aligned*
   (c) *Tighten the inner joint retaining bolts, swivel hub retaining nuts (where fitted) and driveshaft nut to the specified torque and use a new split pin to secure the driveshaft nut.* **Do not attempt to fully tighten the driveshaft nut until the weight of the car is on its wheels**

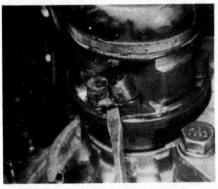

3.5A Remove the protective caps ...

3.5B ... to expose the inner constant velocity joint retaining bolts (arrowed)

3.10 Undo the screws and lift off the access panel

3.12 Withdraw the outer constant velocity joint from the wheel hub

3.13 Position the bearing water shield on the outer joint flange before refitting

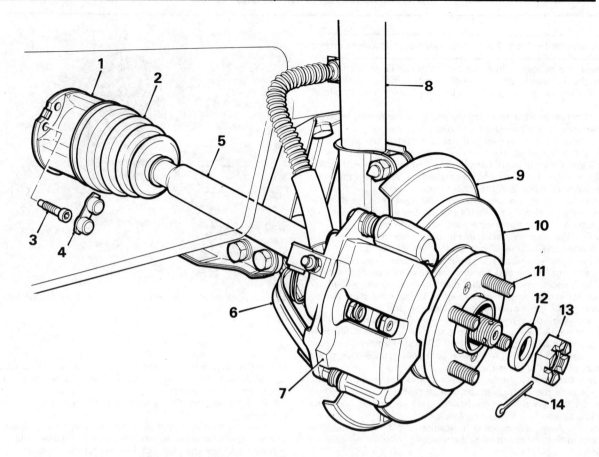

Fig. 8.1 Left-hand driveshaft and related components (Sec 3)

1	Inner constant velocity joint	5	Driveshaft
2	Rubber boot	6	Lower suspension arm
3	Retaining bolt	7	Brake caliper
4	Protective caps	8	Suspension strut

9	Disc shield	12	Flat washer
10	Disc	13	Driveshaft nut
11	Wheel stud	14	Split pin

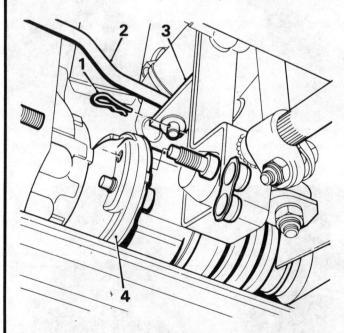

Fig. 8.2 Gearchange linkage disconnection point (Sec 3)

1	Retaining clip	4	Inner constant velocity
2	Linkage arm		joint
3	Pivot		

Fig. 8.3 Exploded view of the constant velocity joints (Secs 4 and 5)

1	Grease retaining plate	8	Driveshaft
2	Circlip	9	Circlip
3	Inner joint inner member	10	Retaining clip
4	Inner joint outer member	11	Rubber boot
5	Retaining clip	12	Retaining clip
6	Rubber boot	13	Outer joint
7	Retaining ring		

4 Inner constant velocity joint – removal, inspection and refitting

1 Remove the driveshaft from the car, as described in Section 3.

2 With the driveshaft on the bench, carefully prise up the tags securing the grease retaining plate to the joint outer member. Lift off the plate.

3 Cut off the outer metal retaining clip and inner rubber ring securing the rubber boot to the driveshaft and joint outer member.

4 Using circlip pliers, extract the circlip securing the joint inner member to the driveshaft.

5 Make an alignment mark between the end of the driveshaft and joint inner member to ensure correct reassembly. Withdraw the inner member, outer member, and the rubber boot.

6 Thoroughly clean the constant velocity joint inner and outer members using paraffin, or a suitable solvent. Dry with a lint-free rag, or compressed air, and then inspect the joint as follows.

7 Examine the bearing tracks in the outer member for signs of scoring, wear ridges or evidence of lack of lubrication. Similarly examine the three bearing caps on the inner member. Check that each bearing cap turns evenly and smoothly on its roller bearings with no trace of tight spots. Insert the inner member into the outer member and check for excessive side movement of the bearing caps in their tracks, and of the inner member in the bearing caps.

8 If any of the above checks indicate wear in the joint, it will be necessary to renew the driveshaft and inner joint as an assembly; they are not available separately. If the joint is in satisfactory condition, obtain a repair kit consisting of a new rubber boot, retaining clips, and the correct quantity of grease (photo).

9 Apply a little rubber grease to the boot inner rubber retaining ring and slide it onto the shaft up to the shoulder (photo).

10 Slide on the new rubber boot (photo) and ease the retaining ring over its end to secure it in position (photo).

11 Place the joint outer member in position and carefully ease the rubber boot over its inner end using a screwdriver, or other suitable flat tool (photo).

12 Position the metal retaining ring over the rubber boot and engage

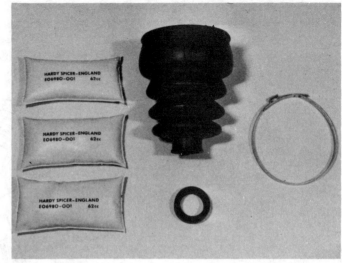

4.8 Components of the inner constant velocity joint repair kit

one of the slots in the clip end over the small tag. Make sure the clip is as tight as possible and, if necessary, use a screwdriver to ease the slot over the tag (photo).

13 Fully tighten the clip by squeezing the raised portion with pliers (photo).

14 Pack the joint inner and outer members with the specified quantity of the grease supplied in the kit (photo) and then slide the inner member onto the shaft (photo). Ensure that the reference marks made during removal are aligned.

4.9 Slide the rubber boot inner retaining ring onto the driveshaft up to the shoulder

4.10A Fit the rubber boot ...

4.10B ... and secure the inner end of the boot with the retaining ring

4.11 Ease the other end of the boot over the joint outer member

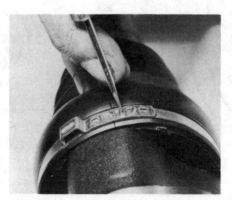

4.12 Hook the retaining clip slot over the clip tag, making sure it is tight

4.13 Tighten the clip fully by squeezing the raised portion

4.14A Pack the outer member with the special grease ...

4.14B ... and then slide on the inner member

4.15 Secure the inner member with the retaining circlip

15 Secure the joint inner member with the retaining circlip (photo).
16 Place the grease retaining plate over the end of the outer member, and bend over the tags to secure it in place (photo).
17 The driveshaft can now be refitted to the car, as described in Section 3.

4.16 Secure the grease retaining plate in position by bending over the tags

5 Outer constant velocity joint – removal, inspection and refitting

1 Remove the driveshaft, as described in Section 3.
2 With the driveshaft on the bench, cut off the two rubber boot retaining clips and fold back the boot to expose the outer joint.
3 Firmly grasp the driveshaft, or support it in a vice. Using a hide, or plastic mallet, sharply strike the outer edge of the joint and drive it off the shaft. The outer joint is retained on the driveshaft by an internal circular section circlip and striking the joint in the manner described forces the circlip to contract into a groove, so allowing the joint to slide off.

4 With the constant velocity joint removed from the driveshaft, thoroughly clean the joint using paraffin, or a suitable solvent, and dry it, preferably using compressed air. Carry out a careful visual inspection of the joint, paying particular attention to the following areas.
5 Move the inner splined driving member from side to side to expose each ball in turn at the top of its track. Examine the balls for cracks, flat spots or signs of surface pitting.
6 Inspect the ball tracks on the inner and outer members. If the tracks have widened, the balls will no longer be a tight fit. At the same time check the ball cage windows for wear or for cracking between the balls. Wear in the balls, ball tracks and ball cage windows will lead to the characteristic clicking noise on full lock described previously.
7 If any of the above checks indicate wear in the joint it will be necessary to renew it complete, as the internal parts are not available separately. If the joint is in a satisfactory condition, obtain a repair kit consisting of a new rubber boot, retaining clips and the correct quantity of grease.
8 The help of an assistant will be necessary whilst refitting the joint to the driveshaft. Ensure that the circlip is undamaged and correctly located in its groove in the driveshaft. Position the new rubber boot over the shaft and locate its end in the shaft groove.
9 Place the retaining clip over the rubber boot and wrap it round until the slot in the clip end can be engaged with the tag. Make sure the clip is as tight as possible using pliers, or a screwdriver, if necessary. Fully tighten the clip by squeezing the raised portion with pliers.
10 Fold back the rubber boot and position the constant velocity joint over the splines on the driveshaft until it abuts the circlip.
11 Using two small screwdrivers placed either side of the circlip, compress the clip and at the same time have your assistant firmly strike the end of the joint with a hide, or plastic, mallet.
12 The joint should slide over the compressed circlip and into position on the shaft. It will probably take several attempts until you achieve success. If the joint does not spring into place the moment it is struck, remove it, reposition the circlip and try again. Do not force the joint, otherwise the circlip will be damaged.
13 With the joint in position against the retaining collar, pack it thoroughly with the specified quantity of the grease supplied in the repair kit. Work the grease well into the ball tracks while twisting the joint, and fill the rubber boot with any excess.
14 Ease the rubber boot over the joint and secure it with the retaining clip, as described in paragraph 9.
15 The driveshaft can now be refitted to the car, as described in Section 3.

6   Fault diagnosis – driveshafts

Symptom	Reason(s)
Vibration and/or noise on turns	Worn constant velocity outer joint(s)
Vibration when accelerating	Worn constant velocity inner joint(s) Bent or distorted driveshaft
Noise on taking up drive	Worn driveshaft or constant velocity joint splines Loose driveshaft nut Worn constant velocity joints

See also Fault Diagnosis – suspension and steering (page 203)

Chapter 9 Braking system

For modifications, and information applicable to later models, see Supplement at end of manual

Contents

Specifications

System type

Diagonally split dual circuit hydraulic with pressure regulating valve in rear hydraulic circuit. Cable-operated handbrake on rear wheels. Servo assistance on all models

Front brakes

Type	Disc with single piston sliding calipers
Disc diameter	9.5 in (241.3 mm)
Disc thickness	0.497 to 0.507 in (12.623 to 12.878 mm)
Maximum disc run-out	0.006 in (0.15 mm)
Minimum pad thickness	0.125 in (3.2 mm)

Rear brakes

Type	Single leading shoe drum, self-adjusting
Drum diameter	8.0 in (203.0 mm)
Lining width	1.5 in (38.1 mm)
Minimum lining thickness	0.0625 in (1.6 mm)
Wheel cylinder diameter:	
Saloon	0.687 in (17.45 mm)
Estate	0.750 in (19.05 mm)
Handbrake lever stop endfloat	0 to 0.080 in (0 to 2.03 mm)

General

Master cylinder bore diameter	0.81 in (20.57 mm)
Servo unit boost ratio	3:1
Brake fluid type/specification	Hydraulic fluid to FMVSS 116 DOT 4 or SAE J1703C (Duckhams Universal Brake and Clutch Fluid)

Torque wrench settings

	lbf ft	Nm
Bleed screws	7	9
Brake caliper anchor bracket-to-swivel hub bolts	53	72
Twin GP valve mounting bolts	9	12
Guide pin bolts	24	33
Master cylinder mounting nuts	9	12
Servo unit mounting nuts	9	12
Wheel cylinder-to-backplate bolts	5	7
Rear hub retaining nut*	50	68
Wheel nuts	53	72
Brake disc to drive flange	8	11

*Tighten to torque then align next split pin hole

1 General description

The braking system is of the servo-assisted, dual circuit hydraulic type with disc brakes at the front and drum brakes at the rear. A diagonally split dual circuit hydraulic system is employed in which each circuit operates one front and one diagonally opposite rear brake from a tandem master cylinder. Under normal conditions both circuits operate in unison; however, in the event of hydraulic failure in one circuit, full braking force will still be available at two wheels. A pressure regulating device or 'twin GP' (Gravity Pressure) valve is incorporated in the rear brake hydraulic circuit. This valve regulates the pressure applied to each rear brake and reduces the possibility of the rear wheels locking under heavy braking.

The front disc brakes are operated by single piston sliding type calipers. At the rear, leading and trailing brake shoes are operated by twin piston wheel cylinders and are self-adjusting by footbrake application.

Driver warning lights are provided for brake pad wear, low brake hydraulic fluid level and handbrake applied.

2 Maintenance and inspection

1 At the intervals given in Routine Maintenance at the beginning of this manual the following service operations should be carried out on the braking system components.

2 Check the brake hydraulic fluid level and if necessary top up with the specified fluid to the MAX mark on the reservoir. Any need for frequent topping-up indicates a fluid leak somewhere in the system which must be investigated and rectified immediately.

3 Check the front disc pads and the rear brake shoe linings for wear and inspect the condition of the discs and drums. Details will be found in Sections 3 and 6 respectively.

4 Check the condition of the hydraulic pipes and hoses as described

in Section 14. At the same time check the condition of the handbrake cables, lubricate the exposed cables and linkages and, if necessary, adjust the handbrake, as described in Section 16.

5 The three braking system warning lights should be tested as follows. Check the operation of the handbrake warning light by applying the handbrake with the ignition switched on. The light should illuminate when the handbrake is applied. To test the low brake hydraulic fluid level warning light, place the car in gear (or P if automatic transmission is fitted), release the handbrake and switch on the ignition. The light should illuminate when the flexible contact cover in the centre of the brake fluid reservoir filler cap is depressed. To check the disc pad wear warning indicator, locate the twin terminal black plastic socket which is in the wiring harness adjacent to the right-hand brake caliper. Switch on the ignition and connect a bridging wire between the terminals of one socket; the pad wear warning light should be illuminated on the instrument panel when the bridging wire is earthed. If any of the lights fail to illuminate in the test condition, then either the bulb is blown, a fuse is at fault or there is a fault in the circuit.

6 Renew the brake hydraulic fluid at the specified intervals by draining the system and refilling with fresh fluid, as described in Section 15.

7 The flexible brake hoses and rubber seals in the brake calipers, wheel cylinders and master cylinder should also be renewed at the less frequent intervals given, as should the air filter in the vacuum servo unit. Details of these operations will be found in the relevant Sections of this Chapter.

3 Front disc pads – inspection and renewal

1 Apply the handbrake, prise off the front wheel trim and slacken the wheel nuts. Jack up the front of the car and support it securely on axle stands. Remove the roadwheels.

2 The thickness of the disc pads can now be checked by viewing

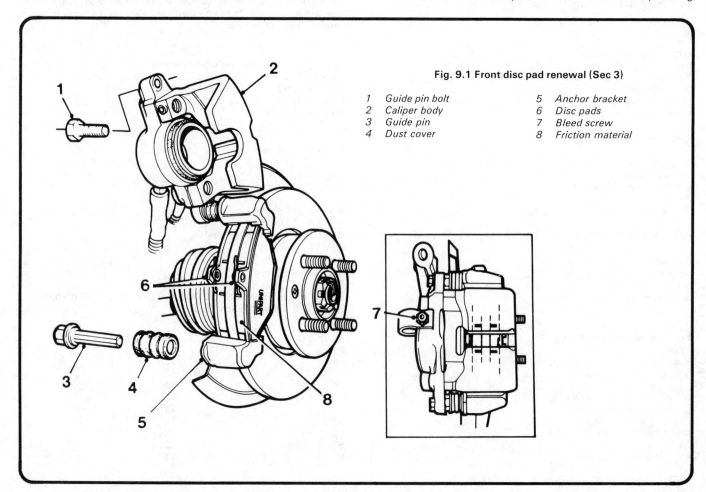

Fig. 9.1 Front disc pad renewal (Sec 3)

1 Guide pin bolt	5 Anchor bracket
2 Caliper body	6 Disc pads
3 Guide pin	7 Bleed screw
4 Dust cover	8 Friction material

3.3 Front caliper bleed screw (A) and pad wear indicator wiring connector (B)

3.5 Unscrew the lower caliper guide pin bolt

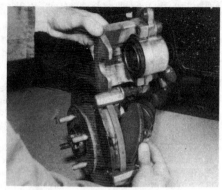

3.6 Pivot the caliper body upwards and withdraw the disc pads

through the slot in the front of the caliper body. If the lining on any of the pads is at, or below, the minimum specified thickness all four pads must be renewed as a complete set.

3   To renew the pads, first remove the protective cap over the caliper bleed screw (photo). Obtain a plastic or rubber tube of suitable diameter to fit snugly over the bleed screw and submerge the free end in a jar containing a small quantity of brake fluid.

4   Open the bleed screw half a turn and pull the caliper body toward you. This will push the piston back into its bore to facilitate removal and refitting of the pads. When the piston has moved in as far as it will go, close the bleed screw, remove the tube and refit the protective cap.

5   Disconnect the pad wear indicator wiring connector (right-hand caliper only) and, using a suitable spanner, unscrew the lower guide pin bolt while holding the guide pin with a second spanner (photo).

6   Pivot the caliper body upwards (photo), withdraw the two disc pads and, where fitted, the anti-squeal shim(s).

7   Brush the dust and dirt from the caliper, piston, disc and pads, but **do not inhale,** as it is injurious to health.

8   Rotate the brake disc by hand and scrape away any rust and scale. Carefully inspect the entire surface of the disc and if there are any signs of cracks, deep scoring or severe abrasions, the disc must be renewed. Also inspect the caliper for signs of fluid leaks around the piston, corrosion, or other damage. Renew the piston seals or the caliper body as necessary.

9   To refit the pads, first attach any anti-squeal shim(s) to them. If a single shim is fitted, it must be fitted to the outer pad. Later models are fitted with an anti-squeal shim behind both pads; in this instance the inner pad's shim can be identified by the cut-out section to facilitate the location of the pad wear warning leads. Very lightly smear the pad and caliper contact areas with a silicone grease, ensuring that no grease is allowed to come into contact with the friction material.

10 Place the pads in position against the disc, noting that the pad with the wear indicator lead is fitted to the inner position on the right-hand caliper.

11 Place the caliper body over the pads and refit the guide pin bolt. Tighten the bolt to the specified torque.

12 Reconnect the wear indicator wiring connector (where applicable), and refit the roadwheels – do not tighten the wheel nuts fully until the weight of the car is on the wheels.

13 Depress the brake pedal several times to bring the piston into contact with the pads and then lower the car to the ground. Check and, if necessary, top up the fluid in the master cylinder reservoir.

4   Front brake caliper – removal, overhaul and refitting

1   Apply the handbrake, prise off the front wheel trim and slacken the wheel nuts. Jack up the front of the car and support it securely on axle stands. Remove the roadwheel.

2   Using a suitable spanner, unscrew the guide pin bolts while holding the guide pins with a second spanner.

3   Disconnect the pad wear indicator wiring connector (right-hand side caliper only) and lift away the caliper body, leaving the disc pads and anchor bracket still in position. It is not necessary to remove the anchor bracket unless it requires renewal because of accident damage or severe corrosion.

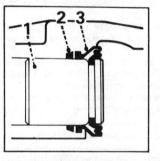

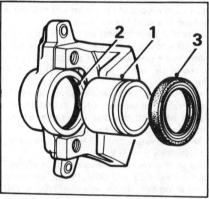

Fig. 9.2 Front brake caliper components (Sec 4)

1 Piston	3 Dust cover
2 Piston seal	

4   With the flexible brake hose still attached to the caliper body, very slowly depress the brake pedal until the piston has been ejected just over halfway out of its bore.

5   Using a brake hose clamp, or self-locking wrench with protected jaws, clamp the flexible brake hose. This will minimise brake fluid loss during subsequent operations.

6   Slacken the brake hose-to-caliper body union, and then, while holding the hose, rotate the caliper to unscrew it from the hose. Lift away the caliper and plug or tape over the end of the hose to prevent dirt entry.

7   With the caliper on the bench wipe away all traces of dust and dirt, but *avoid inhaling the dust as it is injurious to health.*

8   Withdraw the partially ejected piston from the caliper body and remove the dust cover.

9   Using a suitable blunt instrument, such as a knitting needle, carefully extract the piston seal from the caliper bore.

10 Clean all the parts in methylated spirit, or clean brake fluid, and wipe dry using a lint free cloth. Inspect the piston and caliper bore for signs of damage, scuffing or corrosion and if these conditions are

evident renew the caliper body assembly. Also renew the guide pins if bent or damaged.

11 If the components are in a satisfactory condition, a repair kit consisting of new seals and dust cover should be obtained.

12 Thoroughly lubricate the components and new seals with clean brake fluid and carefully fit the seal to the caliper bore.

13 Position the dust cover over the innermost end of the piston so that the caliper bore sealing lip protrudes beyond the base of the piston. Using a blunt instrument, if necessary, engage the sealing lip of the dust cover with the groove in the caliper. Now push the piston into the bore until the other sealing lip of the dust cover can be engaged with the groove in the piston. Having done this, push the piston fully into its bore. Ease the piston out again slightly, and make sure that the cover lip is correctly seating in the piston groove.

14 Remove the guide pins from the anchor bracket and smear them with a high melting-point brake grease. Fit new dust covers to the guide pins and refit them to the anchor bracket.

15 Hold the flexible brake hose and screw the caliper body back onto the hose.

16 With the piston pushed fully into its bore, refit the caliper and secure it with the guide pin bolts. Tighten the bolts to the specified torque.

17 Fully tighten the brake hose union and remove the clamp. If working on the right-hand caliper, reconnect the pad warning light wiring connector.

18 Refer to Section 15 and bleed the brake hydraulic system; noting that, if precautions were taken to minimise fluid loss, it should only be necessary to bleed the relevant front wheel.

19 Refit the roadwheel and lower the car to the ground before fully tightening the wheel nuts and refitting the wheel trim.

5 Front brake disc – removal and refitting

1 Apply the handbrake, remove the front wheel trim and slacken the wheel nuts. Jack the front of the car up and support it securely on axle stands. Remove the roadwheel.

2 Rotate the disc by hand and examine it for deep scoring, grooving or cracks. Light scoring is normal, but if excessive the disc must be renewed. Any loose rust and scale around the outer edge of the disc can be removed by lightly tapping it with a small hammer while rotating the disc.

3 To remove the disc undo and remove the two bolts securing the brake anchor bracket to the swivel hub. Withdraw the caliper assembly complete with pads, off the disc and support it to one side. Avoid straining the flexible brake hose.

4 Undo and remove the two screws securing the disc to the drive flange and withdraw the disc.

5 Refitting is the reverse sequence to removal. Ensure that the mating face of the disc and drive flange are thoroughly clean and tighten all retaining bolts to the specified torque. Prior to refitting the caliper unit, check that the run-out of the brake disc is within the specified limits. If not, the disc is warped and must be renewed.

6 Rear brake shoes – inspection and renewal

1 Chock the front wheels, remove the rear wheel trim and slacken the rear wheel nuts. Jack up the rear of the car and support it securely on axle stands. Remove the roadwheel and release the handbrake.

2 By judicious tapping and levering remove the hub cap and extract the split pin from the hub retaining nut.

3 Using a large socket and bar, undo and remove the hub retaining nut and flat washer (photo). *Note that the left-hand nut has a left-hand thread and the right-hand nut has a conventional right-hand thread.* **Take care not to tip the car from the axle stands.** If the hub nuts are particularly tight, temporarily refit the roadwheel and lower the car to the ground. Slacken the nut in this more stable position and then raise and support the car before removing the nut.

4 Withdraw the hub and brake drum assembly from the stub axle (photo). If it is not possible to withdraw the hub due to the brake drum binding on the brake shoes, the following procedure should be adopted. Refer to Section 16, if necessary, and slacken off the handbrake cable at the adjuster. From the rear of the brake backplate, prise out the handbrake lever stop, which will allow the brake shoes to retract sufficiently from the hub assembly to be removed. It will, however, be necessary to remove the brake shoes and fit a new handbrake lever stop to the backplate.

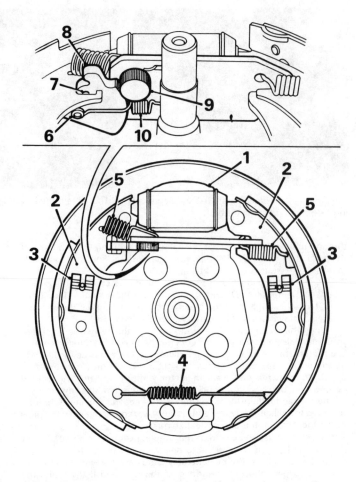

Fig. 9.3 Rear brake assembly – left-hand side (Sec 6)

1	Wheel cylinder	6	Self-adjust quadrant
2	Brake shoes		pivot pin
3	Hold-down springs	7	Hollow pin
4	Lower return spring	8	Quadrant
5	Upper return springs	9	Ratchet wheel
		10	Operating spring

6.3 Rear hub retaining nut and flat washer

6.4 Remove the hub and brake drum assembly from the stub axle

5 With the brake drum assembly removed, brush or wipe the dust from the brake drum, brake shoes and backplate. *Take great care not to inhale the dust, as it is injurious to health.*

6 Measure the brake shoe lining thickness. If it is worn down to the specified minimum amount, or if it is nearly worn down to the rivets, renew all four rear brake shoes. The shoes must also be renewed if any are contaminated with brake fluid or grease, or show signs of cracking or glazing. If contamination is evident, the cause must be traced and cured before fitting new brake shoes.

7 If the brake shoes are in a satisfactory condition proceed to paragraph 19; if removal is necessary, proceed as follows.

8 First make a careful note of the location and position of the various springs and linkages as an aid to refitting (photo).

9 Depress the brake shoe hold-down springs (photo) while supporting the hold-down pin from the rear of the backplate with your finger. Turn the springs through 90° and lift off. Withdraw the hold-down pins.

10 Using a screwdriver, if necessary, release the shoes from their lower pivots and disengage the lower return spring (photo).

11 Using long-nosed pliers, release the upper return spring from the leading shoe and the arm of the self-adjust mechanism (photo). Detach the leading shoe from the wheel cylinder and self-adjust mechanism, and remove it from the backplate.

12 Withdraw the trailing shoe from the backplate, move the handbrake operating lever away from the shoe and unhook the handbrake cable (photo). Release the upper return spring and detach the self-adjust mechanism from the trailing brake shoe.

13 Before fitting the new brake shoes clean off the brake backplate with a rag and apply a trace of silicone grease to the brake shoe contact areas (photo). Clean the self-adjust mechanism and make sure that it is free to move in its elongated slot (photo).

14 Refit the self-adjust mechanism to the trailing brake shoe and secure with the return spring.

15 Position the self-adjust mechanism in the fully retracted position by moving the quadrant away from the ratchet wheel and turning it until the hollow pin locates the inner cutaway (photo).

16 Connect the handbrake cable to the operating lever and position the trailing shoe on the backplate.

17 Refit the leading shoe to the backplate, ensuring that the shoe is properly located around the pin on the self-adjust mechanism (photo).

18 Refit the upper and lower return springs and the hold-down springs and pins.

19 Refit the brake drum and hub assembly, flat washer and hub retaining nut. Tighten the hub retaining nut to the specified torque and then tighten further until a split pin hole is aligned. Fit a new split pin and tap on the hub cap.

6.8 Layout and position of the rear brake components

6.9 Brake shoe hold-down spring and pin

6.10 Brake shoe lower return spring

6.11 Brake shoe upper return spring removal

6.12 Detach the handbrake cable from the trailing shoe operating lever

6.13A Apply a trace of silicone grease to the brake shoe contact areas (arrowed)

6.13B Brake shoe self-adjust mechanism

6.15 Self-adjust mechanism pin (arrowed) in fully retracted position

6.17 Brake shoe correctly located around self-adjust mechanism hollow pin (arrowed)

20 Refit the roadwheel, but do not fully tighten the wheel nuts.

21 If it was necessary to slacken the handbrake cable to enable the brake drum to be removed, adjust the cable as described in Section 16.

22 Lower the car to the ground, tighten the wheel nuts to the specified torque and refit the wheel trim. Depress the footbrake two or three times to operate the brake adjusters.

7 Rear wheel cylinder – removal and refitting

1 Begin by removing the rear hub and brake drum assembly, as described in Section 6, paragraphs 1 to 5 inclusive.

2 Using a screwdriver, or other suitable tool, carefully ease the upper end of the leading brake shoe (the one nearest the front of the car) away from the wheel cylinder piston. Take care not to damage the wheel cylinder rubber boot as you do this. As the leading shoe is moved away from the wheel cylinder, the self-adjust mechanism will expand and hold both brake shoes in the expanded position. This will provide sufficient clearance to allow removal of the wheel cylinder.

3 Using a brake hose clamp, or self-locking wrench with protected jaws, clamp the flexible brake hose just in front of the rear axle pivot mounting. This will minimise brake fluid loss during subsequent operations.

4 At the rear of the brake backplate, unscrew the union nut securing the brake pipe to the wheel cylinder. Carefully ease the pipe out of the cylinder and plug or tape over its end to prevent dirt entry.

5 Undo and remove the two bolts securing the wheel cylinder to the backplate and withdraw the cylinder from between the brake shoes.

6 To refit the wheel cylinder, place it in position on the backplate and engage the brake pipe and union. Screw in the union nut two or three turns to ensure the thread has started.

7 Refit the wheel cylinder retaining bolts and tighten them to the specified torque. Now fully tighten the brake pipe union nut.

8 Using a screwdriver or other suitable tool, as before, ease the leading brake shoe away from the wheel cylinder and retract the self-adjust mechanism. To do this move the quadrant, in its elongated slot, away from the ratchet wheel and turn it until the hollow pin contacts the cutaway in the operating lever. As you do this, ease the brake shoes back into their upper locations in the wheel cylinder pistons.

9 Refit the brake drum and hub assembly, flat washer and hub retaining nut. Tighten the hub retaining nut to the specified torque and then tighten further until a split pin hole is aligned. Fit a new split pin and tap on the hub cap.

10 Remove the clamp from the brake hose and bleed the brake hydraulic system, as described in Section 15. Providing suitable precautions were taken to minimise loss of fluid, it should only be necessary to bleed the relevant rear brake.

11 If it was necessary to slacken the handbrake cable to remove the brake drum, adjust the cable, as described in Section 16.

12 Refit the roadwheel, but do not tighten the wheel nuts fully.

13 Lower the car to the ground and tighten the wheel nuts. Depress the footbrake two or three times to operate the brake adjusters.

8 Rear wheel cylinder – overhaul

1 Remove the wheel cylinder from the car, as described in the previous Section.

2 With the wheel cylinder on the bench, remove the dust cover retainers, where fitted, and withdraw the dust covers from the ends of the piston and cylinder body.

3 Withdraw the pistons and piston spring. Remove the seals from the pistons and unscrew the bleed screw from the cylinder body.

4 Thoroughly clean all the components in methylated spirits or clean brake fluid and dry with a lint-free rag.

5 Carefully examine the surfaces of the pistons and cylinder bore for wear, score marks or corrosion and, if evident, renew the complete wheel cylinder. If the components are in a satisfactory condition, obtain a repair kit consisting of new seals and dust covers.

6 Dip the new seals and pistons in clean brake fluid and assemble the components wet, as follows.

7 Using your fingers, fit the new seals to the pistons with their sealing lips facing inwards.

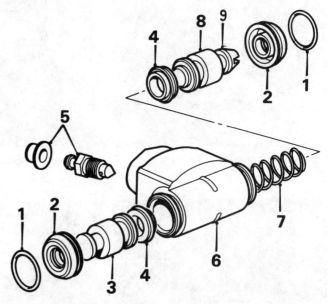

Fig. 9.4 Exploded view of the right-hand rear wheel cylinder (Sec 8)

1 Dust cover retainers (where fitted)	6 Wheel cylinder body
2 Dust cover	7 Piston spring
3 Trailing shoe piston	8 Leading shoe piston
4 Piston seal	9 Leading shoe tappet
5 Bleed screw and dust cap	

8   Lubricate the cylinder bore with clean brake fluid and insert the
spring, followed by the two pistons. Note that the piston with the slot
in its end must face toward the front of the car when fitted.
9   Place the dust covers over the pistons and cylinder and, where
applicable, refit the dust cover retainers.
10 Screw the bleed screw into the cylinder body and then refit the
wheel cylinder to the car, as described in the previous Section.

9   Rear brake backplate – removal and refitting

The rear brake backplate is removed in conjunction with the stub
axle and details of this procedure will be found in Chapter 11, Section
12.

10 Master cylinder – removal and refitting

1   Working under the front of the car, remove the dust cover from the
bleed screw on each front brake caliper. Obtain two plastic or rubber
tubes of suitable diameter to fit snugly over the bleed screws, and
place the other ends in a suitable receptacle.
2   Open the bleed screws half a turn and operate the brake pedal
until the master cylinder reservoir is empty. Tighten the bleed screws
and remove the tubes. Discard the expelled brake fluid.
3   Disconnect the wiring connectors from the reservoir filler cap
terminals.
4   Unscrew the two brake pipe union nuts and carefully withdraw the
pipes from the master cylinder. Plug or tape over the ends of the pipes
to prevent dirt entry, and place rags beneath the master cylinder to
protect the surrounding paintwork (photo).
5   Undo and remove the two retaining nuts and washers and
withdraw the master cylinder from the servo unit.
6   Refitting the master cylinder is the reverse sequence to removal.

10.4 Brake pipe union nuts (A) and master cylinder retaining nuts (B)

Tighten the retaining nuts to the specified torque and, on completion,
bleed the brake hydraulic system, as described in Section 15.

11 Master cylinder – overhaul

1   Remove the master cylinder from the car, as described in the
previous Section. Drain any fluid remaining in the reservoir and

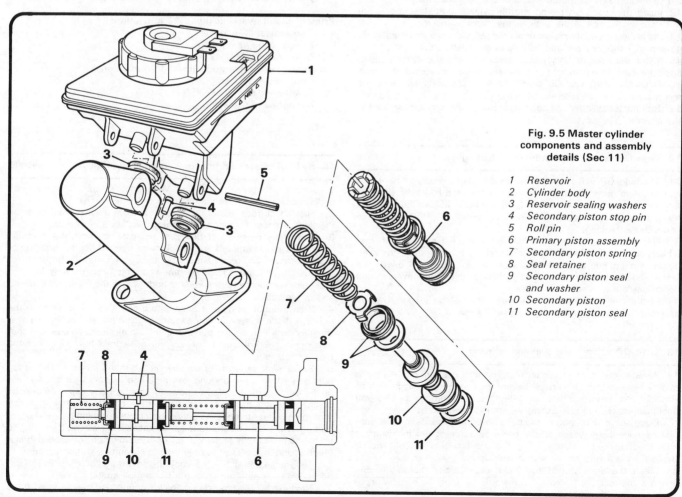

Fig. 9.5 Master cylinder components and assembly details (Sec 11)

1   Reservoir
2   Cylinder body
3   Reservoir sealing washers
4   Secondary piston stop pin
5   Roll pin
6   Primary piston assembly
7   Secondary piston spring
8   Seal retainer
9   Secondary piston seal
    and washer
10 Secondary piston
11 Secondary piston seal

prepare a clean uncluttered working surface, ready for dismantling.

2   Remove the primary piston assembly and then mount the cylinder horizontally in a soft-jawed vice. Using a parallel pin punch, drive out the two roll pins securing the reservoir to the cylinder body.

3   Lift off the reservoir and withdraw the two reservoir sealing washers from the master cylinder inlet ports.

4   Using a blunt instrument, push the secondary piston in as far as it will go and withdraw the secondary piston stop pin from its location in the secondary inlet port.

5   Slowly release the secondary piston, remove the cylinder from the vice and tap it on a block of wood to release the secondary piston from the cylinder bore.

6   Note the location and position of the components on the secondary piston and then remove the piston spring. Withdraw the seal retainer, followed by the seal and washer. Now remove the remaining seal from the other end of the secondary piston.

7   The primary piston should not be dismantled, as parts are not available separately. If the master cylinder is in a serviceable condition, and is to be reused, a complete new primary piston assembly is included in the repair kit.

8   With the master cylinder completely dismantled, clean all the components in methylated spirits, or clean brake fluid, and dry with a lint-free rag.

9   Carefully examine the cylinder bore and secondary piston for signs of wear, scoring or corrosion, and if evident renew the complete cylinder assembly.

10 If the components are in a satisfactory condition, obtain a repair kit consisting of new seals, springs and primary piston assembly.

11 Lubricate the master cylinder bore, pistons and seals thoroughly in clean brake fluid, and assemble them wet.

12 Fit the washer and seal onto the inner end of the secondary piston using your fingers only. Use the notes made during dismantling and the accompanying illustrations as a guide to the direction of fitment of the seal. Place the seal retainer over the seal and refit the spring. Fit the remaining seal to the other end of the secondary piston.

13 Insert the secondary piston into the cylinder bore, taking care not to turn over the lips of the seals as the piston is inserted.

14 Push the secondary piston down the cylinder bore as far as it will go and refit the stop pin into its hole in the inlet port.

15 Insert the new primary piston assembly into the cylinder bore, again taking care not to turn over the seal lips as they enter the bore.

16 Place the two reservoir seals into the inlet ports and refit the reservoir. Secure the reservoir with the two roll pins.

17 The master cylinder can now be refitted to the car, as described in the previous Section.

12 Twin GP valve – description and testing

1   The twin GP valve is mounted in the engine compartment and contains two gravity/pressure valves, one for each hydraulic circuit.

2   The purpose of the valve is to distribute brake fluid to the front and rear brakes, and to limit the fluid pressure supplied to the rear brakes under heavy braking.

3   The operation of the valve may be suspect if one or both rear wheels continually lock during normal braking. It is essential, however, before condemning the valve to ensure the fault does not lie with the brake shoe assemblies or wheel cylinders, and that adverse road conditions are also not responsible.

4   In the event of failure of the valve it must be renewed as an assembly, as parts are not available separately.

13 Twin GP valve – removal and refitting

1   Remove the master cylinder reservoir filler cap and place a piece of polythene over the filler neck. Secure the polythene in place with an elastic band ensuring that an airtight seal is obtained. This will minimise brake fluid loss during subsequent operations.

2   Place some rags around the valve to protect the paintwork from brake fluid spillage. Wash off any brake fluid that comes into contact with the body immediately using copious amounts of cold water. Brake fluid is a very effective paint stripper.

3   Clean the area around the brake pipe unions thoroughly and unscrew the union nuts. Note that the nuts securing the primary brake

pipes are of a smaller diameter than the secondary nuts. Very carefully withdraw the brake pipes from the valve.

4   Undo and remove the two bolts securing the valve to the inner member and withdraw the unit from the engine compartment.

5   Refitting is the reverse sequence to removal. Refer to Fig. 9.6 if in any doubt about the pipe locations, and bleed the brake hydraulic system, as described in Section 15, after refitting.

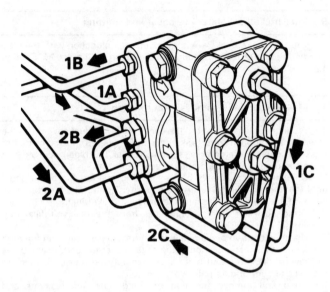

Fig. 9.6 Hydraulic pipe arrangement at twin GP valve (Sec 13)

Primary circuit hydraulic pipes (11 mm unions)
1A From master cylinder
1B To right-hand front brake
1C To left-hand rear brake
Secondary circuit hydraulic pipes (13 mm unions)
2A From master cylinder
2B To left-hand front brake
2C To right-hand rear brake

14 Hydraulic pipes and hoses – inspection, removal and refitting

1   At intervals given in Routine Maintenance carefully examine all brake pipes, hoses, hose connections and pipe unions.

2   First check for signs of leakage at the pipe unions. Then examine the flexible hoses for signs of cracking, chafing and fraying.

3   The brake pipes must be examined carefully and methodically. They must be cleaned off and checked for signs of dents, corrosion or other damage. Corrosion should be scraped off and, if the depth of pitting is significant, the pipes renewed. This is particularly likely in those areas underneath the vehicle body where the pipes are exposed and unprotected.

4   If any section of pipe or hose is to be removed, first unscrew the master cylinder reservoir filler cap and place a piece of polythene over the filler neck. Secure the polythene with an elastic band ensuring that an airtight seal is obtained. This will minimise brake fluid loss when the pipe or hose is removed.

5   Brake pipe removal is usually quite straightforward. The union nuts at each end are undone, the pipe and union pulled out and the centre section of the pipe removed from the body clips. Where the unions nuts are exposed to the full force of the weather they can sometimes be quite tight. As only an open ended spanner can be used, burring of the flats on the nuts is not uncommon when attempting to undo them. For this reason a self-locking wrench, or special brake union spanner, is often the only way to separate a stubborn union.

6   To remove a flexible hose, wipe the unions and brackets free of dirt and undo the union nut from the brake pipe end(s).

7   Next extract the hose retaining clip, or unscrew the nut, and lift the

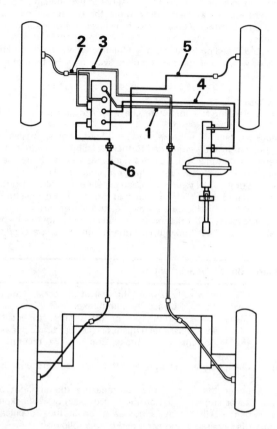

Fig. 9.7 Layout of brake hydraulic pipes and hoses (Sec 14)

Secondary circuit
1   *Master cylinder to GP valve*
2   *GP valve to left-hand front hose*
3   *GP valve to right-hand rear hose*

Primary circuit
4   *Master cylinder to GP valve*
5   *GP valve to right-hand front hose*
6   *GP valve to left-hand rear hose*

The pipe is then bent to shape, using the old pipe as a guide, and is ready for fitting to the car.

9   Refitting the pipes and hoses is a reverse of the removal procedure. Make sure that the hoses are not kinked when in position and also make sure that the brake pipes are securely supported in their clips. After refitting, remove the polythene from the reservoir and bleed the brake hydraulic system, as described in Section 15.

15 Hydraulic system – bleeding

1   The correct functioning of the brake hydraulic system is only possible after removal of all air from the components and circuit; this is achieved by bleeding the system. Note that only clean unused brake fluid, which has remained unshaken for at least 24 hours, must be used.

2   If there is any possibility of incorrect fluid being used in the system, the brake lines and components must be completely flushed with uncontaminated fluid and new seals fitted to the components.

3   *Never reuse* brake fluid which has been bled from the system.

4   During the procedure, do not allow the level of brake fluid to drop more than halfway down the reservoir.

5   Before starting work, check that all pipes and hoses are secure, unions tight and bleed screws closed. Take great care not to allow brake fluid to come into contact with the car paintwork, otherwise the finish will be seriously damaged. Wash off any spilled fluid immediately with cold water.

6   There is a number of one-man, do-it-yourself, brake bleeding kits currently available from motor accessory shops. If one of these kits is not available, it will be necessary to gather together a clean jar and a suitable length of clear plastic tubing which is a tight fit over the bleed screw, and also to engage the help of an assistant.

7   If brake fluid has been lost from the master cylinder due to a leak in the system, ensure that the cause is traced and rectified before proceeding further.

8   If the hydraulic system has only been partially disconnected and suitable precautions were taken to prevent further loss of fluid it should only be necessary to bleed that part of the system (ie primary or secondary circuit).

9   If the complete system is to be bled then it should be done in the following sequence.

   Secondary circuit: Left-hand front then right-hand rear
   Primary circuit: Right-hand front then left-hand rear

10   To bleed the system, first clean the area around the bleed screw (photo) and fit the bleed tube. If necessary, top up the master cylinder reservoir with brake fluid.

14.7 Front brake hose and retaining nut (arrowed)

15.10 Front brake caliper bleed screw (arrowed)

end of the hose out of its bracket (photo). If a front hose is being removed, it can now be unscrewed from the brake caliper.

8   Brake pipes can be obtained individually, or in sets, from most accessory shops or garages with the end flares and union nuts in place.

11 If the system incorporates a vacuum servo, destroy the vacuum by giving several applications of the brake pedal in quick succession.

Bleeding – two-man method

12 Gather together a clean jar and a length of rubber or plastic tubing which will be a tight fit on the brake bleed screws.

13 Engage the help of an assistant.

14 Push one end of the bleed tube onto the first bleed screw and immerse the other end in the jar which should contain enough hydraulic fluid to cover the end of the tube (photo).

15 Open the bleed screw one half turn and have your assistant depress the brake pedal fully then slowly release it. Tighten the bleed screw at the end of each pedal downstroke to obviate any chance of air or fluid being drawn back into the system.

16 Repeat this operation until clean hydraulic fluid, free from air bubbles, can be seen coming through into the jar.

17 Tighten the bleed screw at the end of a pedal downstroke and remove the bleed tube. Bleed the remaining screws in a similar way.

15.14 Bleed tube attached to front caliper bleed screw

Bleeding – using one-way valve kit

18 There is a number of one-man, one-way brake bleeding kits available from motor accessory shops. It is recommended that one of these kits is used wherever possible as it will greatly simplify the bleeding operation and also reduce the risk of air or fluid being drawn back into the system, quite apart from being able to do the work without the help of an assistant.

19 To use the kit, connect the tube to the bleed screw and open the screw one half turn.

20 Depress the brake pedal fully and slowly release it. The one-way valve in the kit will prevent expelled air from returning at the end of each pedal downstroke. Repeat this operation several times to be sure of ejecting all air from the system. Some kits include a translucent container which can be positioned so that the air bubbles can be seen being ejected from the system.

21 Tighten the bleed screw, remove the tube and repeat the operations on the remaining brakes.

22 On completion, depress the brake pedal. If it still feels spongy repeat the bleeding operations, as air must still be trapped in the system.

Bleeding – using a pressure bleed kit

23 These kits are also available from motor accessory shops and are usually operated by air pressure from the spare tyre.

24 By connecting a pressurised container to the master cylinder fluid reservoir, bleeding is then carried out by simply opening each bleed screw in turn and allowing the fluid to run out, rather like turning on a tap, until no air is visible in the expelled fluid.

25 By using this method, the large reserve of hydraulic fluid provides a safeguard against air being drawn into the master cylinder during bleeding which often occurs if the fluid level in the reservoir is not maintained.

26 Pressure bleeding is particularly effective when bleeding 'difficult' systems or when bleeding the complete system at the time of routine fluid renewal.

All methods

27 When bleeding is completed, check and top up the fluid level in the master cylinder reservoir.

28 Check the feel of the brake pedal. If it feels at all spongy, air must still be present in the system and further bleeding is indicated. Failure to bleed satisfactorily after a reasonable repetition of the bleeding operations may be due to worn master cylinder seals.

29 Discard brake fluid which has been expelled. It is almost certain to be contaminated with moisture, air and dirt making it unsuitable for further use. Clean fluid should always be stored in an airtight container as it is hygroscopic (absorbs moisture readily) which lowers its boiling point and could affect braking performance under severe conditions.

16 Handbrake cable – adjustment

1   Chock the front wheels, jack up the rear of the car and support it securely on axle stands. Release the handbrake.

2   Apply the footbrake firmly two or three times to ensure full movement of the self-adjust mechanism on the rear brake shoes. This is particularly important if the brake drums have recently been removed.

3   Apply the handbrake sharply and then release it to equalise the cable loads.

4   From under the rear of the car measure the endfloat of the handbrake lever stops located on the rear of each brake backplate. If the movement of the stops is not as given in the Specifications, adjustment is necessary and is carried out as follows.

5   Turn the cable adjuster, located just in front of the rear axle (photo), clockwise to increase the cable tension and anti-clockwise to decrease it. The direction of rotation assumes the adjuster is being viewed from the rear of the car, ie facing forwards.

6   Turn the cable adjuster until the endfloat of the handbrake lever stops is as specified and the travel of the handbrake lever necessary to lock the wheels is between 5 and 8 clicks of the ratchet. Do not overtighten the cable.

7   After adjustment, ensure that the wheels are free to turn, without binding when the handbrake is released, and then lower the car to the ground.

16.5 Location of the handbrake cable adjuster (arrowed) in front of the rear axle

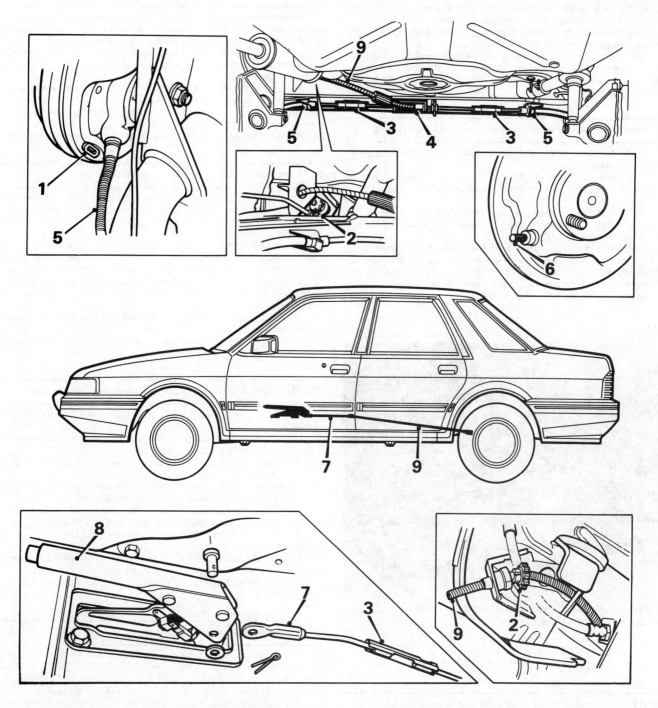

Fig. 9.8 Layout of the handbrake mechanism (Secs 16 to 20)

*1   Brake shoe inspection
     aperture
2   Handbrake cable adjuster*

*3   Cable connector
4   Compensator assembly
5   Handbrake cable – rear*

*6   Handbrake lever stop
7   Handbrake cable – front
8   Handbrake lever*

*9   Handbrake cable –
     intermediate*

17 Handbrake cable (front) – removal and refitting

1   Chock the front wheels, jack up the rear of the car and support it securely on axle stands.

2   From under the rear of the car slacken the handbrake cable adjuster to remove all tension from the cable. Refer to Section 16, if necessary.

3   From inside the car remove the centre console, as described in Chapter 12.

4   Slide both the front seats fully forward then undo and remove the rear bolts securing the inner seat rails to the floor. Now remove the centre console rear mounting bracket.

5   Undo and remove the seat belt centre stalk mounting bolts and remove the two stalks.

6   Undo and remove the bolt securing the left-hand seat belt to the inner sill. Remove the carpet retaining screws amd lift up the rear carpet sufficiently to gain access to the handbrake cable cover plate. Remove the cover plate and seal.

7   Extract the split pin and withdraw the clevis pin securing the

handbrake cable to the handbrake lever. Remove the cable from the lever and feed it through the cover plate aperture.
8   Disconnect the front cable from the intermediate cable at the connector and withdraw it from under the car.
9   Refitting the cable is the reverse sequence to removal. Ensure that the seat belt mounting bolts are tightened to the specified torque (Chapter 12) and adjust the cable, as described in Section 16, after refitting.

18 Handbrake cable (intermediate) – removal and refitting

1   Chock the front wheels, jack up the rear of the car and support it securely on axle stands.
2   From under the rear of the car, slacken the handbrake cable adjuster to remove all tension from the cable. Refer to Section 16 if necessary.
3   Disconnect the intermediate cable from the front cable at the connector (photo), release the cable adjuster from its bracket and pull the cable through the bracket.
4   Disconnect both rear cables from the connectors located to the rear of the rear axle transverse member. Release the intermediate cable from its mounting bushes and withdraw it from under the car.
5   Refitting is the reverse sequence to removal but adjust the cable, as described in Section 16, after fitting.

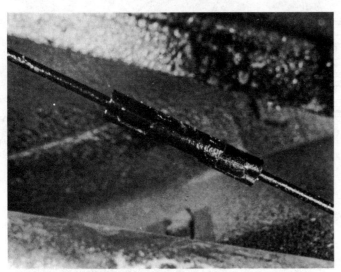

18.3 Handbrake intermediate cable-to-front cable connector located just forward of the fuel tank

19 Handbrake cable (rear) – removal and refitting

1   Chock the front wheels, prise off the rear wheel trim and slacken the wheel nuts. Jack the car up and support it securely on axle stands. Remove the roadwheel.
2   Disconnect the rear cable at the connector just behind the rear axle transverse member.
3   Using circlip pliers, extract the retaining circlip and withdraw the rear cable from its mounting bracket on the rear cable.
4   Refer to Section 6, paragraphs 2 to 4 inclusive, and remove the rear hub and brake drum assembly.
5   With the drum removed, ease the handbrake operating lever on the trailing brake shoe forward and disengage the inner cable nipple from the elongated slot on the lever.
6   Withdraw the cable from the brake backplate and remove the spring collar and felt seal.
7   Refitting is the reverse sequence to removal. Adjust the cable, as described in Section 16, after refitting.

20 Handbrake lever and switch – removal and refitting

1   Refer to Chapter 12, and remove the centre console.
2   If only the warning light switch is to be removed, disconnect the wires from the switch terminals, remove the securing screws and withdraw the switch.
3   To remove the lever assembly, extract the split pin, withdraw the clevis pin and detach the handbrake cable from the lever.
4   Undo and remove the two mounting bolts and lift away the handbrake lever assembly.
5   Refitting is a straightforward reversal of the removal sequence.

21 Footbrake pedal – removal and refitting

1   From inside the car extract the split pin and withdraw the clevis pin and washer securing the servo unit pushrod to the brake pedal.
2   Undo and remove the nut and washer at the clutch pedal end of the pedal pivot bolt.
3   Draw the pivot bolt toward the side of the car to release the clutch pedal. Recover the washer from the pivot shaft (where fitted), unhook the top of the clutch pedal from the cable and withdraw the pedal.
4   Fully remove the pivot bolt and withdraw the brake pedal after detaching the return spring. Where fitted, recover the pivot bolt washers.
5   With the pedal removed, examine the pivot bushes and, if worn, renew them.
6   Refitting the pedals is the reverse sequence to removal. Smear the pivot bolt and bushes with a little lithium-based grease prior to fitting.

22 Stop-light switch – removal, refitting and adjustment

1   To remove the switch, first disconnect the battery negative terminal.
2   Undo and remove the bolts on the right-hand side of the pedal securing the switch bracket in position.
3   Disconnect the wiring connector and remove the switch. Unscrew the switch from the plastic insert.
4   To refit the switch, screw it fully into the plastic insert and position the switch with the terminal in the vertical position.
5   Refit the wiring connector and secure the switch bracket to the pedal bracket with the bolts, finger tight only at this stage.
6   To adjust the switch, reconnect the battery negative terminal and switch on the ignition.
7   Depress the brake pedal by 0.25 in (6 mm) and adjust the switch position so that the stop-lights just come on.
8   Tighten the switch bracket retaining bolts and recheck the stop-light operation.

23 Vacuum servo unit – description

A vacuum servo unit is fitted into the brake hydraulic circuit in series with the master cylinder, to provide assistance to the driver when the brake pedal is depressed. This reduces the effort required by the driver to operate the brakes under all braking conditions.
The unit operates by vacuum obtained from the inlet manifold and comprises basically a booster diaphragm, control valve, and a non-return valve.
The servo unit and hydraulic master cylinder are connected together so that the servo unit piston rod acts as the master cylinder pushrod. The driver's braking effort is transmitted through another pushrod to the servo unit piston and its built-in control system. The servo unit piston does not fit tightly into the cylinder, but has a strong diaphragm to keep its edges in constant contact with the cylinder wall, so ensuring an airtight seal between the two parts. The forward chamber is held under vacuum conditions created in the inlet manifold of the engine and, during periods when the brake pedal is not in use, the controls open a passage to the rear chamber so placing it under vacuum conditions as well. When the brake pedal is depressed, the vacuum passage to the rear chamber is cut off and the chamber opened to atmospheric pressure. The consequent rush of air pushes the servo piston forward in the vacuum chamber and operates the main pushrod to the master cylinder.

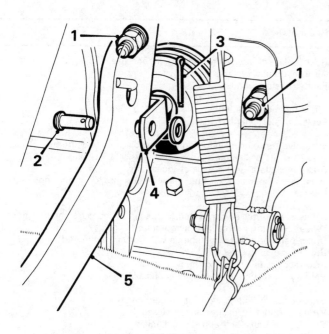

Fig. 9.9 Vacuum servo unit attachments (Sec 24)

1   Servo unit retaining nuts   4   Servo pushrod
2   Clevis pin   5   Brake pedal
3   Split pin

24 Vacuum servo unit – removal and refitting

1   Undo and remove the two nuts and washers securing the brake master cylinder to the servo unit. Carefully withdraw the master cylinder, taking great care not to strain the brake pipes, and tie it to one side, just clear of the servo.
2   Using a wide-bladed screwdriver, carefully prise the vacuum hose elbow out of the grommet on the front face of the servo.
3   From inside the car, extract the split pin and withdraw the clevis pin securing the servo pushrod to the brake pedal.
4   Undo and remove the two nuts and washers securing the servo to the bulkhead and withdraw the unit from the engine compartment.
5   Refitting the servo unit is the reverse sequence to removal. Tighten the servo and master cylinder retaining nuts to the specified torque and use a new split pin in the servo pushrod clevis pin.

25 Vacuum servo unit – servicing

1   At the intervals given in Routine Maintenance the servo unit air filter should be renewed as follows. Note that this is the only work that can be carried out on the servo and no attempt should be made to dismantle the unit or alter the setting of the domed nut in the output rod (where fitted).
2   From inside the car, extract the split pin and withdraw the clevis pin securing the servo pushrod to the brake pedal.
3   Pull back the rubber dust cover, release the end cap and extract the old filter. Cut the new filter, as shown in Fig. 9.10, place it over the pushrod and push the filter into the neck of the servo. Refit the end cap and dust cover.
4   To test the operation of the servo, depress the footbrake and then start the engine. As the engine starts there should be a noticeable 'give' in the brake pedal. Allow the engine to run for at least two minutes and then switch it off. If the brake pedal is now depressed again, it should be possible to detect a hiss from the servo when the pedal is depressed. After about four or five applications no further hissing will be heard and the pedal will feel considerably firmer.

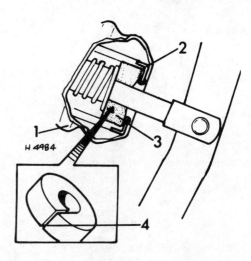

Fig. 9.10 Servo unit air filter renewal (Sec 25)

1   Dust cover   3   Filter
2   End cap   4   Position of cut in filter

26 Fault diagnosis – braking system

Symptom	Reason(s)
Excessive pedal travel	Rear brake self-adjust mechanism inoperative Air in hydraulic system Faulty master cylinder
Brake pedal feels spongy	Air in hydraulic system Faulty master cylinder
Judder felt through brake pedal or steering wheel when braking	Excessive run-out or distortion of front discs or rear drums Brake pedals or linings worn Brake backplate or disc caliper loose Wear in suspension or steering components or mountings – see Chapter 11
Excessive pedal pressure required to stop car	Faulty servo unit, disconnected, damaged or insecure vacuum hose Wheel cylinder(s) or caliper piston seized Brake pads or brake shoe linings worn or contaminated Brakes shoes incorrectly fitted Incorrect grade of pads or linings fitted Primary or secondary hydraulic circuit failure
Brakes pull to one side	Brake pads or linings worn or contaminated Wheel cylinder or caliper piston seized Seized rear brake self-adjust mechanism Brake pads or linings renewed on one side only Faulty twin GP valve Tyre, steering or suspension defect – see Chapter 11
Brakes binding	Wheel cylinder or caliper piston seized Handbrake incorrectly adjusted Faulty master cylinder
Rear wheels locking under normal braking	Rear brake shoe linings contaminated Faulty twin GP valve

Chapter 10 Electrical system

For modifications, and information applicable to later models, see Supplement at end of manual

Contents

Specifications

System type .... 12 volt, negative earth

Battery
Type .... Unipart – 'Sealed for life'
Performance – cold start current (amps)/reserve capacity (minutes)
    1.3 litre models 280/50
    1.6 litre models 360/70

Alternator
Type .... Lucas 127/55
Maximum output .... 55 amps
Brush length:
    New .... 0.8 in (20.0 mm)
    Minimum .... 0.4 in (10.0 mm)
Brush spring tension (brush face flush with brushbox) .... 4.7 to 9.8 ozf (1.3 to 2.7 N)

Starter motor
Type:
    1.3 litre models .... Lucas 8M90 pre-engaged
    1.6 litre models .... Lucas 9M90 pre-engaged
Minimum brush length .... 0.4 in (10.0 mm)
Brush spring tension .... 32 to 36 ozf (9 to 10 N)
Commutator minimum skimmed thickness .... 0.08 in (2.0 mm)

Wiper blades .... Champion X-5103 (front) and X-3603 (rear)

Wiper arms .... Champion CCA6

Relays and control units

Component:	Location
Direction indicator/hazard warning flasher unit	In fusebox on right-hand side of facia
Starter solenoid relay	In fusebox on right-hand side of facia
Ignition auxiliary relay	In fusebox on right-hand side of facia
Heated rear screen relay	In fusebox on right-hand side of facia
Headlamp relay	In fusebox on right-hand side of facia
Inlet manifold heater (if fitted)	In fusebox on right-hand side of facia
Fuel system electronic mixture control unit – ECU	Attached to panel on glovebox roof
Wipe/wash program control unit	Attached to panel on glovebox roof
Programmed ignition system control unit (if fitted)	Attached to engine compartment left-hand valance
Electric window lift relay and control unit (if fitted)	Behind the driver's door inner trim panel
Central locking control unit (if fitted)	Behind the driver's door inner trim panel

Fuses

Fuse colour coding:	Current rating
Violet	3 amp
Tan	5 amp
Red	10 amp
Blue	15 amp
Natural	25 amp
Green	30 amp
Fusible links:	
Link A	150 mm long, 28/0.3 wire
Link B	200 mm long, 28/0.3 wire
Link C	450 mm long, 14/0.3 wire

Bulbs

	Wattage
Headlamps	60/55
Sidelamps	4
Direction indicators	21
Reverse lamps	21
Stop/tail lamps	21/5
Tail lamps	5
Rear foglamps	21
Number plate lamp	5
Interior and luggage compartment lamps	10
Glovebox lamp	5
Switch illumination	0.36
Selector illumination lamp (automatic transmission models)	3
Instrument panel illumination and warning lamps	1.2
Ignition warning lamp	2
Heater control illumination	0.36 and 1.2

Torque wrench settings

	lbf ft	Nm
Alternator adjustment arm to alternator	18	24
Alternator adjustment arm to engine:		
1.3 litre models	28	38
1.6 litre models	67	91
Alternator pivot mounting bolt:		
1.3 litre models	16	22
1.6 litre models	13	18
Alternator pulley nut	28	38
Battery tray to body	18	24
Starter motor retaining bolts:		
Manual gearbox models	38	52
Automatic transmission models	18	24
Wiper arm retaining nuts	7	9
Wiper motor spindle nuts:		
M18	3	4
Plastic nut	2	3

1 General description

The electrical system is of the 12 volt negative earth type, and consists of a 12 volt battery, alternator, starter motor and related electrical accessories, components and wiring. The battery is of the maintenance free, 'sealed for life' type and is charged by an alternator which is belt-driven from the crankshaft pulley. The starter motor is of the pre-engaged type incorporating an integral solenoid. On starting, the solenoid moves the drive pinion into engagement with the flywheel ring gear before the starter motor is energised. Once the engine has started, a one-way clutch prevents the motor armature being driven by the engine until the pinion disengages from the flywheel.

Further details of the major electrical systems are given in the relevant Sections of this Chapter.

Caution: *Before carrying out any work on the vehicle electrical system, read through the precautions given in Safety First! at the beginning of this manual and in Section 2 of this Chapter.*

2 Electrical system – precautions

It is necessary to take extra care when working on the electrical system to avoid damage to semi-conductor devices (diodes and transistors), and to avoid the risk of personal injury. In addition to the precautions given in Safety First! at the beginning of this manual, observe the following items when working on the system.

1 *Always remove rings, watches, etc before working on the electrical system.* Even with the battery disconnected, capacitive discharge could occur if a component live terminal is earthed through a metal object. This could cause a shock or nasty burn.

2 *Do not reverse the battery connections.* Components such as the alternator or any other having semi-conductor circuitry could be irreparably damaged.

3 If the engine is being started using jump leads and a slave battery, connect the batteries *positive to positive* and *negative to negative.* This also applies when connecting a battery charger.

4 Never disconnect the battery terminals, or alternator multi-plug connector, when the engine is running.

5 The battery leads and alternator multi-plug must be disconnected before carrying out any electric welding on the car.

6 Never use an ohmmeter of the type incorporating a hand cranked generator for circuit or continuity testing.

7 1.6 litre models are equipped with an inlet manifold heater which will operate whenever the ignition is switched on and the coolant temperature is below 90°F (32°C). Due to the high current consumption of this unit, always disconnect the wiring connector if the ignition is to be left on for any length of time with the engine stopped. Further details will be found in Chapter 3.

3 Maintenance and inspection

1 At regular intervals (see Routine Maintenance) carry out the following maintenance and inspection operations on the electrical system components.

2 Check the operation of all the electrical equipment, ie wipers, washers, lights, direction indicators, horn etc. Refer to the appropriate Sections of this Chapter if any components are found to be inoperative.

3 Visually check all accessible wiring connectors, harnesses and retaining clips for security, or any signs of chafing or damage. Rectify any problems encountered.

4 Check the alternator drivebelt for cracks, fraying or damage. Renew the belt if worn or, if satisfactory, check and adjust the belt tension. These procedures are covered in Chapter 2.

5 Check the condition of the wiper blades and if they are cracked or show signs of deterioration, renew them, as described in Section 33. Check the operation of the windscreen and headlamp washers (if fitted). Adjust the nozzles using a pin, if necessary.

6 Check the battery terminals and, if there is any sign of corrosion, disconnect and clean them thoroughly. Smear the terminals and battery posts with petroleum jelly before refitting the plastic covers. If there is any corrosion on the battery tray, remove the battery, clean the deposits away and treat the affected metal with an anti-rust preparation. Repaint the tray in the original colour after treatment.

7 Top up the windscreen washer reservoir and check the security of the pump wires and water pipes.

8 It is advisable to have the headlight aim adjusted using optical beam setting equipment.

9 While carrying out a road test check the operation of all the instruments and warning lights, and the operation of the direction indicator self-cancelling mechanism.

10 At less frequent intervals (see Routine Maintenance) renew the alternator drivebelt, as described in Chapter 2.

4 Battery – removal and refitting

1 The sealed for life battery is located on the left-hand side of the engine compartment.

2 To remove the battery, slacken the negative (-) terminal clamp bolt and lift the terminal off the battery post (photo).

3 Lift the plastic cover from the positive (+) terminal, loosen the clamp bolt and lift the terminal off the battery post.

4 Undo and remove the retaining bolt and lift off the battery clamp plate (photo).

5 Withdraw the battery from the carrier tray.

6 Refitting is the reverse sequence to removal, but make sure that the polarity is correct before connecting the leads, and do not overtighten the clamp bolts.

5 Battery – charging

1 In winter when a heavy demand is placed on the battery, such as when starting from cold and using more electrical equipment, it may be necessary to have the battery fully charged from an external source.

2 Charging is best done overnight at a 'trickle' rate of 1 to 1.5 amps. Due to the design of certain maintenance-free batteries, rapid or boost charging is not recommended. If in any doubt about the suitability of certain types of charging equipment for use with maintenance-free batteries, consult a BL dealer or automotive electrical specialist.

3 Battery charging must be carried out with the battery removed from the vehicle and in a well-ventilated area.

4 The terminals of the battery and the leads of the charger must be connected positive to positive and negative to negative.

6 Alternator – removal and refitting

1 Disconnect the battery negative terminal.

2 Release the retaining clip and remove the multi-plug from the rear of the alternator (photo).

3 Slacken the adjustment arm bolt and the pivot mounting bolt and move the unit in towards the engine. Slip the drivebelt off the pulleys.

4 Remove the adjustment arm bolt and washers. Remove the pivot bolt, nuts and washers, taking note of the position of any additional spacers. Withdraw the alternator from the engine.

5 Refitting is a reverse of the removal procedure, but before tightening the mounting bolts and the adjustment arm bolt tension the drivebelt, as described in Chapter 2.

7 Alternator – fault tracing and rectification

Due to the specialist knowledge and equipment required to test or repair an alternator, it is recommended that, if the performance is suspect, the car be taken to an automobile electrician who will have the facilities for such work. Because of this recommendation, information is limited to the inspection and renewal of the brushes. Should

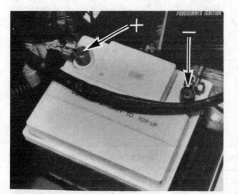

4.2 Battery terminal identification

4.4 Battery clamp retaining bolt

6.2 Alternator wiring multi-plug

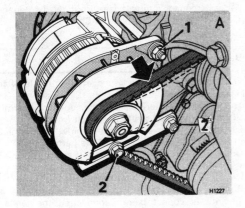 

Fig. 10.1 Alternator mounting details (Sec 6)

A   *1.3 litre models*   B   *1.6 litre models*   1   *Pivot mounting bolt*   2   *Adjustment arm nuts and bolts*

Arrows indicate belt tension checking point

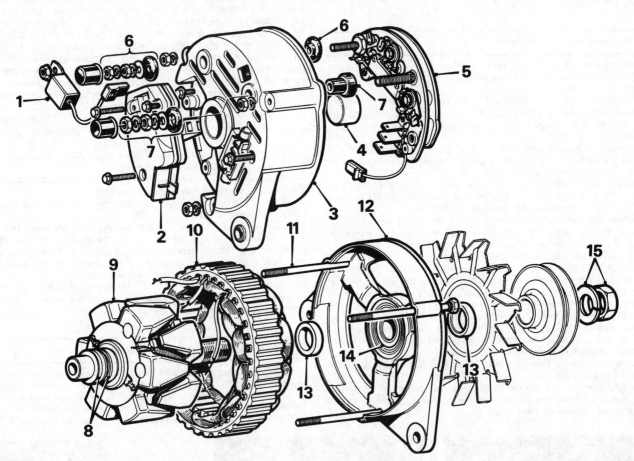

Fig. 10.2 Exploded view of the alternator (Sec 8)

1 *Suppression capacitor*	5 *Rectifier*	8 *Slip rings*	13 *Spacers*
2 *Regulator and brushbox assembly*	6 *Phase terminal and washer assembly*	9 *Rotor assembly*	14 *Bearing*
3 *Slip ring end bracket*	7 *Main terminal and washer assembly*	10 *Stator*	15 *Pulley retaining nut and washer*
4 *Bearing*		11 *Through-bolts*	
		12 *Drive end bracket*	

the alternator not charge, or the system be suspect, the following points should be checked before seeking further assistance:

(a)   *Check the drivebelt condition and tension*
(b)   *Ensure that the battery is fully charged*
(c)   *Check the ignition warning light bulb, and renew it if blown*

8   Alternator brushes – removal, inspection and refitting

1   Remove the alternator, as described in Section 6.
2   Disconnect the electrical lead and remove the suppression capacitor from the rear of the alternator.

3   Undo the retaining screws, lift off the regulator and brushbox assembly and disconnect the electrical lead.
4   Check the brush length and the brush spring tension against the figures given in the Specifications. New brushes are supplied as an assembly complete with brushbox and regulator.
5   Refitting is the reverse sequence to removal.

9   Starter motor – testing in the car

1   If the starter motor fails to operate, first check the condition of the battery by switching on the headlamps. If they glow brightly then gradually dim after a few seconds, the battery is in an uncharged condition.
2   If the battery is satisfactory, check the starter motor main terminal and the engine earth cable for security. Check the terminal connections on the starter solenoid – located on top of the starter motor.
3   If the starter still fails to turn use a voltmeter, or 12 volt test lamp and leads, to ensure that there is battery voltage at the solenoid main terminal (containing the cable from the battery positive terminal).
4   With the ignition switched on and the ignition key in position III, check that voltage is reaching the solenoid terminal with the Lucar connector, and also the starter main terminal.
5   If there is no voltage reaching the Lucar connector there is a wiring or ignition switch fault. If voltage is available, but the starter does not operate, then the starter or solenoid is likely to be at fault.

10   Starter motor – removal and refitting

1   Disconnect the battery negative terminal.
2   Disconnect the electrical leads at the solenoid terminals noting their respective locations (photo). **Note:** *Take care when disconnecting the leads from the Lucar type terminal connections as they are easily broken. Pull back the cover along the lead to expose the connector, then using a small screwdriver, depress the lock tag in the centre of the connector to release it.*

Manual gearbox models
3   Using a suitable Allen key, undo and remove the two bolts, nuts and washers securing the starter to the gearbox flange and lift the unit off. Note that on 1.6 litre models the starter motor bolts also retain the engine front snubber bracket (photo) and ignition system crankshaft sensor bracket.
4   Refitting the starter motor is the reverse sequence to removal, but tighten the retaining bolts to the specified torque.

Automatic transmission models
5   Jack up the front of the car and securely support it on axle stands.
6   From under the car undo and remove the three bolts securing the starter motor to the transmission and manipulate the motor out from under the car.
7   Refitting is the reverse sequence to removal, but tighten the retaining bolts to the specified torque.

10.2 Wiring connections at the starter motor solenoid

10.3 Removing the front snubber bracket

11   Starter motor – overhaul

1   Remove the starter motor from the car, as described in the previous Section.
2   At the rear of the solenoid, unscrew the nut and lift away the lead from the solenoid STA terminal.
3   Undo and remove the two screws securing the solenoid to the drive end housing. Disengage the solenoid plunger from the engaging lever and withdraw the solenoid.
4   Withdraw the end cap from the commutator end housing and then prise off the 'spire' retaining washer from the armature shaft.
5   Undo and remove the two through-bolts and withdraw the end housing sufficiently to gain access to the brushes.
6   Release the field brushes from their brushbox locations and then remove the end housing.
7   On models fitted with automatic transmission, slacken the locknut and unscrew the eccentric pivot pin from the drive end housing.
8   Mark the relationship of the field coil assembly to the drive end housing, unscrew the two housing retaining screws and withdraw the drive end housing.
9   Slide the armature and drive assembly out of the field coil assembly.
10 Mount the armature in a vice and, using a suitable tubular drift, tap the thrust collar on the end of the shaft toward the pinion to expose the jump ring. Prise the jump ring out of its groove and slide it off the shaft. Withdraw the thrust collar and the drive assembly.
11 With the starter motor now completely dismantled, clean all the components in paraffin or a suitable solvent and wipe dry.
12 Check the length of the brushes and the tension of the brush springs. If the length and tension are not as given in the Specifications renew the brushes and springs. Note that the field brushes must be soldered in place.
13 Check that the brushes move freely in their holders, and clean the holders and brushes with a petrol-moistened rag if they show any tendency to stick. If necessary use a fine file on the brushes if they still stick after cleaning.
14 Check the armature shaft for distortion, and examine the commutator for excessive scoring, wear or burns. If necessary, the commutator may be skimmed in a lathe and then polished with fine glass paper. Make sure that it is not reduced below the specified minimum thickness and do not undercut the mica insulation.
15 Check the taping of the field coils, check all joints for security and check the coils and commutator for signs of burning.
16 Check the drive pinion assembly, drive end housing, engaging lever and solenoid for wear or damage. Make sure that the drive pinion one-way clutch permits movement of the pinion in one direction only and renew the complete assembly, if necessary.
17 Check the condition of the bush in the commutator end housing and, if necessary, renew it. Note that on manual gearbox models the

170

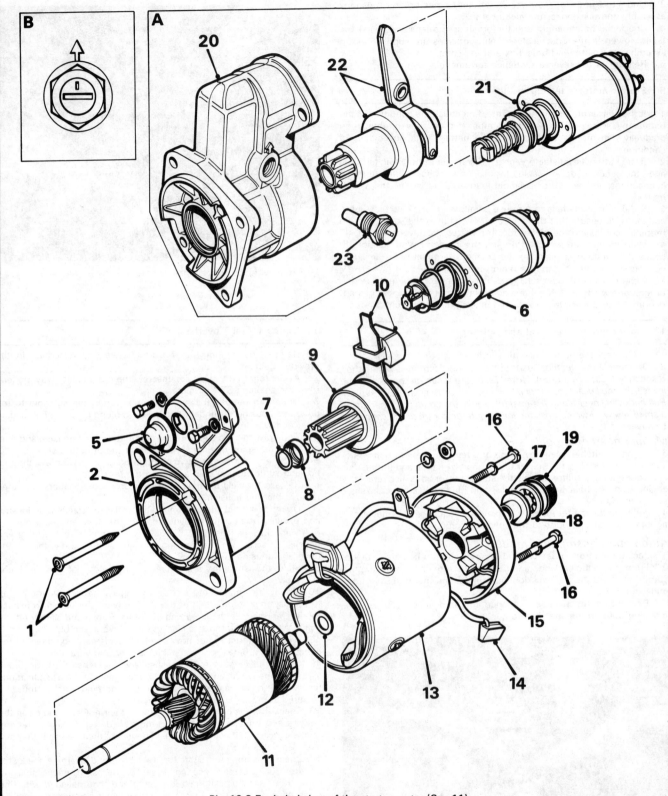

Fig. 10.3 Exploded view of the starter motor (Sec 11)

1 Housing retaining screws	10 Engaging lever and bush	17 Bush	23 Eccentric pivot pin
2 Drive end housing	11 Armature	18 'Spire' washer	A Alternative components fitted
5 End cover	12 Thrust washer	19 End cap	to automatic transmission models
6 Solenoid	13 Field coil assembly	20 Drive end housing	B Eccentric pivot pin mark and
7 Jump ring	14 Field brush	21 Solenoid	housing arrow alignment
8 Thrust collar	15 Commutator end housing	22 Drive and engaging lever	
9 Drive assembly	16 Through-bolt	assembly	

armature is supported at the drive end by a bush in the gearbox housing, and that access to this bush entails removal of the gearbox from the car unless the BL special removing tool can be obtained. Soak the new bushes in engine oil for 24 hours before fitting.

18 Accurate checking of the armature, commutator and field coil windings and insulation requires the use of special test equipment. If the starter motor was inoperative when removed from the car and the previous checks have not highlighted the problem, then it can be assumed that there is a continuity or insulation fault and the unit should be renewed.

19 If the starter is in a satisfactory condition, or if a fault has been traced and rectified, the unit can be reassembled using the reverse of the dismantling procedure. When refitting the eccentric pivot pin (automatic transmission models) screw the pin fully into the housing, unscrew it one full turn then align the mark with the arrow on the housing. Hold the pin in this position and tighten the locknut.

12 Fuses, relays and control units – general

Fuses
1 The fusebox is situated below the facia on the right-hand side, adjacent to the steering column. To gain access to the fusebox, release the two turnbuckles using a coin and lift off the fusebox cover (photo). The fuse locations, current rating and circuits protected are shown on the fusebox cover (photo). Each fuse is colour-coded and has its rating stamped on it.

2 To remove a fuse from its location, hook it out using the small plastic removal tool provided. This is located, together with the spare fuses, in a holder on the left-hand side of the fusebox. Hook the tool over the fuse and pull up to remove. Refit the fuse by pressing it firmly into its location.

3 Always renew a fuse with one of an identical rating. Never renew a fuse more than once without finding the source of the trouble.

4 Circuits protected by fuses C1, C2 and C3 operate when the ignition switch is at positions I or II. Circuits A1, A2, C5 and C6 only operate when the ignition switch is at position II. All other circuits operate irrespective of ignition switch position.

Relays and control units
5 The relays are located in the fusebox under the right-hand side of the facia.

6 The relays can be removed by simply pulling them from their respective locations. If a system controlled by a relay becomes inoperative, and the relay is suspect, operate the system and if the relay is functioning it should be possible to hear it click as it is energized. If the relay is not being energized then the relay is not receiving a main supply voltage, a switching voltage, or the relay itself is faulty.

7 Control units for the wiper delay and fuel system electronic mixture control are located on a panel attached to the roof of the glovebox. Release the two turnbuckles using a coin and lower the panel to gain access (photo).

8 Where applicable, the programmed ignition system control unit will be found on the left-hand valance in the engine compartment. The

electric window lift relay and control unit together with the central locking control unit are located inside the driver's door behind the trim panel.

13 Fusible link – general

1 Certain electrical circuits on Montego models are protected by fusible links which form the first part of the battery positive (+) cable. The links consist of fusible cable of different thicknesses and lengths according to the circuits they protect. The wire size and fusible link lengths are given in the Specifications.

2 If power to a number of related electrical circuits is lost, the relevant fusible link may be suspect. The circuits protected by the fusible links are as follows.

> Link A   Ignition circuit
> Link B   Window lift circuit
> Link C   Interior lights, lighter, clock circuit

3 To renew a fusible link, disconnect the battery leads then remove the leads from the battery positive (+) terminal.

4 Carefully cut back the binding and pull the sleeve from the cables. Identify the link to be renewed by referring to the Specifications.

5 Remove the joint cover and unsolder the fusible link at both ends.

6 Solder a new fusible link into place and seal the joints with insulating tape. Retape the sleeve to the cables and reconnect the battery terminals.

14 Direction indicator and hazard flasher system – general

1 The flasher unit is located in the fusebox situated under the right-hand side of the facia.

2 Should the flashers become faulty in operation, check the bulbs for security and make sure that the contact surfaces are not corroded. If one bulb blows or is making a poor connection due to corrosion, the system will not flash on that side of the car.

3 If the flasher unit operates in one direction and not the other, the fault is likely to be in the bulbs, or wiring to the bulbs. If the system will not flash in either direction, operate the hazard flashers. If these function, check for a blown fuse in position C6. If the fuse is satisfactory renew the flasher unit.

15 Ignition switch/steering column lock – removal and refitting

The ignition switch can be separated from the steering column lock (see Chapter 11, Section 19), after extracting the grub screw.

16 Steering column switches – removal and refitting

1 Disconnect the battery negative terminal.

2 Remove the steering wheel, as described in Chapter 11.

3 Undo the retaining screws and lift off the left- and right-hand steering column cowls.

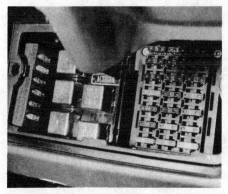

12.1A Fuse and relay locations in the fusebox ...

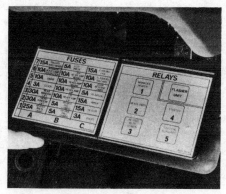

12.1B ... and circuit details on the fusebox cover

12.7 Control unit locations on glovebox roof panel

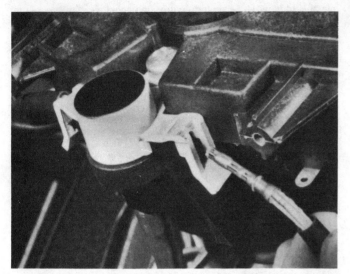

16.4 Removing the fibre optic guide from the steering column switch bulb housing

17 Facia switches – removal and refitting

1   Disconnect the battery negative terminal.
2   Undo the retaining screws and withdraw the right-hand steering column cowl from the column (photo). Position the steering wheel so that the spokes are facing away from the cowl to provide sufficient clearance for removal.
3   Undo the two retaining screws and withdraw the switch panel from the facia (photos).
4   Disconnect the wiring multi-plug and push the switch out of the panel.
5   Refitting is the reverse sequence to removal.

18 Instrumentation – description

Electro-mechanical instrumentation is used on Montego models covered by this manual. On all versions the instrumentation consists of a speedometer, fuel gauge, temperature gauge and driver warning lights. Additionally, according to model, a tachometer, digital clock and additional warning lights for low fuel and high engine temperature may be fitted. A multi-function unit attached to the rear of the instrument panel provides a stabilized 10 volt supply for the instruments and warning lights and controls the tachometer operation, where applicable.

4   Carefully lift up the retaining clip and slip the fibre optic guide out of the bulb housing (photo).
5   Depress the retainers at the top and bottom of the switch then pull the switch out of the steering column boss (photo).
6   Disconnect the wiring multi-plug(s) as applicable and remove the switch (photo).
7   Refitting is the reverse sequence to removal. Position the switch striker bush with the arrow pointing towards the direction indicator switch before refitting the steering wheel.

19 Instrument panel and instruments – removal and refitting

1   Disconnect the battery negative terminal.
2   Undo the two retaining screws and ease the instrument panel surround from its location (photo). Disconnect the clock wiring harness plug and remove the surround (photo).
3   Undo the five screws securing the instrument panel mounting brackets to the facia. Ease the panel down slightly and tip it towards the steering wheel at the top.

16.5 Depress the retainers to release the switch ...

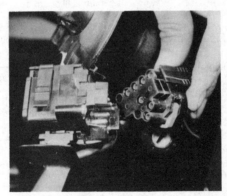

16.6 ... then disconnect the wiring multi-plugs

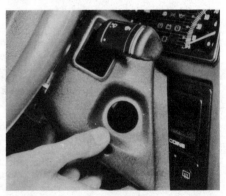

17.2 Remove the steering column cowl ...

17.3A ... undo the two screws (arrowed) ...

17.3B ... then withdraw the switch panel

19.2A Undo the retaining screws ...

19.2B ... then withdraw the instrument panel surround and disconnect the clock wiring

19.4 Disconnect the instrument panel wiring multi-plugs

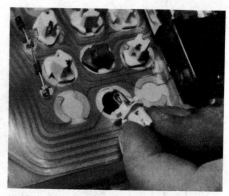

20.2 Remove the instrument panel bulb holders by turning them anti-clockwise

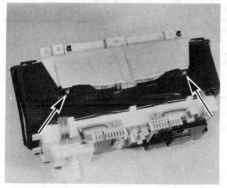

20.3 Panel mounting bracket retaining nuts (arrowed)

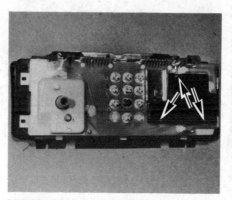

20.5 Multi-function unit cover retaining nuts (arrowed)

20.15 Speedometer retaining screws (arrowed)

4   Disconnect the wiring harness multi-plugs (photo) then reach round behind the panel and disconnect the speedometer cable by pulling the retainer straight from the speedometer head. It may be necessary to push the cable through the bulkhead grommet from the engine compartment side to facilitate removal.
5   Withdraw the instrument panel from the facia.
6   Refitting is the reverse sequence to removal.

20 Instrument panel and instruments – dismantling and reassembly

1   Remove the instrument panel from the car, as described in the previous Section.

Panel illumination and warning lamp bulbs
2   The bulb holders are secured to the rear of the instrument panel by a bayonet fitting and are removed by turning the holders anti-clockwise (photo). Note that only the no-charge warning lamp bulb can be separated from its bulb holder (coloured red). All the other illumination and warning lamp bulbs are renewed complete with their bulb holders.
3   To gain access to the panel front illumination lamp bulb, remove the upper instrument panel mounting bracket (photo).
4   Refit the bulb holders by turning clockwise to lock.

Multi-function unit
5   Undo the three nuts and lift off the multi-function unit cover (photo).
6   Undo the four nuts securing the fuel and temperature gauges and, where fitted, the three nuts securing the tachometer.
7   Undo the multi-function unit retaining screws and withdraw the unit from the instrument panel.
8   Refit the multi-function unit using the reverse sequence to removal.

Printed circuit
9   Remove all the instrument panel warning and illumination bulb holders as previously described.
10   Undo the three nuts and remove the multi-function unit cover.
11   Carefully extract the pegs securing the printed circuit to the instrument panel. Remove the tape, ease the printed circuit off the pegs and studs and lift off.
12   Refitting the printed circuit is the reverse sequence to removal.

Speedometer
13   Release the retaining wires and lift off the instrument panel shroud and window assembly on the front illuminated panel. Note that the two parts will be connected by the printed circuit.
14   Ease the face plate from the case noting that it is secured between the fuel and temperature gauges with double-sided tape.
15   Undo the retaining screws and withdraw the speedometer from the instrument panel (photo).
16   Refitting is the reverse sequence to removal, but use new double-sided tape to secure the face plate.

Fuel gauge, temperature gauge and tachometer
17   Carry out the operations described in paragraphs 13 and 14.
18   Undo the retaining nuts and remove the cover from the multi-function unit.
19   Undo the nuts securing the fuel and temperature gauges and, where fitted, the tachometer to the printed circuit and instrument panel.
20   With the panel held face down, carefully remove the instruments. Recover the wavy washers from the gauge studs.
21   Refitting is the reverse sequence to removal bearing in mind the following points:

(a)   Ensure that an equal number of wavy washers are positioned on each of the fuel and temperature gauge studs

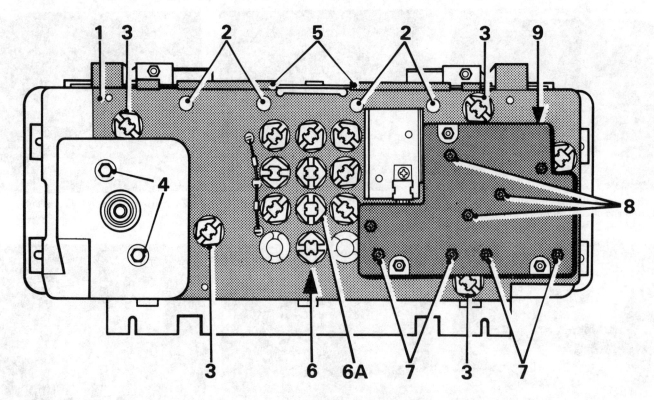

Fig. 10.4 Instrument panel connections and component locations (Sec 20)

1 Printed circuit
2 Printed circuit retaining pegs
3 Panel illumination bulbs
4 Speedometer retaining screws
5 Wiring multi-plug connections
6 Warning lamp bulb holders
6A Ignition no-charge warning lamp bulb holder
7 Fuel and temperature gauge retaining nuts
8 Tachometer retaining nuts (where fitted)
9 Multi-function unit

(b) Position all the instruments in the panel before tightening the retaining nuts; where fitted, tighten the tachometer retaining nuts first.

21 Clock – removal and refitting

1 Disconnect the battery negative terminal.
2 Undo the two retaining screws and ease the instrument panel surround from its location.
3 Disconnect the clock wiring harness plug and remove the surround.
4 Undo the two screws securing the clock to the surround and carefully lift the clock from its location (photo).

5 The clock time set push rods should not be disturbed. If they become dislodged, the longest legs must be positioned horizontally.
6 Refitting the clock is the reverse sequence to removal. Reset the hour and minute display by depressing first the hour then the minute control buttons using a pencil or ballpoint pen. To zero the display depress both control buttons simultaneously.

22 Headlamp and sidelamp bulbs – renewal

1 From within the engine compartment pull off the wiring connector at the rear of the headlamp bulb (photo).
2 Slip off the rubber cover to gain access to the bulbs (photo).

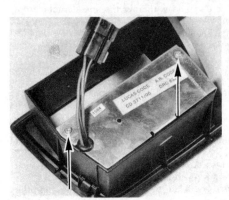

21.4 Clock retaining screws (arrowed)

22.1 Disconnect the headlamp bulb wiring connector ...

22.2 ... then slip off the rubber cover

22.3A Release the headlamp bulb retaining clip

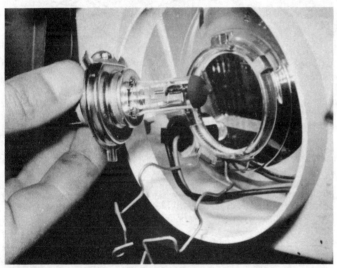

22.3B ... and withdraw the bulb

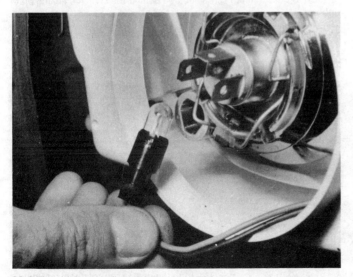

22.6 The sidelamp bulb holder is a push fit in the lens assembly

Headlamp bulb

3   Release the bulb retaining clip and withdraw the bulb from its location in the headlamp (photos). Take care not to touch the bulb glass with your fingers; if touched, clean the bulb with methylated spirits.
4   Fit the bulb ensuring that the lugs on the bulb engage with the slots in the headlamp.
5   Refit the retaining clip, rubber cover and wiring connector.

Sidelamp bulb

6   Withdraw the bulb holder from the headlamp and remove the bulb from the holder by turning anti-clockwise (photo).
7   Fit the new bulb to the holder, push the holder into the headlamp and refit the rubber cover.

23   Headlamp lens assembly – removal and refitting

1   From within the engine compartment, pull off the wiring connector at the rear of the headlamp bulb.
2   Slip off the rubber cover then withdraw the sidelamp bulb holder.
3   Undo the two nuts securing the top of the lens unit to the body, release the two lower plastic ball pegs from their sockets and remove the assembly from the car.
4   Remove the headlamp bulb from the lens assembly if required, as described in the previous Section.
5   Refitting is the reverse sequence to removal, but have the headlamp aim adjusted using optical beam setting equipment (see Section 32).

24   Front direction indicator bulb – renewal

1   Working in the engine compartment, release the spring retainer (photo) and ease the lamp body and seal from the body panel (photo). Release the bulb holder by turning anti-clockwise.
2   Remove the bulb from the holder by turning anti-clockwise.
3   Fit the bulb and holder to the lamp body, locate the lamp flange behind the panel and push the unit into position. Make sure the seal seats correctly.
4   Secure the lamp body with the spring retainer.

25   Rear lamp cluster bulbs – renewal

1   Working in the luggage compartment, press and disengage the retainers then withdraw the bulb panel from the lens unit (photos).
2   The bulb can now be removed from the panel as required. All the bulbs except the tail lamp bulb are removed by turning them anti-clockwise. The tail lamp bulb is a push fit.
3   Fit the bulbs and bulb panel using the reverse sequence to removal.

26   Number plate lamp bulb – renewal

1   Push the lens housing forward and lift up to release it from the bumper (photo).
2   Turn the bulb holder anti-clockwise and withdraw it from the lens housing. The bulb is a push fit in the holder.
3   Refit using the reverse sequence to removal.

27   Interior courtesy lamp bulbs – renewal

1   Carefully ease the lamp lens from its location and then withdraw the bulb by turning anti-clockwise.
2   Fit the bulb and push the unit into place.

24.1A Release the direction indicator lamp
body spring retainer ...

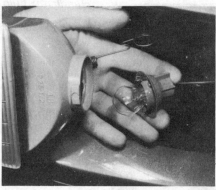

24.1B ... to gain access to the bulb holder and
bulb

25.1A Depress the plastic retainers ...

25.1B ... then withdraw the rear lamp cluster
bulb panel

26.2 Push the lens housing forward and lift to
release from bumper

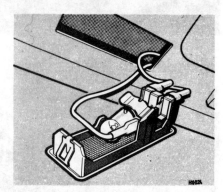

Fig. 10.5 Interior courtesy lamp bulb and
holder arrangement (Sec 27)

**28 Instrument panel illumination and warning lamp bulbs –
renewal**

   Refer to Section 20.

---

29 Switch illumination bulbs – renewal

1   Remove the facia switch panel, as described in Section 17, to gain
access to the illumination bulbs.
2   With the wiring multi-plug disconnected, remove the relevant bulb
which is a push fit in the holder (photo).

3   To renew the hazard warning switch bulb, lift off the switch lens
and remove the push-fit bulb (photo).
4   Refitting is the reverse sequence to removal.

---

30 Heater control illumination bulbs – removal

1   Carefully prise off the switch blanking plate on the right-hand side
of the heater control knobs.
2   Undo the retaining screw located behind the switch blanking plate
(photo).
3   Centralize the heater control knobs and ease the control panel,
complete with knobs, off the facia.

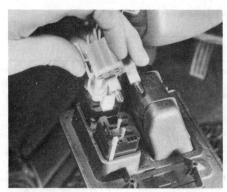

29.2 Switch panel bulbs located in wiring
multi-plug

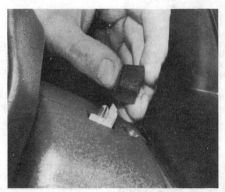

29.3 Lift off the switch lens to renew the
hazard warning switch bulb

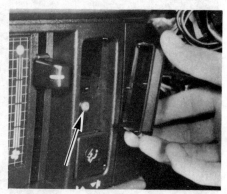

30.2 Undo the heater control panel retaining
screw (arrowed)

4   Disconnect the switch multi-plug to gain access to the push-fit bulbs.
5   The panel illumination bulb holder and bulb are removed by turning anti-clockwise (photo).
6   Refitting is the reverse sequence to removal.

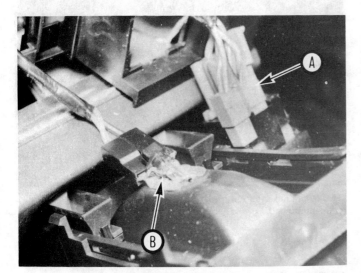

30.5 Heater panel illumination bulbs located in wiring multi-plug (A) and at rear of panel (B)

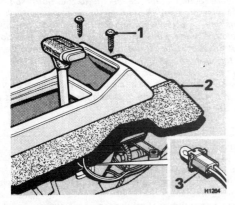

Fig. 10.6 Selector lever illumination bulb location (Sec 31)

1   Console retaining screws   *3   Illumination bulb holder*
2   Centre console

31 Selector lever illumination bulbs (automatic transmission models) – renewal

1   Remove the centre console sufficiently to gain access to the bulbs, using the procedure given in Chapter 12.
2   Pull the relevant bulb holder from the side of the selector housing and remove the push-fit bulb.
3   Refitting is the reverse sequence to removal.

32 Headlamp aim – adjustment

1   At regular intervals (see Routine Maintenance) headlamp aim should be checked and, if necessary, adjusted.
2   Due to the light pattern of the homofocal headlamp lenses fitted to Montego models, optical beam setting equipment must be used to achieve satisfactory aim of the headlamps. It is recommended therefore that this work be entrusted to a BL dealer.
3   Holts Amber Lamp is useful for temporarily changing the headlight colour to conform with the normal usage on Continental Europe.

33 Wiper blades and arms – removal and refitting

Wiper blades
1   The wiper blades should be renewed when they no longer clean the windscreen effectively.
2   Lift the wiper arm away from the window.
3   Release the spring retaining catch and separate the blade from the wiper arm (photo).
4   Insert the new blade into the arm, making sure that the spring retainer catch is engaged correctly.

Wiper arms
5   To remove a wiper arm, open the bonnet, lift the hinged cover and unscrew the retaining nut (photo).
6   Carefully prise the arm off the spindle.
7   Before refitting the arm switch the wipers on and off, allowing the mechanism to return to the 'park' position.
8   Noting that the longer arm is fitted to the driver's side, refit the arm to the spindle with the blade positioned on the finisher at the bottom edge of the windscreen.
9   Refit and tighten the retaining nut then close the hinged cover.

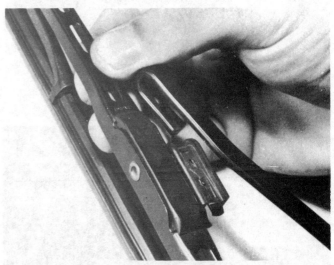

33.3 Release the catch and separate the wiper blade from the arm

33.5 Lift the hinged cover to gain access to the wiper arm retaining nut

34 Windscreen wiper motor – removal and refitting

1   Disconnect the battery negative terminal.
2   Undo the screws securing the washer reservoir in place and move the reservoir to one side.
3   Pull the rubber edge seal off the top of the closure panel, undo the retaining screws and lift away the panel (photo).

34.3A Undo the retaining screws ...

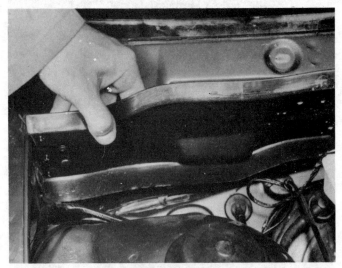

34.3B ... and lift out the closure panel

4   Remove the wiper arms, as described in Section 33.
5   Undo the retaining nut and remove the wiper linkage rotary link from the motor spindle (photo).
6   Undo the three bolts securing the wiper motor mounting bracket to the bulkhead.
7   Ease the motor assembly out of its fitted position and prise the wiring harness grommet out of the bulkhead.
8   From inside the car, open the glovebox lid and lower the relay panel from the roof of the glovebox.
9   Identify the wiper motor wiring harness (coloured grey) and disconnect the multi-plugs.
10 Pull the wiring harness through the bulkhead and remove the wiper motor assembly from its location.

34.5 Wiper linkage rotary link retaining nut (A) and motor bracket retaining bolts (B)

11 Refitting is the reverse sequence to removal, bearing in mind the following points:

(a)   Before reconnecting the linkage to the motor spindle, switch the wipers on and then off again to set the motor in the 'park' position. Apply silicone grease to the rotary link and gimbal, hold the primary link and turn the rotary link anti-clockwise to set the cam in its extended position. Hold the primary link and rotary link in a straight line, as shown in Fig. 10.7, and refit the link to the wiper motor spindle

(b)   Switch the wipers on and off once more so that the linkage is in the 'park' position then refit the wiper arms, as described in Section 33

35 Windscreen wiper linkage – removal and refitting

1   Disconnect the battery negative terminal.
2   Undo the screws securing the washer reservoir in place and move the reservoir to one side.
3   Pull the rubber edge seal off the top of the closure panel, undo the retaining screws and lift away the panel.
4   Remove the wiper arms, as described in Section 33.
5   Carefully prise the primary link arm socket off the rotary link assembly.
6   Remove the wiper spindle sealing covers followed by the retaining nuts, washers and upper spacers.
7   Push the wiper spindles through the body then withdraw the linkage assembly and central water shield from the engine compartment.
8   Remove the central water shield and lower spindle spacers.
9   Refitting is the reverse sequence to removal, bearing in mind the following points:

(a)   Apply silicone grease to the primary link ball and socket then press the two parts together, using a lightly adjusted self-locking wrench

(b)   Switch the wipers on and off to set the linkage in the 'park' position then refit the wiper arms as described in Section 33

36 Windscreen wiper primary link – removal and refitting

1   Disconnect the battery negative terminal.
2   Undo the screws securing the washer reservoir in place and move the reservoir to one side.
3   Pull the rubber edge seal off the top of the closure panel, undo the retaining screws and lift away the panel.

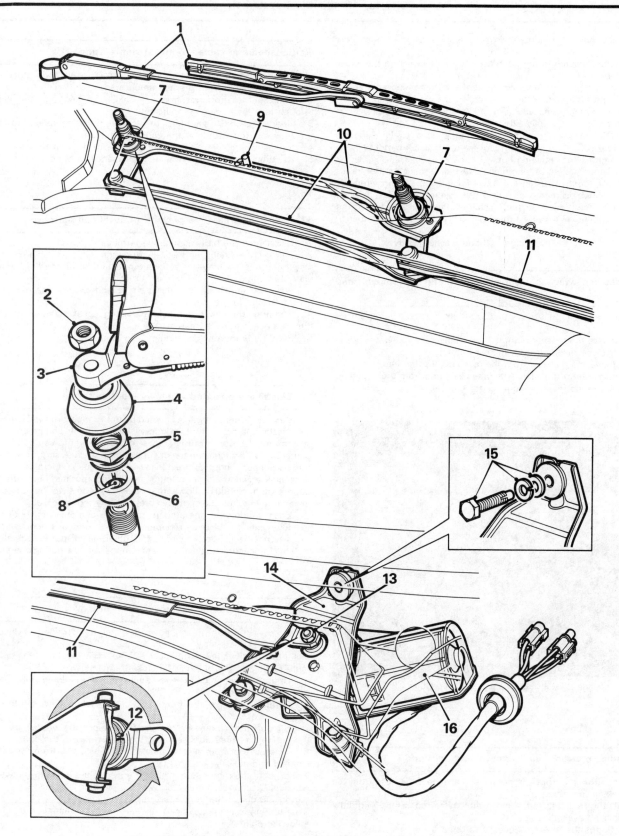

Fig. 10.7 Windscreen wiper motor and linkage layout (Secs 34, 35 and 36)

1	Wiper blade and arm	6	Upper spacer
2	Wiper arm retaining nut	7	Lower spacer
3	Wiper arm	8	Wiper spindle
4	Sealing cover	9	Wiper arm stop
5	Spindle retaining nut and washer	10	Drive link assembly
		11	Primary link arm
		12	Rotary link assembly
		13	Rotary link retaining nut
14	Wiper motor mounting bracket		
15	Mounting bracket retaining bolt and washers		
16	Wiper motor		

4   Carefully prise the primary link arm socket off the rotary link assembly.
5   Undo the retaining nut and remove the rotary link from the wiper motor spindle.
6   To refit the primary link, reconnect the battery then switch the wipers on and off again to set the motor in the 'park' position.
7   Hold the primary link and turn the rotary link anti-clockwise to position the cam in the fully extended position.
8   Hold the primary and rotary link together in a straight line, as shown in Fig. 10.7, then refit the rotary link to the wiper motor spindle. Refit and tighten the retaining nut.
9   Apply silicone grease to the primary link ball and socket then press the two parts together, using a lightly adjusted self-locking wrench.
10   With the windscreen wet, check the operation of the wipers; ensuring that they 'park' on the finisher at the bottom edge of the windscreen. If necessary adjust the wiper arm position, as described in Section 33.
11   Refit the closure panel, edge seal and washer reservoir.

37 Windscreen washer reservoir and pump – removal and refitting

1   Undo the retaining screws and withdraw the reservoir from its location.
2   Disconnect the washer pump wiring connectors and water hoses then remove the reservoir from the car (photo).
3   To remove the pump(s) from the reservoir, lift the pump and pull the inlet nozzle from the seal. Withdraw the pump from the reservoir.
4   Refitting is the reverse sequence to removal.

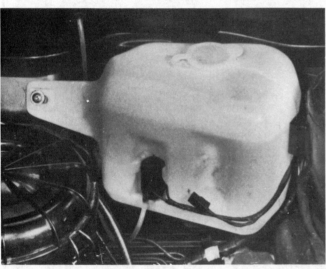

37.2 Windscreen washer reservoir and pump locations

38 Headlamp washer jet – removal and refitting

1   From within the front bumper panel disconnect the hose from the washer extension tube.
2   Undo the retaining nut, remove the seating collar and withdraw the washer jet assembly.
3   Refitting is the reverse sequence to removal.

39 Horn – removal and refitting

1   Disconnect the battery negative terminal.
2   Working under the left-hand front wing, disconnect the electrical leads, undo the retaining nut and remove the horn from its bracket.
3   The horn is a sealed unit which cannot be dismantled or adjusted. In the event of horn failure, renewal will be necessary.

40 Speedometer cable – removal and refitting

1   Refer to Section 19, paragraphs 1 to 4 inclusive and detach the speedometer cable from the instrument panel.
2   Release the grommet from the bulkhead and pull the cable through into the engine compartment.
3   Undo the bolt securing the cable retaining plate to the transmission and withdraw the cable and pinion.
4   Release the cable clips and remove the cable from the car.
5   Refitting the cable is the reverse sequence to removal.

41 Radio, radio/cassette player – removal and refitting

1   Disconnect the battery negative terminal.
2   Pull off the knobs and bezels from the radio or radio/cassette player controls, unscrew the retaining nuts and lift off the finisher and masking plate.
3   Push the unit back into its aperture and remove the mounting plate from inside the panel.
4   Withdraw the unit, disconnect the wiring, speaker and aerial leads then remove it from the car.
5   Refitting is the reverse sequence to removal.

42 Electrically-operated windows – description

Certain Montego models covered by this manual are available with electrically-operated front windows as optional equipment. The system enables both front windows to be raised or lowered independently by two switches on the driver's door armrest. A single switch on the passenger's door armrest allows independent operation of the passenger's window. A 'one-touch' facility incorporated in the driver's door switch circuitry allows the driver's window to be raised or lowered fully when the switch is pressed fully then released.
Each window is operated by an electric motor acting on the window regulator. A relay mounted within the driver's door controls the electrical supply to the circuit and a thermal cut-out in each motor isolates the circuit should any object jam the window during operation.

43 Window lift motor – removal and refitting

1   Remove the front door inner trim panel, as described in Chapter 12.
2   If the door window is not closed, temporarily reconnect the switch multi-plugs and the battery and close the window.
3   Undo the retaining screws and remove the inner trim panel mounting bracket.
4   Carefully remove the polythene condensation barrier from the door.
5   Support the window in the raised position using wooden wedges then undo the three nuts securing the motor unit to the door.
6   Release the three nylon wheels of the lift arms from the lifting channel and auxiliary channel.
7   Operate the motor so that the lift arms are set in the fully lowered position. Disconnect the switch wiring multi-plug and manoeuvre the motor unit out of the door aperture.
8   Refitting is the reverse sequence to removal, bearing in mind the following points.

  (a) Lubricate the lifting and auxiliary channels with graphite grease
  (b) Install the motor unit with the lifting arms in the lowered position then engage the nylon wheels of the arms into the channels with the arms in the raised position
  (c) With the motor unit installed, check the operation of the window and adjust the lower ends of the window channels if necessary
  (d) Ensure that all wiring is neatly secured in its clips

44 Window lift 'one-touch' control unit – removal and refitting

1   Remove the front door inner trim panel, described in Chapter 12. Carefully peel back the polythene condensation barrier as necessary, starting at the lower front corner of the door.
2   Disconnect the wiring multi-plug, undo the retaining screws and withdraw the control unit from inside the door.
3   Refitting is the reverse sequence to removal.

45 Central door locking – description

A central door locking system is available as an option on certain Montego models covered by this manual. The system enables the passenger's front door lock, both rear door locks and the boot lid lock to be operated simultaneously by the action of the driver's door interior lock button or exterior private lock.

The passenger's door, both rear doors and the boot lid each incorporate a solenoid which is connected to the lock mechanism.

The driver's door is equipped with the normal private lock and interior lock button arrangement and is also fitted with a control unit.

Operation of the driver's door lock causes the control unit to supply an electric current to each of the door lock solenoids, thus locking or unlocking the doors in unison with the driver's door.

46 Central locking solenoid – removal and refitting

1   To remove a passenger door solenoid or driver's door control unit, first remove the inner trim panel from the relevant door, as described in Chapter 12.
2   If the door window is not closed, temporarily reconnect the switch multi-plugs and battery if electrically-operated windows are fitted, and close the window.
3   Undo the retaining screws and remove the inner trim panel mounting bracket.
4   Carefully remove the polythene condensation barrier from the door.
5   Undo the retaining screws, disconnect the wiring multi-plug and cable clips then lift the solenoid or control unit off the door panel. Detach the lock lever link and remove the unit.
6   Refitting is the reverse sequence to removal.

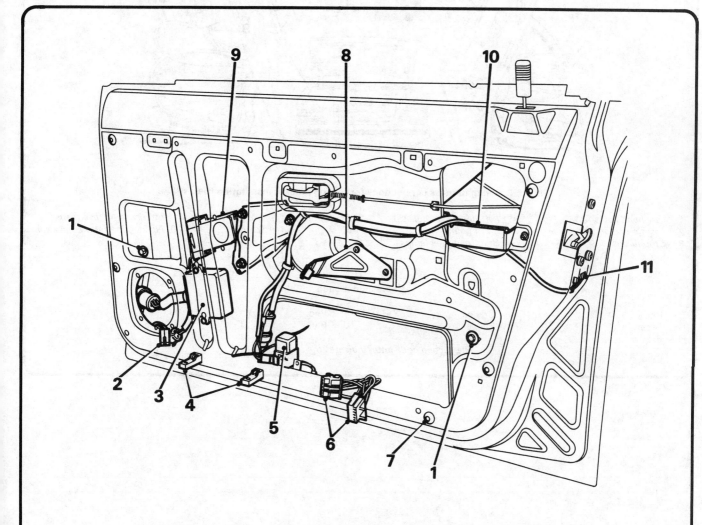

Fig. 10.8 Electric window and central locking door compartment locations (Secs 43 to 46)

1 Window auxiliary channel retaining screws	4 Inner trim panel retainers	8 Inner trim panel mounting bracket	10 Central locking switch unit
2 Speaker wiring connector	5 Window lift relay	9 Window lift motor	11 Interior light door switch
3 Window lift 'one-touch' control unit	6 Window lift multi-plugs		
	7 Inner trim panel retaining studs		

47 Boot lock solenoid – removal and refitting

1   Disconnect the battery negative terminal.
2   Open the boot lid and undo the nut securing the solenoid mounting plate to the lock barrel. Remove the lock barrel.
3   Disconnect the wiring multi-plug then position the solenoid mounting plate in the left-hand boot lid aperture.
4   Undo the retaining screws and withdraw the solenoid from behind the mounting plate.
5   Refitting is the reverse sequence to removal.

48 Accessory wiring – general

1   If an electrical accessory is to be fitted, electrical connections should be made at the fusebox, on the feed side of the following fuses.

(a) If the accessory is to operate through the ignition switch, connect to fuse C1, C2 or C3 (light green/white wire)
(b) If the accessory is to operate independently of the ignition switch, connect to fuse A6 (brown wire)

2   Always use a separate line fuse of the appropriate rating to protect the accessory being fitted.
3   To avoid the risk of damage resulting from overloading the vehicle electrical circuits, do not plug any accessory into the cigarette lighter socket unless it is approved by BL for such fitment.
4   For towing bracket installations, connection for the towing socket can be made at the wiring harness behind the rear light clusters. To gain access, fold back the floor covering and release the rear of the side panel covers. The cable colours, locations and their respective circuits are shown in Fig. 10.9. Note: a special plug-in type relay is also required. Always disconnect the battery negative terminal before making any wiring conections.

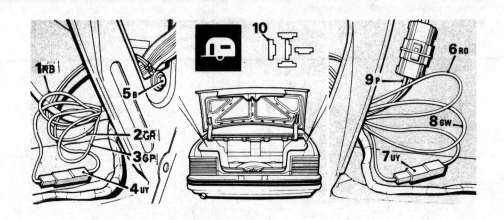

Fig. 10.9 Wiring harness connections for towing bracket installation (Sec 48)

1 Left-hand tail lamps	4 Left-hand rear foglamp	7 Right-hand rear foglamp	9 Interior lamp (live feed)
2 Left-hand indicator	5 Earth point	8 Right-hand indicator	10 Flasher unit relay connector
3 Stop lamps	6 Right-hand tail lamps		

Cable colour code

B	Black	O	Orange	R	Red	W	White
G	Green	P	Purple	U	Blue	Y	Yellow
N	Brown						

The second code letter indicates the tracer colour

Fault diagnosis appears overleaf

49 Fault diagnosis – electrical system

Symptom	Reason(s)
Starter fails to turn engine	Battery discharged or defective Battery terminal and/or earth leads loose Starter motor connections loose Starter solenoid faulty Starter brushes worn or sticking Starter commutator dirty or worn Starter field coils earthed Starter inhibitor switch faulty (automatic transmission only)
Starter turns engine very slowly	Battery discharged Starter motor connections loose Starter brushes worn or sticking
Starter spins but does not turn engine	Pinion or flywheel ring gear teeth broken or badly worn Starter pinion sticking
Starter noisy	Pinion or flywheel ring gear teeth badly worn Mounting bolts loose
Battery will not hold charge for more than a few days	Battery defective internally Battery terminals loose Alternator drivebelt slipping Alternator or regulator faulty Short circuit
Ignition light stays on	Alternator faulty Alternator drivebelt broken
Ignition light fails to come on	Warning bulb blown Indicator light open circuit Alternator faulty
Instrument readings increase with engine speed	Faulty instrument multi-function unit
Fuel or temperature gauge gives no reading	Wiring open circuit Sender or thermistor faulty Faulty instrument multi-function unit
Fuel or temperature gauge gives continuous maximum reading	Wiring short circuit Sender or thermistor faulty Faulty instrument multi-function unit Faulty gauge
Lights inoperative	Bulb blown Fuse blown Fusible link blown Battery discharged Switch faulty Relay faulty Wiring open circuit Bad connection due to corrosion
Failure of component motor	Commutator dirty or burnt Armature faulty Brushes sticking or worn Armature bearings dry or misaligned Field coils faulty Fuse blown Relay faulty Fusible link blown Poor or broken wiring connections
Failure of an individual component	Fuse blown Relay faulty Fusible link blown Poor or broken wiring connections Switch faulty Component faulty

Chapter 11 Suspension and steering

For modifications, and information applicable to later models, see Supplement at end of manual

Contents

Specifications

Front suspension

Type	Independent by MacPherson struts with coil springs and integral telescopic shock absorbers. Anti-roll bar on 1.6 litre models only
Coil spring free length:	
1.3 litre models (white spring)	14.48 in (368.0 mm)
1.6 litre models:	
Manual gearbox (purple spring)	14.72 in (373.9 mm)
Automatic transmission (yellow spring)	15.16 in (385.1 mm)
Trim height (measured from the centre of the front hub to the edge of the wheel arch)	14.7 to 16.0 in (375.0 to 405.0 mm))
Trim height permissible side difference	1.02 in (26.0 mm)

Rear suspension

Type	Trailing twist axle with coil springs and telescopic shock absorbers
Coil spring free length	13.5 in (343.0 mm)
Trim height (measured from the centre of the rear hub to the edge of the wheel arch):	
Up to VIN 131623	13.9 to 15.1 in (352.0 to 382.0 mm)
From VIN 131624	14.5 to 15.7 in (368.0 to 398.0 mm)
Trim height permissible side difference	1.02 in (26.0 mm)
Rear wheel toe setting	0°20′ to 0°40′ toe-in
Rear wheel camber angle	0° to 1° negative

Steering

Type	Rack and pinion
Turns lock to lock	4.3
Steering wheel diameter	15 in (381.0 mm)
Camber angle	0°05′ negative to 0°26′ negative
Castor angle	0°30′ positive to 1°30′ positive
Steering axis inclination	12°0′ to 13°0′
Toe setting	Parallel ± 0°8′
Steering gear lubricant	Semi-fluid grease (Duckhams Adgear 00)

Roadwheels

Wheel size:	
1.3 litre models	5.00J x 13
1.6 litre models	120 x 365 mm 'TD' type

Tyres

Tyre size:	
1.3 litre models	165 SR 13 steel braced radial ply
1.6 litre models	180/65 R365 'TD' type steel braced radial ply

Tyre pressures (cold):

	Front	Rear
Saloon	26 lbf/in² (1.8 bar)	28 lbf/in² (1.9 bar)
Estate	28 lbf/in² (1.9 bar)	31 lbf/in² (2.1 bar)

When fully loaded, or for high speed use, increase the front pressure by 3.0 lbf/in² (0.2 bar) and the rear pressure by 2.0 lbf/in² (0.1 bar) for Saloons or 7.0 lbf/in² (0.5 bar) for Estates

Torque wrench settings

	lbf ft	Nm
Front suspension		
Anti-roll bar bush to lower arm:		
10 mm nut	13	18
12 mm nut	30	41
Anti-roll bar clamp bolts	33	45
Balljoint-to-lower arm (service replacement bolts)	22	30
Balljoint-to-swivel hub clamp nut	33	45
Driveshaft nut*	150	203
Suspension strut-to-swivel hub nuts	66	90
Suspension strut upper retaining nut	41	56
Suspension strut bearing retaining nut	33	45
Lower arm rear mounting bolts	30	41
Lower arm front mounting bolt	53	72
Subframe to body	67	91
Subframe supports to body	55	75
Subframe supports to subframe	67	91

Refer to Chapter 13, Section 10

	lbf ft	Nm
Rear suspension		
Rear axle pivot bolts (marked '8.8')	41	56
Rear axle pivot bolts (marked '10.9')	63	85
Rear hub retaining nut	50	68
Stub axle-to-trailing arm nuts	39	53
Suspension strut lower mounting	55	75
Suspension strut upper mounting nut	33	45
Suspension strut spring retainer nut	18	24

	lbf ft	Nm
Steering		
Steering column mounting bolts	18	24
Intermediate shaft clamp nuts	18	24
Rack and pinion assembly mounting bolts	33	45
Steering wheel nut	35	47
Steering wheel bolt	24	33
Tie-rod inner balljoint to rack	60	81
Tie-rod outer balljoint to steering arm	22	30

	lbf ft	Nm
Roadwheels		
Wheel nuts (all models)	53	72

1 General description

The independent front suspension is of the MacPherson strut type, incorporating coil springs and integral telescopic shock absorbers. Lateral and longitudinal location of each strut assembly is by pressed steel lower suspension arms utilizing rubber inner mounting bushes and incorporating a balljoint at their outer ends. On 1.6 litre models both lower suspension arms are interconnected by an anti-roll bar. The front swivel hubs, which carry the wheel bearings, brake calipers and the hub/disc assemblies, are bolted to the MacPherson struts and connected to the lower arms via the balljoints.

The rear suspension is of the trailing twist axle type, whereby the two trailing arms are welded to an 'L' section transverse member. This arrangement allows considerable independent up-and-down movement of each trailing arm whilst providing lateral rigidity and anti-roll capability in a structure of minimum unsprung weight. Suspension and damping is by strut assemblies containing coil springs and integral telescopic shock absorbers.

The steering gear is of the conventional rack and pinion type, located behind the front wheels. Movement of the steering wheel is transmitted to the steering gear by an intermediate shaft containing two universal joints. The front wheels are connected to the steering gear by tie-rods, each having an inner and outer balljoint.

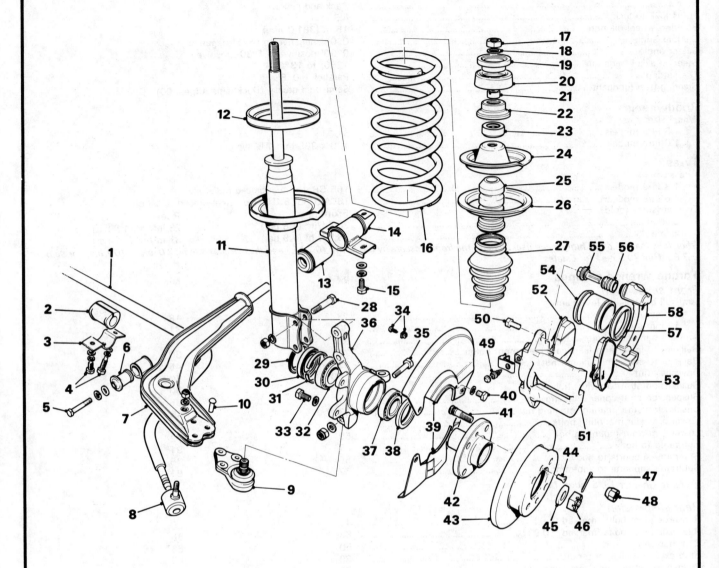

Fig. 11.1 Exploded view of the front suspension (Sec 1)

1	Anti-roll bar	15	Mounting bolt
2	Mounting block	16	Coil spring
3	Clamp	17	Upper mounting nut
4	Clamp retaining bolts	18	Washer
5	Lower arm front mounting bolt	19	Cup and washer assembly
6	Rubber bush	20	Upper mounting rubber
7	Lower suspension arm	21	Bearing retainer nut
8	Anti-roll bar mounting bush	22	Bearing housing
9	Lower suspension arm balljoint	23	Upper bearing
10	Rivet (replaced by nut and bolt after balljoint renewal)	24	Spring seat
		25	Spring aid
11	Front suspension strut	26	Upper insulator ring
12	Lower insulator ring	27	Rubber boot
13	Rubber bush	28	Strut-to-swivel hub retaining bolt
14	Rear mounting bracket	29	Bearing water shield
		30	Inner oil seal

31	Oil seal spacer
32	Hub inner bearing
33	Brake caliper mounting bolt
34	Disc shield retaining screw and clip
35	Lower balljoint clamp bolt
36	Swivel hub
37	Hub outer bearing
38	Outer oil seal
39	Disc shield halves
40	Disc shield retaining bolt
41	Wheel stud
42	Drive flange
43	Brake disc

44	Disc retaining screw
45	Flat washer
46	Driveshaft nut
47	Split pin
48	Wheel nut
49	Bleed screw
50	Caliper retaining (guide pin) bolt
51	Caliper body
52	Inner brake pad
53	Outer brake pad
54	Piston
55	Guide pin
56	Dust cover
57	Piston dust cover
58	Anchor bracket

187

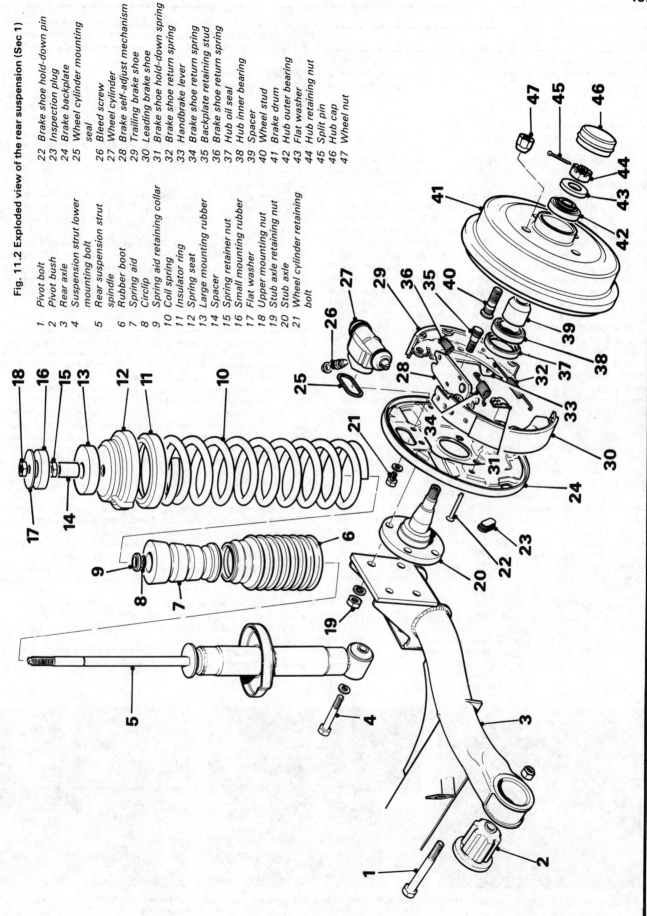

Fig. 11.2 Exploded view of the rear suspension (Sec 1)

1 Pivot bolt
2 Pivot bush
3 Rear axle
4 Suspension strut lower mounting bolt
5 Rear suspension strut spindle
6 Rubber boot
7 Spring aid
8 Circlip
9 Spring aid retaining collar
10 Coil spring
11 Insulator ring
12 Spring seat
13 Large mounting rubber
14 Spacer
15 Spring retainer nut
16 Small mounting rubber
17 Flat washer
18 Upper mounting nut
19 Stub axle retaining nut
20 Stub axle
21 Wheel cylinder retaining bolt
22 Brake shoe hold-down pin
23 Inspection plug
24 Brake backplate
25 Wheel cylinder mounting seal
26 Bleed screw
27 Wheel cylinder
28 Brake self-adjust mechanism
29 Trailing brake shoe
30 Leading brake shoe
31 Brake shoe hold-down spring
32 Brake shoe return spring
33 Handbrake lever
34 Brake shoe return spring
35 Backplate retaining stud
36 Brake shoe return spring
37 Hub oil seal
38 Hub inner bearing
39 Spacer
40 Wheel stud
41 Brake drum
42 Hub outer bearing
43 Flat washer
44 Hub retaining nut
45 Split pin
46 Hub cap
47 Wheel nut

2 Maintenance and inspection

1   At the intervals given in Routine Maintenance at the beginning of this manual a thorough inspection of all suspension and steering components should be carried out using the following procedure as a guide.

Front suspension and steering

2   Apply the handbrake, jack up the front of the car and support it securely on axle stands.
3   Visually inspect the lower balljoint dust covers and the steering rack and pinion gaiters for splits, chafing, or deterioration. Renew the rubber gaiters or the balljoint assembly, as described in Sections 21 and 8 respectively, if any damage is apparent.
4   Grasp the roadwheel at the 12 o'clock and 6 o'clock positions and try to rock it. Very slight free play may be felt, but if the movement is appreciable further investigation is necessary to determine the source. Continue rocking the wheel while an assistant depresses the footbrake. If the movement is now eliminated or significantly reduced, it is likely that the hub bearings are at fault. If the free play is still evident with the footbrake depressed, then there is wear in the suspension joints or mountings. Pay close attention to the lower balljoint and lower arm mounting bushes. Renew any worn components, as described in the appropriate Sections of this Chapter.
5   Now grasp the wheel at the 9 o'clock and 3 o'clock positions and try to rock it as before. Any movement felt now may again be caused by wear in the hub bearings or the steering tie-rod inner or outer balljoints. If the outer balljoint is worn the visual movement will be obvious. If the inner joint is suspect, it can be felt by placing a hand over the rack and pinion rubber gaiter and gripping the tie-rod. If the wheel is now rocked, movement will be felt at the inner joint if wear has taken place. Repair procedures are described in Sections 20 and 23.
6   Using a large screwdriver or flat bar check for wear in the anti-roll bar mountings (where fitted) and lower arm mountings by carefully levering against these components. Some movement is to be expected, as the mountings are made of rubber, but excessive wear should be obvious. Renew any bushes that are worn.
7   With the car standing on its wheels, have an assistant turn the steering wheel back and forth about one eighth of a turn each way. There should be no lost movement whatever between the steering wheel and roadwheels. If this is not the case, closely observe the joints and mountings previously described, but in addition check the intermediate shaft universal joints for wear and also the rack and pinion steering gear itself. Any wear should be visually apparent and must be rectified, as described in the appropriate Sections of this Chapter.

Rear suspension

8   Chock the front wheels, jack up the rear of the car and support it securely on axle stands.
9   Visually inspect the rear suspension components, attachments and linkages for any visible signs of wear or damage.
10   Grasp the roadwheel at the 12 o'clock and 6 o'clock positions and try to rock it. Any excess movement here indicates wear in the hub bearings which may also be accompanied by a rumbling sound when the wheel is spun. Repair procedures are described in Section 11.

Wheels and tyres

11   Carefully inspect each tyre, including the spare, for signs of uneven wear, lumps, bulges or damage to the sidewalls or tread face. Refer to Section 25 for further details.
12   Check the condition of the wheel rims for distortion, damage and excessive run-out. Also make sure that the balance weights are secure with no obvious signs that any are missing. Check the torque of the wheel nuts and check the tyre pressures.

Shock absorbers

13   Check for any signs of fluid leakage around the shock absorber body or from the rubber boot around the piston rod. Should any fluid be noticed the shock absorber is defective internally and renewal is necessary.
14   The efficiency of the shock absorber may be checked by bouncing the car at each corner. Generally speaking the body will return to its normal position and stop after being depressed. If it rises and returns on a rebound, the shock absorber is probably suspect. Examine also the shock absorber upper and lower mountings for any sign of wear. Renewal procedures are contained in Sections 6 and 14.

3 Front swivel hub assembly – removal and refitting

1   Securely apply the handbrake, chock the rear wheels and remove the wheel trim from the front roadwheel.
2   Extract the split pin, then, using a socket and long bar, slacken, but do not remove, the driveshaft nut.
3   Slacken the roadwheel retaining nuts then jack up the front of the car and support it securely on axle stands. Remove the roadwheel. Remove the driveshaft retaining nut and washer (photo).
4   Undo and remove the two bolts securing the disc brake caliper to the anchor bracket (photo). Slide the caliper, complete with brake pads, off the anchor bracket (photo), and suspend it from a convenient place under the wheel arch using string or wire. Take care not to strain the flexible hydraulic hose.
5   Undo and remove the nut securing the tie-rod balljoint to the steering arm on the swivel hub. Release the balljoint from the steering arm using a suitable extractor (see Section 20).
6   At the base of the swivel hub unscrew and remove the nut and washer, withdraw the lower balljoint clamp bolt (photo).
7   Undo and remove the nuts and washers, then withdraw the two bolts securing the suspension strut to the upper part of the swivel hub (photo).
8   Release the swivel hub from the strut and then lift the hub, while pushing down on the suspension arm, to disengage the lower balljoint. Withdraw the swivel hub assembly from the driveshaft and remove it from the car (photo).
9   Secure the swivel hub in a vice, undo the two bolts and remove the brake caliper anchor bracket. Separate the drive flange and disc

3.3 Remove the driveshaft nut and washer

3.4A Brake caliper-to-anchor bracket upper retaining bolt (arrowed)

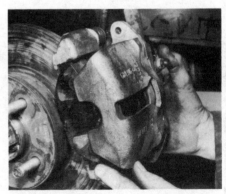

3.4B Remove the caliper and brake pads from the anchor bracket

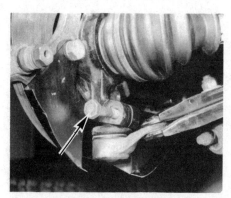

3.6 Undo the nut and withdraw the lower balljoint clamp bolt (arrowed)

3.7 Remove the two nuts and bolts (arrowed) securing the swivel hub to the suspension strut

3.8 Withdraw the hub assembly from the driveshaft

3.9A Remove the brake caliper anchor bracket

3.9B Use a hammer and tube to drift out the drive flange if tight ...

3.9C ... then withdraw the drive flange and disc assembly from the hub

from the hub using a hammer and suitable tube or socket, if necessary (photos).

10 If required, the two halves of the disc shield can now be removed after unscrewing the retaining bolt and screw.

11 Refitting the swivel hub is the reverse of the removal sequence, bearing in mind the following points:

(a) Ensure that the bearing water shield is in position on the driveshaft before fitting the swivel hub. Fill the groove in the water shield with a general purpose grease

(b) Where applicable, tighten all retaining nuts and bolts to the specified torque. **Do not attempt to fully tighten the driveshaft nut until the weight of the car is on the roadwheels**

(c) Tighten the driveshaft retaining nut to the specified torque, with reference to Chapter 13, Section 10

4 Front hub bearings – removal and refitting

1 Remove the swivel hub assembly from the car, as described in the previous Section.

2 With the hub assembly on the bench, prise out the inner and outer oil seals using a screwdriver or suitable flat bar. Note that there is a spacer fitted between the inner oil seal and the bearing.

3 With the hub supported on blocks use a hammer and drift to drive out one of the bearing inner races from the centre of the hub. Take care not to lose the balls which will be dislodged as the inner race is released. Turn the hub over and repeat the procedure for the other inner race. The outer races can now be driven out in the same way.

4 Wipe away any surplus grease from the bearings and swivel hub and then clean these components thoroughly using paraffin, or a suitable solvent. Dry with a lint-free rag. Remove any burrs or score marks from the hub bore with a fine file or scraper.

5 Carefully examine the bearing inner and outer races, the balls and ball cage for pitting, scoring or cracks, and if at all suspect renew the

bearings as a pair. It will also be necessary to renew the oil seals as they will have been damaged during removal.

6 If the old bearings are in satisfactory condition and are to be reused, reassemble the ball cage, holding the balls in position with grease, and then place this assembly in the outer race. Lay the inner race over the balls and push it firmly into place.

7 Before refitting the bearings to the hub, pack them thoroughly with a high melting-point grease.

8 Place one of the bearings in position on the hub with the word THRUST, or the markings stamped on the edge of the inner race, facing away from the centre of the hub (photo).

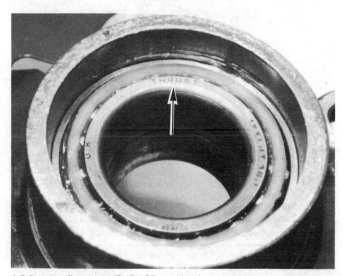

4.8 Bearing fitted with THRUST marking (arrowed) facing away from the hub centre

4.11A Fit the outer oil seal with its sealing lip facing inwards

4.11B Place the spacer in position over the bearing ...

4.11C ... then fit the inner oil seal, with its sealing lip facing inwards

9   Using a tube of suitable diameter, a large socket or a soft metal drift, drive the bearing into the hub bore until it contacts the shoulder in the centre of the hub. Ensure that the bearing does not tip slightly and bind as it is being fitted. *If this happens the outer race may crack, so take great care to keep it square.*

10 Turn the hub over and repeat this procedure for the other bearing

11 Dip the new oil seals in oil and carefully fit them to the hub using a tube of suitable diameter or one of the old seals to drive them in. Note that both oil seals are fitted with their sealing lips inward (photo) and that the inner seal has a second lip on its outer face. Don't forget to fit the spacer between the bearing and inner oil seal (photos).

12 The swivel hub can now be refitted to the car, as described in the previous Section.

3   From within the engine compartment prise off the small plastic cap in the centre of the strut upper mounting (photo), and remove the upper mounting cover.

4   Insert an Allen key of the appropriate size into the centre of the strut spindle and, while holding the key to prevent the spindle turning, unscrew the mounting nut (photo). Lift off the spring washer and the cup and washer assembly.

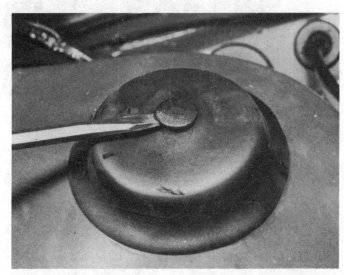
5.3 Prise off the plastic cap and remove the front strut upper mounting cover

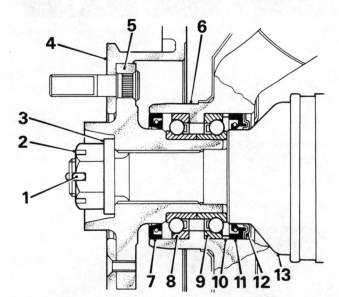

Fig. 11.3 Cross-sectional view of the front hub (Sec 4)

1 Split pin	8 Outer bearing
2 Driveshaft nut	9 Inner bearing
3 Flat washer	10 Spacer
4 Brake disc	11 Inner oil seal
5 Drive flange	12 Bearing water shield
6 Hub	13 Driveshaft
7 Outer oil seal	

5   Front suspension strut – removal and refitting

1   Securely apply the handbrake, chock the rear wheels and remove the trim from the front roadwheel.

2   Slacken the roadwheel retaining nuts, then jack up the front of the car and support it securely on axle stands. Remove the roadwheel.

5.4 Undo the strut upper mounting nut whilst holding the spindle

5   Undo and remove the two nuts, bolts and spring washers securing the suspension strut to the swivel hub (photo). Release the strut from the hub and withdraw it from under the wheel arch (photo).
6   Refitting is the reverse sequence to removal. Ensure that all retaining nuts and bolts are tightened to the specified torque, where applicable.

5.5A Undo the strut-to-swivel hub retaining nuts and bolts (arrowed) ...

5.5B ... then remove the strut assembly from under the wheel arch

6.2 Compressors in position on the front coil spring

6.3A Lift off the upper mounting rubber ...

6.3B ... and bearing housing (arrowed)

6   Front suspension strut – dismantling and reassembly

Note: *Before attempting to dismantle the front suspension strut, a suitable tool to hold the coil spring in compression must be obtained. Adjustable coil spring compressors are readily available and are recommended for this operation. Any attempt to dismantle the strut without such a tool is likely to result in damage or personal injury.*
1   Proceed by removing the front suspension strut, as described in the previous Section.
2   Position the spring compressors on either side of the spring (photo) and compress the spring evenly until there is no tension on the upper spring seat or upper mounting.
3   Lift off the upper mounting rubber (photo) and bearing housing (photo).

4   To remove the bearing retainer nut it will be necessary to make up a suitable tool which will engage in the slots of the nut enabling it to be unscrewed. A tool can be made out of a large nut with one end suitably shaped by cutting or filing so that two projections are left which will engage with the slots in the retainer nut (Fig. 11.4).

5   While holding the strut spindle with an Allen key, unscrew the bearing retainer nut. Lift off the upper bearing, spring seat and insulator ring.

6   The coil spring can now be removed, with compressors still in position if desired, followed by the spring aid, rubber boot and lower insulator ring.

7   With the strut completely dismantled, the components can be examined as follows.

8   Examine the strut for signs of fluid leakage. Check the strut spindle for signs of wear or pitting along its entire length and check the strut body for signs of damage or elongation of the mounting bolt holes. Test the operation of the strut, while holding it in an upright position, by moving the spindle through a full stroke and then through short strokes of 2 to 4 in (50 to 100 mm). In both cases the resistance felt should be smooth and continuous. If the resistance is jerky or uneven, or if there is any visible sign of wear or damage to the strut, renewal is necessary.

9   If any doubt exists about the condition of the coil spring, remove the spring compressors and check the spring for distortion. Also measure the free length of the spring and compare the measurement with the figure given in the Specifications. Renew the spring if it is distorted or outside the specified length.

10 Begin assembly by fitting the rubber boot and spring aid to the strut.

11 Place the lower insulator ring in position, followed by the spring, ensuring that the end of the bottom coil locates in the step of the spring seat.

12 With the spring suitably compressed, withdraw the strut spindle as far as it will go and refit the upper insulator ring and spring seat. Make sure that the end of the spring upper coil locates in the step of the spring seat.

13 Place the bearing, with the small internal diameter downwards, over the spindle and refit the bearing retainer nut. Tighten the nut to the specified torque and remove the spring compressors.

14 Finally refit the bearing housing and the upper mounting rubber, large opening uppermost.

15 The strut can now be refitted to the car, as described in the previous Section.

7   Front lower suspension arm – removal and refitting

1   Apply the handbrake, chock the rear wheels and remove the front wheel trim. Slacken the wheel nuts. Jack up the front of the car, support it securely on axle stands and remove the roadwheel.

2   Undo and remove the nut and washer and then withdraw the clamp bolt securing the lower balljoint to the swivel hub.

3   Lever the lower suspension arm downwards to release the balljoint from the hub.

4   Undo and remove the three bolts securing the suspension arm rear mounting to the underbody (photo).

5   Undo and remove the nut and bolt securing the front mounting to the crossmember (photo).

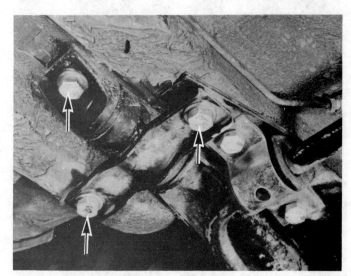

7.4 Front suspension arm rear mounting bolts (arrowed)

Fig. 11.4 Suspension strut bearing retainer nut removal tool (Sec 6)

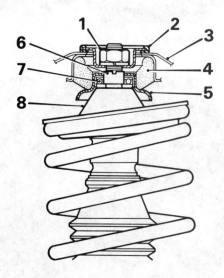

Fig. 11.5 Front suspension strut upper mounting details (Sec 6)

1	Mounting nut	5	Bearing housing
2	Cup and washer assembly	6	Bearing retainer nut
3	Inner front wing valance	7	Upper bearing
4	Upper mounting rubber	8	Spring seat

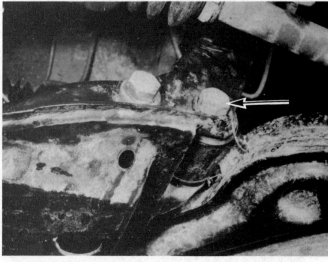

7.5 Front suspension arm front mounting retaining bolt (arrowed)

6   On models equipped with an anti-roll bar, undo and remove the nut and washer securing the anti-roll bar mounting bush to the suspension arm.

7   Ease the front mounting out of the subframe, lift the suspension arm to disengage the anti-roll bar (where fitted) and manoeuvre the arm out from beneath the vehicle.

8   With the arm removed from the car, carefully examine the rubber mounting bushes for swelling or deterioration, the balljoint for slackness, and check for damage to the rubber boot. Renewal of the balljoint is described in Section 8. Renewal of the suspension arm mounting bushes may be carried out as follows.

9   Using a two-legged puller, or similar tool, draw the rear bush assembly off its spigot. The rear bush can be removed from its mounting bracket by pressing it out in a vice with the aid of tubes of suitable length and diameter. Removal of the front bush in the arm follows the same procedure.

10   To refit the new bushes first lubricate them with rubber grease and then press them fully into place using a vice. Slide the rear bush assembly onto its spigot on the suspension arm. Note that the rear bush housings are handed and must now be interchanged from side to side. When fitted the small hole in the bush housing must be toward the centre of the car.

11   Refitting the lower suspension arm is the reverse sequence to removal. Ensure that all retaining nuts and bolts are tightened to the specified torque.

8   Front lower suspension arm balljoint – removal and refitting

1   Remove the lower suspension arm, as described in the previous Section.

2   Drill out the heads of the three rivets securing the balljoint to the arm and drive out the rivets with a punch.

3   Withdraw the balljoint from the suspension arm.

4   Replacement balljoint kits are supplied with nuts and bolts to secure the joint in place of the factory fitted rivets. Slide the new balljoint into position and fit the bolts so that the bolt heads are located above the arm. Screw on the nuts and tighten them to the specified torque.

5   The lower suspension arm can now be fitted to the arm as described in the previous Section.

9   Anti-roll bar – removal and refitting

Note: *For this operation the front suspension must be kept in a laden condition. If a vehicle lift or inspection pit are not available, it will be necessary to drive the front of the car up on ramps.*

1   Undo and remove the nuts and washers securing the anti-roll bar mounting bush, one each side, to the lower suspension arms.

2   Undo and remove the four bolts securing the anti-roll clamps to the subframe. Lift away the clamps, disengage the mounting bushes from the lower suspension arms and lower the anti-roll bar to the ground.

3   If the mounting blocks require renewal, slip them off the bar and place new blocks in position after lubricating liberally with rubber grease.

4   To renew the mounting bushes first measure and record the distance from the outer face of the bush to the end of the anti-roll bar. Now draw off the bushes using a two-legged puller.

5   Lubricate the new bushes with rubber grease and drive them onto the anti-roll bar using a hammer and tube of suitable diameter. Use the measurement recorded during removal as a setting dimension.

6   To refit the anti-roll bar to the car, first locate the bush assemblies in the lower suspension arms and fit the washers and retaining nuts finger tight.

7   Place the clamps over the mounting blocks and refit the retaining bolts, noting that the long bolts locate in the rear holes.

8   First tighten the clamp bolts to the specified torque, followed by the bush retaining nuts.

9   Lower the car to the ground (if applicable).

10   Rear hub assembly – removal and refitting

1   Chock the front wheels, remove the rear wheel trim and slacken the wheel nuts. Jack up the rear of the car and support it securely on axle stands. Remove the roadwheel and release the handbrake.

2   By judicious tapping and levering, extract the hub cap and withdraw the retaining split pin from the hub retaining nut.

3   Using a large socket and bar, undo and remove the hub retaining nut and flat washer. *Note that the left-hand nut has a left-hand thread and the right-hand nut has a conventional right-hand thread.* **Take care not to tip the car from the axle stands.** If the hub nuts are particularly tight, temporarily refit the roadwheel and lower the car to the ground. Slacken the nut in this more stable position and then raise and support the car before removing the nut.

4   Withdraw the hub and brake drum assembly from the stub axle (photo). If it is not possible to withdraw the hub due to the brake drum binding on the brake shoes, the following procedure should be adopted. Refer to Chapter 9, if necessary, and slacken off the handbrake cable at the cable adjuster. From the rear of the brake backplate, prise out the handbrake lever stop, which will allow the brake shoes to retract sufficiently for the hub assembly to be removed. It will, however, be necessary to remove the brake shoes and fit a new handbrake lever stop to the backplate.

5   Refitting the hub assembly is the reverse sequence to removal. Tighten the hub retaining nut to the specified torque and then align the next split pin hole. Always use a new split pin.

10.4 Withdraw the hub and brake drum from the stub axle

11   Rear hub bearings – removal and refitting

1   Remove the rear hub asembly from the car, as described in the previous Section.

2   With the hub on the bench, prise out the rear oil seal using a stout screwdriver or suitable flat bar.

3   Support the hub on blocks and, using a soft metal drift, tap out the two bearing inner races. Take care not to lose any of the balls which will be released from the ball cage as the inner races are removed.

4   Withdraw the spacer located between the two bearings and then drive the two outer races from the centre of the hub.

5   Wipe away any surplus grease and then thoroughly clean all the parts in paraffin or a suitable solvent. Dry with a lint-free rag.

6   Carefully examine the bearing inner and outer races, the ball cage and the balls for scoring, pitting or wear ridges. Renew both bearings as a set if any of these conditions are apparent. The hub oil seal must be renewed as it will have been damaged during removal. If the bearings are in a satisfactory condition, reassemble the balls and ball cage on the outer race, holding them in place with grease, and then press the inner race back into position (photo).

7   Before refitting the bearings, remove any burrs that may be present in the bore of the hub using a fine file or scraper.

11.6 Reassemble the balls in the hub bearing cage then press the inner race into position

11.10 Fit the spacer between the bearings, smaller diameter outward

11.11A Fit the bearings with the word THRUST or the markings facing away from the hub centre

11.11B Fit the inner bearing to the hub ...

11.11C ... followed by the oil seal

8   Pack the bearings with a high melting-point grease and place the outer bearing in position with the word THRUST, or the markings stamped on the edge of the inner race, facing outwards. Drive the bearing into position using a tube of suitable diameter, or a drift, in contact with the bearing outer race. Take great care to keep the bearing square as it is installed, otherwise it will jam in the hub bore which could crack the outer race.

9   Dip a new oil seal in clean engine oil and position it over the outer bearing with its sealing lip facing inwards. Using a hammer and block of wood, tap the seal into the hub until it is flush with the end face of the hub.

10   Turn the hub over and lay the spacer over the installed outer bearing, with the smaller diameter of the spacer facing outwards (photo).

11   Fit the inner bearing and oil seal to the hub, using the same procedure as for the outer bearing, with the word THRUST, or the bearing markings, also facing out from the hub centre (photos).

12   The rear hub assembly can now be refitted to the car, as described in the previous Section.

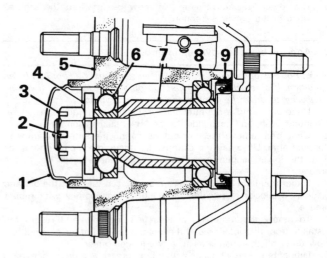

Fig. 11.6 Cross-sectional view of the rear hub (Sec 11)

1	Hub cap	6	Outer bearing
2	Split pin	7	Spacer
3	Hub retaining nut	8	Inner bearing
4	Flat washer	9	Oil seal
5	Hub and brake drum assembly		

12 Rear stub axle – removal and refitting

1   Remove the rear hub assembly, as described in Section 10.

2   Working under the car, disconnect the handbrake inner cable at the cable connector located beneath the rear axle transverse member.

3   Using circlip pliers, extract the circlip securing the handbrake outer cable to the bracket on the rear axle.

4   Using a brake hose clamp, or self-locking wrench with protected jaws, clamp the flexible brake hose located just in front of the rear axle. This will minimise brake fluid loss during subsequent operations.

5   Unscrew the brake pipe union nut at the rear of the wheel cylinder and carefully ease the pipe out of the cylinder. Plug the end of the pipe to prevent dirt entry.

6   Undo and remove the four nuts and spring washers securing the

stub axle to the trailing arm and lift away the stub axle and brake backplate as an assembly.

7   Support the stub axle in a vice and drive out the four mounting studs using a soft metal drift. Take care not to damage the ends of the threads during this operation.

8   With the studs removed, separate the backplate from the stub axle.

9   Refitting the stub axle is the reverse sequence to removal. Tighten all retaining nuts and bolts to the specified torque and, on completion, bleed the brake hydraulic system, as described in Chapter 9; if suitable precautions were taken to minimise fluid loss, as described, it should only be necessary to bleed the relevant wheel and not the entire system.

13 Rear suspension strut – removal and refitting

1   Chock the wheels, prise off the rear wheel trim and slacken the wheel nuts. Jack up the rear of the car and support it securely on axle stands. Remove the roadwheel.

2   As a safety precaution place a jack on suitable blocks beneath, and in contact with, the suspension trailing arm.

3   From inside the luggage compartment release the plastic cap from the strut upper mounting by rotating it whilst at the same time lifting upwards.

4   Engage a suitable small spanner on the flats of the strut spindle to prevent it turning and unscrew the upper mounting nut (photo). Lift off the flat washer and small mounting rubber.

5   From underneath the car, undo and remove the bolt securing the strut lower mounting to the rear suspension trailing arm (photo).

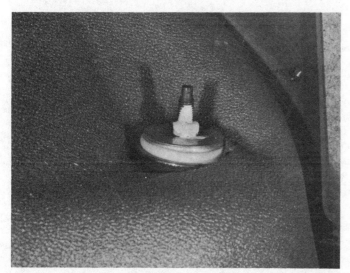

13.4 Rear suspension upper mounting, accessible from inside the luggage compartment

13.5 Rear suspension lower mounting bolt (arrowed)

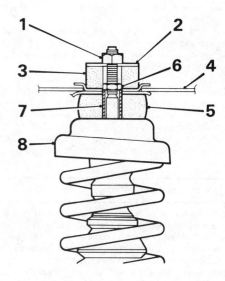

Fig. 11.7 Rear suspension strut upper mounting details (Sec 13)

1 *Upper mounting nut*	5 *Large mounting rubber*
2 *Flat washer*	6 *Spring retainer nut*
3 *Small mounting rubber*	7 *Spacer*
4 *Rear wheel arch*	8 *Spring seat*

Withdraw the strut from under the wheel arch and remove the large mounting rubber.

6   Refitting the suspension strut is the reverse sequence to removal. Ensure that the upper and lower mountings are tightened to the specified torque.

14 Rear suspension strut – dismantling and reassembly

Note: *Before attempting to dismantle the rear suspension strut, a suitable tool to hold the coil spring in compression must be obtained. Adjustable coil spring compressors are readily available and are recommended for this operation. Any attempt to dismantle the strut without such a tool is likely to result in damage or personal injury.*

1   Proceed by removing the rear suspension strut, as described in the previous Section.

2   Position the spring compressors on either side of the spring and compress the spring evenly until all the tension on the spring retainer nut is released.

3   With a spanner engaged with the flats of the strut spindle to stop it turning, unscrew the spring retainer nut. Lift off the spacer, spring seat and insulator ring.

4   The coil spring can now be removed, with the compressors still in position, if desired.

5   Carefully tap the spring aid retaining collar off the strut spindle then lift off the spring aid and the protective rubber boot.

6   With the strut completely dismantled, the components can be examined as follows.

7   Examine the strut body for signs of damage or corrosion, and for any trace of fluid leakage. Check the strut spindle for distortion, wear, pitting, or corrosion along its entire length. Test the operation of the strut, while holding it in an upright position, by moving the spindle through a full stroke and then through short strokes of 2 to 4 in (50 to 100 mm). In both cases the resistance felt should be smooth and continuous. If the resistance is jerky or uneven, or if there is any sign of wear or damage to the strut, renewal is necessary.

8   If the strut is in a satisfactory condition check the condition of the circlip on the strut spindle and, if it is in any way damaged or distorted, fit a new circlip.

9   If any doubt exsts about the condition of the coil spring, remove the spring compressors and check the spring for distortion. Also measure the free length of the spring and compare the measurement

with the figure given in the Specifications. Renew the spring if it is distorted or not of the specified length.

10 Begin reassembly by fitting the rubber boot to the strut, followed by the spring aid and retaining collar.

11 Place the spring in position, ensuring that the end of the bottom coil locates in the step of the retainer.

12 With the spring suitably compressed, withdraw the strut spindle as far as it will go and refit the insulator ring and spring seat, ensuring that the end of the coil locates in the step of the spring seat.

13 Refit the spacer, followed by the spring retainer nut. Tighten the nut to the specified torque.

14 The spring compressors can now be removed and the strut assembly refitted to the car, as described in the previous Section.

15 Rear axle – removal and refittng

1 Chock the wheels, prise off the rear wheel trim and slacken the wheel nuts. Jack up the rear of the car and support it securely on axle stands. Remove the roadwheel and release the handbrake.

2 Disconnect the two rear handbrake inner cables at the connectors located behind the rear axle transverse member.

3 Using two brake hose clamps, or self-locking wrenches with protected jaws, clamp the brake hydraulic flexible hoses, one located on each side of the car, adjacent to the rear axle front pivot bolts. This will minimise brake fluid loss during subsequent operations.

4 Undo and remove the two rear brake pipes unions at their connections with the flexible hoses. Plug or tape over the pipe and hose ends after disconnection to prevent dirt entry.

5 Using pliers extract the retaining clips and release the flexible hoses from the brackets on the rear axle.

6 Place a jack beneath one of the rear axle trailing arms and just take the weight of the axle.

7 Undo and remove the single bolt each side securing the rear suspension strut lower mountings to the rear axle trailing arms.

8 Undo and remove the nuts from the rear axle pivot bolts (one each side) (photo). Suitably support the axle beneath the transverse member and drift the pivot bolts from their locations.

9 Lever the axle out of its pivot mountings, lower it to the ground and withdraw it from under the car.

10 If necessary the axle can be completely dismantled by removing the rear hub assemblies and stub axles, as described in Sections 10 and 12 respectively.

11 If the axle pivot bushes require renewal, the old ones may be prised out and new ones inserted after lubricating them thoroughly with rubber grease.

12 Refitting the axle to the car is a straightforward reversal of the removal sequence. Ensure that all nuts and bolts are tightened to the specified torque and, on completion, bleed the brake hydraulic system, as described in Chapter 9.

16 Steering wheel – removal and refitting

1 Set the front wheels in the straight-ahead position.

2 Ease off the steering wheel pad to provide access to the retaining

15.8 Rear axle pivot bolt retaining nut (arrowed)

nut (photo).

3 Using a socket and bar, undo and remove the retaining nut and lockwasher (photo).

4 Mark the steering wheel and inner column in relation to each other and withdraw the wheel from the inner column splines (photo). If it is tight, tap it upwards near the centre, using the palm of your hand. *Refit the steering wheel retaining nut two turns before doing this, for obvious reasons.*

5 Refitting is the reverse of removal, but align the previously made marks and tighten the retaining nut to the specified torque. Note that there is a small arrow stamped on the striker bush of the multifunction switch and this must point towards the direction indicator switch when fitting the steering wheel.

17 Steering column assembly – removal, overhaul and refitting

1 Disconnect the battery negative terminal.

2 Remove the steering wheel, as described in the previous Section.

3 Undo and remove the screws securing the two steering column cowl halves to the column and lift off both cowls.

4 Disconnect the wiring multi plugs from the steering column multi function switches and from the ignition switch. Release any cable ties securing the wiring harness to the column.

5 From under the facia, remove the steering column cover and seal and then, to assist refitting, mark the inner column in relation to the intermediate shaft.

16.2 Ease off the steering wheel pad ...

16.3 ... undo the nut ...

16.4 ... then pull the steering wheel off the inner column

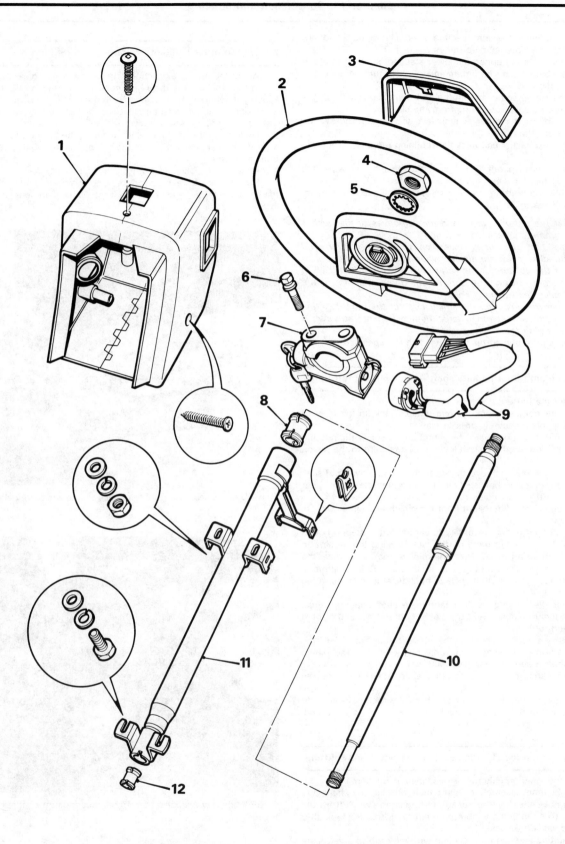

Fig. 11.8 Exploded view of the steering column assembly (Sec 17)

1 Steering column cowl	4 Steering wheel retaining nut	7 Steering lock/ignition switch	10 Inner column
2 Steering wheel	5 Lockwasher	8 Top bush	11 Outer column
3 Steering wheel pad	6 Shear bolt	9 Ignition switch	12 Bottom bush

6   Undo and remove the intermediate shaft clamp bolt and nut securing the intermediate shaft to the inner column.
7   Undo and remove the upper and lower steering column mounting bolts, lift the column off the intermediate shaft and remove the assembly from the car.
8   Slacken the clamp screw and lift the multi-function switch off the steering column. Release the wiring from the clip.
9   Carefully support the outer column in a vice and withdraw the inner column from the top of the outer column.
10 Lift out the top bush, bend back the retaining tag and extract the bottom bush.
11 Clean the parts in paraffin, or a suitable solvent, and wipe dry. Examine the bushes for wear and renew, if necessary.
12 Reassembly and refitting is a reverse of removal and dismantling, bearing in mind the following points:

(a) *Smear the outer surfaces and inner grooves of the bushes with graphite grease before fitting*
(b) *Ensure that the marks made on the inner column and intermediate shaft are aligned when fitting the column*
(c) *Adjust the position of the outer column so that it is 3.25 in (82.6 mm) below the top of the inner column before tightening the retaining bolts*
(d) *Position the arrow on the striker bush of the multi-function switch so that it points towards the direction indicator switch before fitting the steering wheel*
(e) *Tighten all retaining nuts and bolts to the specified torque*

18 Steering column intermediate shaft – removal and refitting

1   Disconnect the battery negative terminal.
2   Apply the handbrake, chock the rear wheels, jack up the front of the car and support it securely on axle stands.
3   From inside the car, pull back the carpet and unscrew the nut and bolt securing the intermediate shaft lower universal joint to the pinion shaft.
4   From under the facia remove the steering column cover and seal.
5   Unscrew the nut and bolt securing the intermediate shaft upper universal joint to the inner column.
6   Release the intermediate shaft from the pinion and inner column, and withdraw it from the car.
7   Before refitting it is first necessary to centralize the steering gear. To do this move the rubber band on the rack housing to one side to expose the centralizing hole. Turn the pinion shaft or move the roadwheels until a corresponding hole in the rack is in alignment. Insert a 6 mm bolt or drill into the hole to hold the steering gear in the central position.
8   From inside the car, engage the intermediate shaft upper universal joint with the inner column. Refit the clamp bolt and nut and tighten to the specified torque.
9   Position the steering wheel in the straight-ahead position and engage the intermediate shaft lower universal joint with the pinion shaft. Refit and tighten the clamp nut and bolt.
10 Refit the steering column cover and seal and place the carpet in position.
11 Remove the centralizing bolt or drill from the rack housing, refit the rubber band over the hole and lower the car to the ground. Reconnect the battery.

19 Steering column lock/ignition switch – removal and refitting

1   Remove the steering column, as described in Section 17.
2   With the column assembly on the bench, drill out the shear bolt heads and remove the clamp plate and lock/ignition switch from the outer column. The ignition switch can be removed from the lock after extracting the small grub screw.
3   Refit the ignition switch and secure it with the grub screw. Locate the new lock body centrally over the slot in the outer column. Lightly bolt the clamp plate into position, but take care not to shear the bolt heads.
4   Refit the steering column, as decribed in Section 17, but before fitting the cowls check that the lock and ignition switch operate correctly.
5   Tighten the shear bolts until the heads break off, and then refit the cowls.

20 Tie-rod outer balljoint – removal and refitting

1   Apply the handbrake, chock the rear wheels, prise off the front wheel trim and slacken the wheel nuts. Jack up the front of the car and support it securely on axle stands. Remove the roadwheel.
2   Using a suitable spanner, slacken the balljoint locknut on the tie-rod by a quarter of a turn.
3   Undo and remove the locknut securing the balljoint to the steering arm and then release the tapered ball-pin using a balljoint separator tool (photos).
4   Unscrew the balljoint from the tie-rod.
5   Refitting is the reverse sequence to removal. Tighten the nuts to the specified torque and check the front wheel alignment, as described in Section 24.

20.3A Using a separator tool to release the tie-rod outer balljoint

20.3B Withdraw the balljoint from the steering arm after releasing the taper

21 Steering rack rubber gaiter – removal and refitting

1   Remove the tie-rod outer balljoint, as described in the previous Section.
2   Release the retaining wire or unscrew the securing clip screws and slide the gaiter off the rack and pinion housing and tie-rod.

3   Slide on a new gaiter and position it over the housing and tie-rod.
4   Using new clips, or two or three turns of soft iron wire, secure the gaiter in position.
5   Refit the outer balljoint, as described in the previous Section.

22 Rack and pinion steering gear – removal and refitting

1   Disconnect the battery negative terminal.
2   Apply the handbrake, chock the rear wheels and remove the trim from both front roadwheels. Slacken the wheel nuts before jacking up the front of the car and then support it securely on axle stands. Remove the roadwheels.
3   Unscrew the nuts securing the tie-rod outer balljoints to the steering arms and release the tapered ballpins using a balljoint separator tool.
4   From inside the car pull back the carpet and unscrew the nut and bolt securing the intermediate shaft lower universal joint to the pinion shaft.
5   Using a sharp knife make diagonal cuts in the sound-deadening material around the pinion shaft to provide access to the pinion cover plate. The pinion cover plate can now be removed.
6   Undo and remove the two nuts and washers securing the gearchange linkage relay lever bracket to the rack housing. Move the linkage clear of the steering gear.
7   Undo and remove the two bolts and washers securing the pinion end of the steering gear to the bulkhead. Now remove the two bolts and washers, clamp plate and plastic seating securing the other end of the rack housing to the bulkhead.
8   Slacken the front crossmember support strut bolts on the driver's side. Remove the two short bolts, but leave the long bolt in position. Pivot the support strut outward at the top.
9   Ease the pinion shaft out of the intermediate shaft splines, manoeuvre the steering gear as necessary and withdraw the assembly out through the driver's side wheel arch.

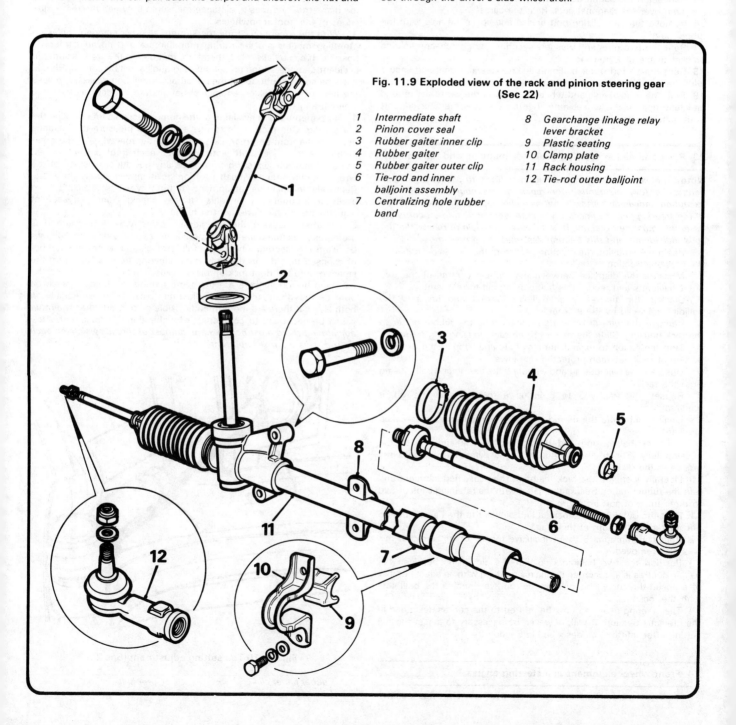

Fig. 11.9 Exploded view of the rack and pinion steering gear (Sec 22)

1   Intermediate shaft
2   Pinion cover seal
3   Rubber gaiter inner clip
4   Rubber gaiter
5   Rubber gaiter outer clip
6   Tie-rod and inner balljoint assembly
7   Centralizing hole rubber band
8   Gearchange linkage relay lever bracket
9   Plastic seating
10 Clamp plate
11 Rack housing
12 Tie-rod outer balljoint

10 To refit the steering gear, manoeuvre it into position through the wheel arch and locate the pinion in the bulkhead aperture.

11 Fit the retaining bolts, finger tight, to the pinion end of the steering gear. At the other end place the plastic seating and clamp in position, refit the retaining bolts and tighten them to the specified torque. Now tighten the pinion end retaining bolts to the specified torque.

12 Refit the gearchange linkage relay lever bracket and the two crossmember support strut bolts. Do not tighten the support strut bolts until the weight of the car is on its wheels.

13 Move the rubber band on the rack housing to one side to expose the centralizing hole. Move the rack as necessary, by turning the pinion, until a 6 mm drill or bolt can be inserted through the hole in the housing and into the corresponding hole in the rack.

14 From inside the car, refit the pinion cover plate, position the steering wheel in the straight-ahead position and refit the intermediate shaft universal joint to the pinion shaft. Refit and tighten the clamp bolt and nut.

15 Reposition the sound-deadening material around the pinion shaft and tape over the diagonal cuts. Refit the carpets.

16 Remove the centralizing bolt or drill and cover the hole with the rubber band.

17 Refit the two nuts and washers securing the gearchange linkage bracket to the rack housing.

18 Refit the tie-rod outer balljoints to the steering arms and secure with the retaining nuts tightened to the specified torque.

19 Refit the roadwheels and wheel nuts, lower the car to the ground and fully tighten the wheel nuts. Tighten the crossmember support strut bolts and refit the wheel trim.

---

23 Rack and pinion steering gear – dismantling and reassembly

Note: *The rack and pinion steering gear fitted to Montego models cannot be fully dismantled for repair or overhaul and, with one exception, individual parts are not available separately. Should repair of the steering gear be necessary, due to wear or damage, a complete assembly must be obtained. It is, however, possible to renew the tie-rods individually and this Section describes the procedure.*

1 Begin by removing the steering gear from the car, as described in the previous Section.

2 Measure the distance between the ball-pin centres of the two outer balljoints and record this figure as an aid to reassembly.

3 Slacken the tie-rod outer balljoint locknut and unscrew the balljoint, followed by the locknut.

4 Remove the wire or retaining clips securing the rubber gaiter to the rack housing. Slide the gaiter off the rack housing and tie-rod.

5 Turn the pinion to extend the rack fully and support the exposed portion of rack between protected vice jaws.

6 Unscrew the ballhousing and withdraw the tie-rod and ballhousing from the rack.

7 Repeat paragraphs 3 to 6 inclusive for the other tie-rod, if required.

8 Liberally lubricate the tie-rod and ballhousing, using the specified lubricant.

9 Refit the ballhousing and tie-rod to the rack and tighten the housing fully. Secure the ballhousing by staking its edge into the groove in the rack.

10 Liberally lubricate the rack teeth with the specified lubricant and refit the rubber gaiter. Secure the gaiter with the retaining clips or two to three turns of soft iron wire.

11 Refit the outer balljoint locknut and balljoint to the tie-rod, but do not tighten the locknut at this stage.

12 Repeat paragraphs 8 to 11 inclusive for the other tie-rod, if this was also removed.

13 Position the two balljoints so that the dimension between the ballpin centres is as recorded during dismantling with an equal number of exposed threads visible on each tie-rod. Now secure the balljoints with the locknuts.

14 The steering gear can now be refitted to the car, as described in the previous Section. It will, however, be necessary to adjust the toe setting after refitting, as described in Section 24.

---

24 Front wheel alignment and steering angles

1 Accurate front wheel alignment is essential to provide positive

steering and prevent excessive tyre wear. Before considering the steering/suspension geometry, check that the tyres are correctly inflated, the front wheels are not buckled and the steering linkage and suspension joints are in good order, without slackness or wear.

2 Wheel alignment consists of four factors:

Camber is the angle at which the front wheels are set from the vertical when viewed from the front of the car. 'Positive camber' is the amount (in degrees) that the wheels are tilted outward at the top from the vertical.

Castor is the angle between the steering axis and a vertical line when viewed from each side of the car. 'Positive castor' is when the steering axis is inclined rearward.

Steering axis inclination is the angle (when viewed from the front of the car) between the vertical and an imaginary line drawn between the suspension strut upper mounting and the lower suspension arm balljoint.

Toe setting is the amount by which the distance between the front inside edges of the roadwheels (measured at hub height) differs from the diametrically opposite distance measured between the rear inside edges of the front roadwheels.

3 With the exception of the toe setting, all other steering angles on Montego models are set during manufacture and no adjustment is possible. It can be assumed, therefore, that unless the car has suffered accident damage all the preset steering angles will be correct. Should there be some doubt about their accuracy it will be necessary to seek the help of a BL dealer, as special gauges are needed to check the steering angles.

4 Two methods are available to the home mechanic for checking the toe setting. One method is to use a gauge to measure the distance between the front and rear inside edges of the roadwheels. The other method is to use a scuff plate in which each front wheel is rolled across a movable plate which records any deviation, or scuff, of the tyre from the straight-ahead position as it moves across the plate. Relatively inexpensive equipment of both types is available from accessory outlets to enable these checks, and subsequent adjustments, to be carried out at home.

5 If, after checking the toe setting using whichever method is preferable, it is found that adjustment is necessary, proceed as follows.

6 Turn the steering wheel onto full left lock and record the number of exposed threads on the right-hand steering tie-rod. Now turn the steering onto full right lock and record the number of threads on the left-hand tie-rod. If there are the same number of threads visible on both tie-rods then subsequent adjustments can be made equally on both sides. If there are more threads visible on one side than the other it will be necessary to compensate for this during adjustment. *After adjustment there must be the same number of threads visible on each tie-rod. This is most important.*

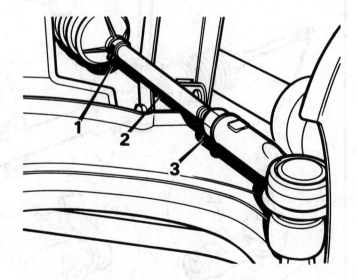

Fig. 11.10 Toe setting adjustment (Sec 24)

1   *Rubber gaiter clip*   3   *Outer balljoint locknut*
2   *Tie-rod*

7   To alter the toe setting slacken the locknut on the tie-rod and turn the rod using a self-grip wrench to achieve the desired setting. When viewed from the side of the car, turning the tie-rod clockwise will increase the toe-in, turning it anti-clockwise will increase the toe-out. Only turn the tie-rods by a quarter of a turn each time and then recheck the setting using the gauges, or scuff plate.

8   After adjustment tighten the locknuts and reposition the steering gear rubber gaiters, if necessary, to remove any twist caused by turning the tie-rods.

25 Wheels and tyres

1   Check the tyre pressures regularly (see Routine Maintenance) when the tyres are cold, and periodically check the tread depth using a depth gauge.

2   Frequently inspect the tyre walls and treads for damage and pick out any stones which have become trapped in the tread pattern.

3   In the interests of extending tread life, the wheels and tyres can be moved between front and rear on the same side of the car and the spare incorporated in the rotational pattern. If the wheels have previously been balanced on the car it will be necessary to have them rebalanced after rotation.

4   Never mix tyres of different construction, or very dissimilar tread patterns. This is particularly important on models equipped with TD

(Total Deflation) wheels and tyres on which two grooves running around the wheel retain the tyre in place in the event of a puncture. TD tyres and wheels are not interchangeable with wheels and tyres of different design and construction, and replacement must be of identical specification to those originally fitted to the car.

5   Always keep the roadwheel nuts tightened to the specified torque and if the wheel stud holes become elongated or flattened, renew the wheel.

6   Occasionally clean the inner faces of the roadwheels and, if there is any sign of rust or corrosion, paint them with metal preservative paint.

7   Before removing a roadwheel which has been balanced on the car, always mark one wheel stud hole and the wheel, so that the roadwheel may be refitted in the same position, to maintain the balance.

8   Should unexpected excessive wear be noticed on any of the tyres its cause must be identified and rectified immediately. Generally speaking the wear pattern can be used as a guide to the cause. If a tyre is worn excessively in the centre of the tread face, but not on the edges, over inflation is indicated. Similarly if the edges are worn, but not the centre, this may be due to under inflation. If both the front or rear tyres are wearing on their inside or outside edges, this is likely to be due to incorrect toe setting. If only one tyre is exhibiting this tendency then there may be a problem with the steering geometry, a worn steering or suspension component, or a faulty tyre. Wheel and tyre imbalance is indicated by irregular and uneven wear patches appearing periodically around the tread face.

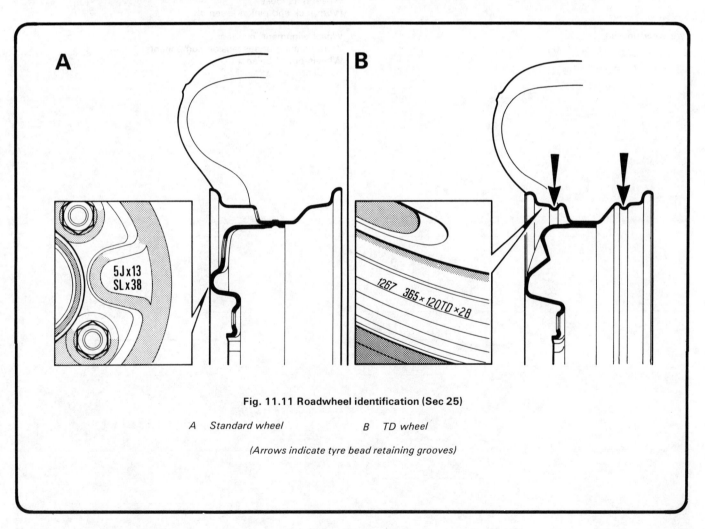

Fig. 11.11 Roadwheel identification (Sec 25)

A   Standard wheel   B   TD wheel

(Arrows indicate tyre bead retaining grooves)

26 Fault diagnosis – suspension and steering

Note: *Before diagnosing steering or suspension faults, be sure that the trouble is not due to incorrect tyre pressures, mixture of tyre types or binding brakes*

Symptom	Reason(s)
Vehicle pulls to one side	Incorrect wheel alignment Wear in front suspension or steering components Accident damage to steering or suspension components
Steering stiff or heavy	Lack of steering gear lubricant Seized balljoint Wheel alignment incorrect Steering rack or column bent or damaged
Excessive play in steering	Worn steering or suspension joints Wear in intermediate shaft universal joints Worn rack and pinion assembly
Wheel wobble and vibration	Roadwheels out of balance Roadwheels buckled or distorted Faulty or damaged tyre Worn steering or suspension joints Wheel nuts loose Worn rack and pinion assembly
Tyre wear uneven	Wheel alignment incorrect Worn steering or suspension components Wheels out of balance Accident damage

Chapter 12 Bodywork

For modifications, and information applicable to later models, see Supplement at end of manual

Contents

Specifications

Torque wrench settings

	lbf ft	Nm
Seat belt mountings	24	32
Bumper bracket retaining bolts	18	24
Front body panel retaining bolts	6	8
Door lock retaining bolts	6	8
Door striker pins	33	45
Bonnet lock retaining bolts	6	8
Boot lock retaining bolts	6	8
Boot lock striker retaining bolts	18	24

1 General description

The bodyshell and underframe is of all-steel welded construction and is of computer-based design. The assembly and welding of the main body unit is completed entirely by computer-controlled robots, and the finished unit is checked for dimensional accuracy using modern computer and laser technology.

The front wings are bolted in position and are detachable should renewal be necessary after a front end collision.

2 Maintenance – bodywork and underframe

The general condition of a vehicle's bodywork is the one thing that significantly affects its value. Maintenance is easy but needs to be regular. Neglect, particularly after minor damage, can lead quickly to

further deterioration and costly repair bills. It is important also to keep watch on those parts of the vehicle not immediately visible, for instance the underside, inside all the wheel arches and the lower part of the engine compartment.

The basic maintenance routine for the bodywork is washing – preferably with a lot of water, from a hose. This will remove all the loose solids which may have stuck to the vehicle. It is important to flush these off in such a way as to prevent grit from scratching the finish. The wheel arches and underframe need washing in the same way to remove any accumulated mud which will retain moisture and tend to encourage rust. Paradoxically enough, the best time to clean the underframe and wheel arches is in wet weather when the mud is thoroughly wet and soft. In very wet weather the underframe is usually cleaned of large accumulations automatically and this is a good time for inspection.

Periodically, except on vehicles with a wax-based underbody protective coating, it is a good idea to have the whole of the underframe of the vehicle steam cleaned, engine compartment included, so that a thorough inspection can be carried out to see what minor repairs and renovations are necessary. Steam cleaning is available at many garages and is necessary for removal of the accumulation of oily grime which sometimes is allowed to become thick in certain areas. If steam cleaning facilities are not available, there are one or two excellent grease solvents available such as Holts Engine Cleaner or Holts Foambrite which can be brush applied. The dirt can then be simply hosed off. Note that these methods should not be used on vehicles with wax-based underbody protective coating or the coating will be removed. Such vehicles should be inspected annually, preferably just prior to winter, when the underbody should be washed down and any damage to the wax coating repaired using Holts Undershield. Ideally, a completely fresh coat should be applied. It would also be worth considering the use of such wax-based protection for injection into door panels, sills, box sections, etc, as an additional safeguard against rust damage where such protection is not provided by the vehicle manufacturer.

After washing paintwork, wipe off with a chamois leather to give an unspotted clear finish. A coat of clear protective wax polish, like the many excellent Turtle Wax polishes, will give added protection against chemical pollutants in the air. If the paintwork sheen has dulled or oxidised, use a cleaner/polisher combination such as Turtle Extra to restore the brilliance of the shine. This requires a little effort, but such dulling is usually caused because regular washing has been neglected. Care needs to be taken with metallic paintwork, as special non-abrasive cleaner/polisher is required to avoid damage to the finish. Always check that the door and ventilator opening drain holes and pipes are completely clear so that water can be drained out. Bright work should be treated in the same way as paint work. Windscreens and windows can be kept clear of the smeary film which often appears, by the use of a proprietary glass cleaner like Holts Mixra. Never use any form of wax or other body or chromium polish on glass.

---

3 Maintenance – upholstery and carpets

Mats and carpets should be brushed or vacuum cleaned regularly to keep them free of grit. If they are badly stained remove them from the vehicle for scrubbing or sponging and make quite sure they are dry before refitting. Seats and interior trim panels can be kept clean by wiping with a damp cloth and Turtle Wax Carisma. If they do become stained (which can be more apparent on light coloured upholstery) use a little liquid detergent and a soft nail brush to scour the grime out of the grain of the material. Do not forget to keep the headlining clean in the same way as the upholstery. When using liquid cleaners inside the vehicle do not over-wet the surfaces being cleaned. Excessive damp could get into the seams and padded interior causing stains, offensive odours or even rot. If the inside of the vehicle gets wet accidentally it is worthwhile taking some trouble to dry it out properly, particularly where carpets are involved. *Do not leave oil or electric heaters inside the vehicle for this purpose.*

---

4 Minor body damage – repair

The colour bodywork repair photographic sequences between pages 32 and 33 illustrate the operations detailed in the following sub-sections.

Note: *For more detailed information about bodywork repair, the Haynes Publishing Group publish a book by Lindsay Porter called The Car Bodywork Repair Manual. This incorporates information on such aspects as rust treatment, painting and glass fibre repairs, as well as details on more ambitious repairs involving welding and panel beating.*

Repair of minor scratches in bodywork

If the scratch is very superficial, and does not penetrate to the metal of the bodywork, repair is very simple. Lightly rub the area of the scratch with a paintwork renovator like Turtle Wax New Color Back, or a very fine cutting paste like Holts Body + Plus Rubbing Compound, to remove loose paint from the scratch and to clear the surrounding bodywork of wax polish. Rinse the area with clean water.

Apply touch-up paint, such as Holts Dupli-Color Color Touch or a paint film like Holts Autofilm, to the scratch using a fine paint brush; continue to apply fine layers of paint until the surface of the paint in the scratch is level with the surrounding paintwork. Allow the new paint at least two weeks to harden; then blend it into the surrounding paintwork by rubbing the scratch area with a paintwork renovator or a very fine cutting paste, such as Holts Body + Plus Rubbing Compound or Turtle Wax New Color Back. Finally, apply wax polish from one of the Turtle Wax range of wax polishes.

Where the scratch has penetrated right through to the metal of the bodywork, causing the metal to rust, a different repair technique is required. Remove any loose rust from the bottom of the scratch with a penknife, then apply rust inhibiting paint, such as Turtle Wax Rust Master, to prevent the formation of rust in the future. Using a rubber or nylon applicator fill the scratch with bodystopper paste like Holts Body + Plus Knifing Putty. If required, this paste can be mixed with cellulose thinners, such as Holts Body + Plus Cellulose Thinners, to provide a very thin paste which is ideal for filling narrow scratches. Before the stopper-paste in the scratch hardens, wrap a piece of smooth cotton rag around the top of a finger. Dip the finger in cellulose thinners, such as Holts Body + Plus Cellulose Thinners, and then quickly sweep it across the surface of the stopper-paste in the scratch; this will ensure that the surface of the stopper-paste is slightly hollowed. The scratch can now be painted over as described earlier in this Section.

Repair of dents in bodywork

When deep denting of the vehicle's bodywork has taken place, the first task is to pull the dent out, until the affected bodywork almost attains its original shape. There is little point in trying to restore the original shape completely, as the metal in the damaged area will have stretched on impact and cannot be reshaped fully to its original contour. It is better to bring the level of the dent up to a point which is about ⅛ in (3 mm) below the level of the surrounding bodywork. In cases where the dent is very shallow anyway, it is not worth trying to pull it out at all. If the underside of the dent is accessible, it can be hammered out gently from behind, using a mallet with a wooden or plastic head. Whilst doing this, hold a suitable block of wood firmly against the outside of the panel to absorb the impact from the hammer blows and thus prevent a large area of the bodywork from being 'belled-out'.

Should the dent be in a section of the bodywork which has a double skin or some other factor making it inaccessible from behind, a different technique is called for. Drill several small holes through the metal inside the area – particularly in the deeper section. Then screw long self-tapping screws into the holes just sufficiently for them to gain a good purchase in the metal. Now the dent can be pulled out by pulling on the protruding heads of the screws with a pair of pliers.

The next stage of the repair is the removal of the paint from the damaged area, and from an inch or so of the surrounding 'sound' bodywork. This is accomplished most easily by using a wire brush or abrasive pad on a power drill, although it can be done just as effectively by hand using sheets of abrasive paper. To complete the preparation for filling, score the surface of the bare metal with a screwdriver or the tang of a file, or alternatively, drill small holes in the affected area. This will provide a really good 'key' for the filler paste.

To complete the repair see the Section on filling and re-spraying.

Repair of rust holes or gashes in bodywork

Remove all paint from the affected area and from an inch or so of the surrounding 'sound' bodywork, using an abrasive pad or a wire brush on a power drill. If these are not available a few sheets of abrasive paper will do the job just as effectively. With the paint removed you will be able to gauge the severity of the corrosion and therefore decide whether to renew the whole panel (if this is possible) or to repair the affected area. New body panels are not as expensive as most people think and it is often quicker and more satisfactory to fit a new panel than to attempt to repair large areas of corrosion.

Remove all fittings from the affected area except those which will act as a guide to the original shape of the damaged bodywork (eg headlamp shells etc). Then, using tin snips or a hacksaw blade, remove all loose metal and any other metal badly affected by corrosion. Hammer the edges of the hole inwards in order to create a slight depression for the filler paste.

Wire brush the affected area to remove the powdery rust from the surface of the remaining metal. Paint the affected area with rust inhibiting paint like Turtle Wax Rust Master; if the back of the rusted area is accessible treat this also.

Before filling can take place it will be necessary to block the hole in some way. This can be achieved by the use of aluminium or plastic mesh, or aluminium tape.

Aluminium or plastic mesh or glass fibre matting, such as the Holts Body + Plus Glass Fibre Matting, is probably the best material to use for a large hole. Cut a piece to the approximate size and shape of the hole to be filled, then position it in the hole so that its edges are below the level of the surrounding bodywork. It can be retained in position by several blobs of filler paste around its periphery.

Aluminium tape should be used for small or very narrow holes. Pull a piece off the roll and trim it to the approximate size and shape required, then pull off the backing paper (if used) and stick the tape over the hole; it can be overlapped if the thickness of one piece is insufficient. Burnish down the edges of the tape with the handle of a screwdriver or similar, to ensure that the tape is securely attached to the metal underneath.

Bodywork repairs – filling and re-spraying

Before using this Section, see the Sections on dent, deep scratch, rust holes and gash repairs.

Many types of bodyfiller are available, but generally speaking those proprietary kits which contain a tin of filler paste and a tube of resin hardener are best for this type of repair, like Holts Body + Plus or Holts No Mix which can be used directly from the tube. A wide, flexible plastic or nylon applicator will be found invaluable for imparting a smooth and well contoured finish to the surface of the filler.

Mix up a little filler on a clean piece of card or board – measure the hardener carefully (follow the maker's instructions on the pack) otherwise the filler will set too rapidly or too slowly. Alternatively, Holts No Mix can be used straight from the tube without mixing, but daylight is required to cure it. Using the applicator apply the filler paste to the prepared area; draw the applicator across the surface of the filler to achieve the correct contour and to level the filler surface. As soon as a contour that approximates to the correct one is achieved, stop working the paste – if you carry on too long the paste will become sticky and begin to 'pick up' on the applicator. Continue to add thin layers of filler paste at twenty-minute intervals until the level of the filler is just proud of the surrounding bodywork.

Once the filler has hardened, excess can be removed using a metal plane or file. From then on, progressively finer grades of abrasive paper should be used, starting with a 40 grade production paper and finishing with 400 grade wet-and-dry paper. Always wrap the abrasive paper around a flat rubber, cork, or wooden block – otherwise the surface of the filler will not be completely flat. During the smoothing of the filler surface the wet-and-dry paper should be periodically rinsed in water. This will ensure that a very smooth finish is imparted to the filler at the final stage.

At this stage the 'dent' should be surrounded by a ring of bare metal, which in turn should be encircled by the finely 'feathered' edge of the good paintwork. Rinse the repair area with clean water, until all of the dust produced by the rubbing-down operation has gone.

Spray the whole repair area with a light coat of primer, either Holts Body + Plus Grey or Red Oxide Primer – this will show up any imperfections in the surface of the filler. Repair these imperfections with fresh filler paste or bodystopper, and once more smooth the surface with abrasive paper. If bodystopper is used, it can be mixed with cellulose thinners to form a really thin paste which is ideal for filling small holes. Repeat this spray and repair procedure until you are satisfied that the surface of the filler, and the feathered edge of the paintwork are perfect. Clean the repair area with clean water and allow to dry fully.

The repair area is now ready for final spraying. Paint spraying must be carried out in a warm, dry, windless and dust free atmosphere. This condition can be created artificially if you have access to a large indoor working area, but if you are forced to work in the open, you will have to pick your day very carefully. If you are working indoors, dousing the floor in the work area with water will help to settle the dust which would otherwise be in the atmosphere. If the repair area is confined to one body panel, mask off the surrounding panels; this will help to minimise the effects of a slight mis-match in paint colours. Bodywork fittings (eg chrome strips, door handles etc) will also need to be masked off. Use genuine masking tape and several thicknesses of newspaper for the masking operations.

Before commencing to spray, agitate the aerosol can thoroughly, then spray a test area (an old tin, or similar) until the technique is mastered. Cover the repair area with a thick coat of primer; the thickness should be built up using several thin layers of paint rather than one thick one. Using 400 grade wet-and-dry paper, rub down the surface of the primer until it is really smooth. While doing this, the work area should be thoroughly doused with water, and the wet-and-dry paper periodically rinsed in water. Allow to dry before spraying on more paint.

Spray on the top coat using Holts Dupli-Color Autospray, again building up the thickness by using several thin layers of paint. Start spraying in the centre of the repair area and then work outwards, with a side-to-side motion, until the whole repair area and about 2 inches of the surrounding original paintwork is covered. Remove all masking material 10 to 15 minutes after spraying on the final coat of paint.

Allow the new paint at least two weeks to harden, then, using a paintwork renovator or a very fine cutting paste such as Turtle Wax New Color Back or Holts Body + Plus Rubbing Compound, blend the edges of the paint into the existing paintwork. Finally, apply wax polish.

5 Major body damage – repair

Where serious damage has occurred or large areas need renewal due to neglect, it means that completely new sections or panels will need welding in, and this is best left to professionals. If the damage is due to impact, it will also be necessary to completely check the alignment of the bodyshell structure. Due to the principle of construction, the strength and shape of the whole car can be affected by damage to one part. In such instances the services of a BL agent with specialist checking jigs are essential. If a body is left misaligned, it is first of all dangerous, as the car will not handle properly, and secondly uneven stresses will be imposed on the steering, engine and transmission, causing abnormal wear or complete failure. Tyre wear may also be excessive.

6 Maintenance – hinges and locks

1 Oil the hinges of the bonnet, boot lid and doors with a drop or two of light oil at regular intervals (see Routine Maintenance).
2 At the same time, lightly oil the bonnet release mechanism and all door locks.
3 Do not attempt to lubricate the steering lock.

7 Door rattles – tracing and rectification

1 Check first that the door is not loose at the hinges, and that the latch is holding the door firmly in position. Check also that the door lines up with the aperture in the body. If the door is out of alignment, adjust it, as described in Sections 22 and 28.
2 If the latch is holding the door in the correct position, but the latch still rattles, the lock mechanism is worn and should be renewed.
3 Other rattles from the door could be caused by wear in the window operating mechanism, interior lock mechanism, or loose glass channels.

8 Bonnet – removal, refitting and adjustment

1 Support the bonnet in its open position and place some cardboard, or rags, beneath the corners by the hinges.
2 Mark the location of the hinges with a soft pencil, then loosen the four retaining nuts and bolts (photo).
3 With the help of an assistant, release the stay, unscrew and remove the retaining bolts and withdraw the bonnet from the car.
4 Refitting is a reversal of removal, but adjust the hinges to their original positions. The bonnet rear edge should be flush with the scuttle, and the gaps at either side equal.

9 Bonnet lock – adjustment

1 Adjustment is possible by screwing the rubber buffers at each end of the front top rail in or out, or by repositioning the lockpin mounted on the bonnet (photo).
2 Slacken the locknut and use a screwdriver to adjust the length of the lockpin so that the bonnet closes easily and is held firmly in place.
3 Tighten the locknut.

8.2 Slacken the bonnet hinge retaining bolts

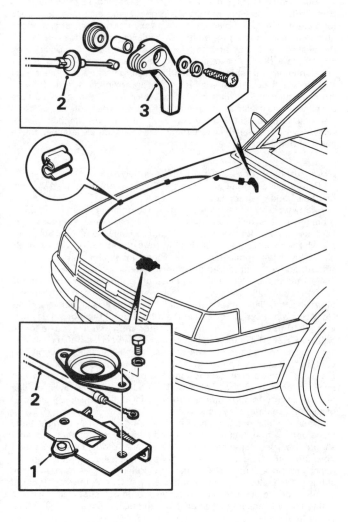

Fig. 12.1 Bonnet lock and release cable details (Sec 10)

1   Bonnet lock   3   Release lever
2   Release cable

9.1 Bonnet lock adjustment points – locknut (A) and lockpin (B)

10 Bonnet lock and release cable – removal and refitting

1   Working inside the car, remove the bonnet release lever and disconnect the cable.
2   Working in the engine compartment, release the outer cable from the support bracket on the lock, and detach the inner cable from the lock lever.
3   Release the cable from its retaining clips, pull the cable through the bulkhead grommet and remove it from the car.
4   To remove the lock, undo and remove the retaining bolts and remove the lock assembly through the aperture in the front body panel.
5   Refitting is the reverse sequence to removal. Adjust the bonnet lock as described in the previous Section, if necessary.

11 Radiator grille – removal and refitting

1   Open the bonnet and support it in the raised position.
2   Very carefully lift the three plastic retaining tags, one at a time, and tip the grille outward at the top.
3   Lift the grille up to release the lower legs (photo) and remove the grille from the car.
4   Refitting is the reverse sequence to removal.

11.3 Lift the grille up to release the lower edge

12 Front body panel – removal and refitting

1   Remove the bonnet lock and cable from the panel, as described in Section 10, and the radiator grille, as described in the previous Section.
2   Detach the air cleaner cold air intake hose from the panel.
3   Undo and remove the three screws each side securing the panel to the body side-members.
4   Release the bonnet stay, lift up the panel and remove it from the car.
5   Refitting is the reverse sequence to removal. Ensure that the lugs on the radiator and cooling fan cowl are correctly located in the body panel as it is fitted.

13 Front wing – removal and refitting

1   Open the bonnet and support it in the raised position.
2   Remove the headlamp and front direction indicator lamp, as described in Chapter 10.
3   Remove the front bumper, as described in Section 32 of this Chapter.
4   Undo the screws and remove the wheel arch liner.
5   Refer to Fig. 12.2 and remove the wing retaining bolts from their locations.
6   Carefully ease the front wing off its mating flanges and remove it from the car.
7   Refitting is the reverse sequence to removal, but use a mastic sealing compound to seal the wing to the body, apply protective coating to the wing under surface, and finish the top surface to match the body colour.

14 Boot lid – removal and refitting

1   Release the retaining clips and plastic rivets securing the hinge covers in position and lift away the covers.
2   Prise off the private lock operating rod retaining clip and slide the rod out of the private lock lever.
3   Undo the three bolts securing the lock mechanism to the boot lid and slide the lock assembly clear of the boot lid.
4   Ease the boot release cable outer sheath out of its support plate and disengage the inner cable end from the lock lever.
5   Disconnect the wiring harness connectors then withdraw the harness and release cable from the boot lid.
6   Using a soft pencil, mark the outline of the hinges on the boot lid.
7   Have an assistant support the boot lid, then undo the hinge retaining bolts and lift the boot lid away (photo).
8   Refitting is the reverse sequence to removal. Align the hinges with the outline marks made during removal. Adjust the hinge position slightly if necessary so that the lid closes easily and is in correct alignment. Further adjustment is possible by slackening the striker plate bolts and repositioning the plate (photo).

15 Boot lock – removal and refitting

Note: *On vehicles equipped with central locking, further information will be found in Chapter 10.*
1   Prise off the private lock operating rod retaining clip and slide the rod out of the private lock lever.
2   Undo the three bolts securing the lock mechanism to the boot lid and slide the lock clear (photo).

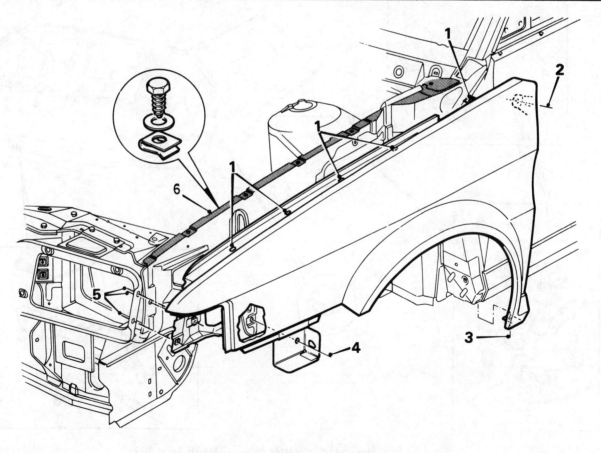

Fig. 12.2 Front wing attachment points (Sec 13)

1 Wing-to-inner valance attachments	*3 Wing-to-sill panel attachment*	*5 Wing-to-headlamp surround attachment*
2 Wing-to-front pillar attachment	*4 Wing-to-lower valance attachment*	*6 Mastic sealant application area*

14.7 Boot lid hinge retaining bolts

14.8 Boot lock striker plate and retaining bolts

15.2 Boot lock retaining bolts (arrowed)

3   Ease the boot release cable outer sheath out of its support plate and disengage the inner cable end from the lock lever. Where applicable disconnect the wiring connectors then remove the lock from the boot lid.

4   Refitting is the reverse sequence to removal. If necessary adjust the striker plate position after refitting the lock so that the boot lid shuts and locks without slamming.

16 Boot private lock – removal and refitting

Note: On vehicles equipped with central locking, further information will be found in Chapter 10.

1   Prise off the private lock operating rod retaining clip and slide the rod out of the private lock lever.

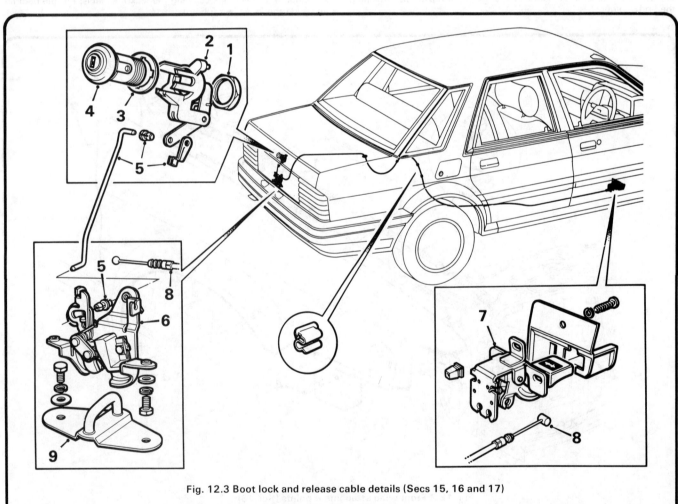

Fig. 12.3 Boot lock and release cable details (Secs 15, 16 and 17)

1 Private lock retaining nut	4 Private lock barrel	7 Release lever
2 Lock retainer	5 Operating rod and clips	8 Release cable
3 Washer	6 Boot lock	9 Striker plate

16.2 Boot private lock retaining nut (arrowed)

method of retention, special tools and equipment are required to remove and refit the screen, and this task is definitely beyond the scope of the home mechanic. If it is necessary to have the windscreen removed, this job should be left to a suitably equipped specialist or BL dealer.

19 Rear window and rear quarter windows – removal and refitting

The rear windows, like the windscreen, also feature direct bonding retention and their removal and refitting should be entrusted to a BL dealer.

20 Front door inner trim panel – removal and refitting

Base and L models
1 Undo the screws securing the door bin to the door, ease the bin away at the bottom and release the upper tags from the armrest (photos).
2 Undo the two armrest retaining screws and remove the armrest from the door (photo).
3 Prise out the finisher cap then undo the window regulator handle retaining screw (photos). Lift off the handle.
4 Prise out the finisher cap and undo the internal door release lever surround retaining screw (photos). Slide the surround off the lever.
5 Starting at the lower rear corner, release the trim panel retaining buttons by carefully levering between the panel and door with a screwdriver or suitable flat bar (photo).
6 Undo and remove the screws securing the upper trim capping to the door, lift the rear of the capping over the lock button and lift off.
7 To gain access to the door components, carefully peel back the polythene condensation barrier as necessary (photo).
8 Refitting is the reverse sequence to removal.

2 Undo the large nut securing the private lock to the lock retainer (photo). Withdraw the private lock, washer and lock retainer from the boot lid.
3 Refitting is the reverse sequence to removal.

17 Boot release cable and lever – removal and refitting

1 Refer to Section 14 and carry out the operations described in paragraphs 1 to 5.
2 Refer to Section 30 and remove the rear seat cushion.
3 Undo the front seat belt lower anchorage attachment and remove the sill interior trim panels.
4 Undo the driver's seat outer rail retaining bolts and withdraw the oddments tray.
5 Undo the screw securing the release lever shield to the body and remove the shield.
6 Undo the two release lever bracket retaining screws, remove the bracket assembly and withdraw the cable.
7 Refitting is the reverse sequence to removal.

HL models
9 Prise out the finisher cap then undo the window regulator handle retaining screw. Lift off the handle.
10 Prise out the finisher cap and undo the internal door release lever surround retaining screw. Slide the surround off the lever.
11 Undo the three screws securing the trim panel to the door; one located in the well of the door pull, and two at the bottom of the door bin.
12 Using a screwdriver or suitable flat bar carefully lever between the panel and door at the front and rear sides to release the panel retaining buttons.
13 Pull out the bottom edge and lift the panel upwards over the lock button. Disconnect the speaker wiring and any additional wiring then remove the panel.
14 To gain access to the door components, carefully peel back the polythene condensation barrier as necessary.
15 Refitting is the reverse sequence to removal.

18 Windscreen – removal and refitting

All Montego models are equipped with a flush-glazed, laminated windscreen, secured to the bodyshell by direct bonding. Due to this

20.1A Undo the door bin retaining screws ...

20.1B ... and disengage the upper tags from the armrest

20.2 Undo the armrest retaining screws

20.3A Prise out the finisher cap ...

20.3B ... then remove the regulator handle retaining screw

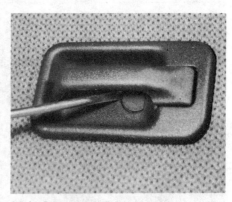

20.4A Prise out the finisher cap ...

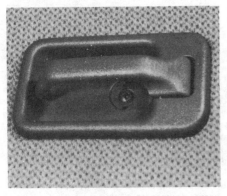

20.4B ... and undo the door release lever surround retaining screw

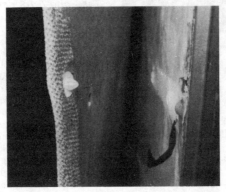

20.5 Release the trim panel retaining buttons

20.7 Peel back the polythene sheet to gain access to the door components

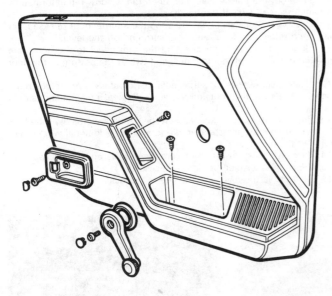

Fig. 12.4 Front door inner trim panel removal – HL models (Sec 20)

21 Front door – removal and refitting

1   Remove the inner trim panel, as described in Sectioon 20.
2   Disconnect the wiring multi-plugs and connectors and, where fitted, at the central locking and electrically-operated window components. Remove the wiring harness from the door.

3   Have an assistant securely support the door then, using a suitable punch, drift out the hinge retaining pins. Withdraw the door from the car.
4   Refitting is the reverse sequence to removal.

22 Front door lock – removal, refitting and adjustment

Note: On vehicles equipped with central locking, further information will be found in Chapter 10.
1   Remove the door inner trim panel, as described in Section 20.
2   Unscrew the interior lock button, temporarily refit the window regulator handle and wind the window fully up.
3   Prise the operating rod out of its retaining bush on the exterior door handle operating lever (photo).
4   Release the retaining clip and withdraw the operating rod from the private lock lever.
5   Undo and remove the screw securing the internal release lever to the door. Release the operating rod from its steady clip and slide the lever rearwards to release it from its location (photo).
6   Undo and remove the screw securing the rear window channel to the door. Release the channel from the felt guide and remove the channel from the door.
7   Undo and remove the three retaining screws (photo) and withdraw the lock assembly, complete with operating rods, from the door aperture. Disconnect the courtesy lamp switch plug.
8   The external handle can also be removed after unscrewing the two nuts and washers.
9   Refitting is the reverse sequence to removal, but adjust the lock striker pin as follows.
10 Check that the latch disc is in the open position then loosen the striker pin nut and position the pin so that the door can be closed easily and is held firmly. Close the door gently, but firmly, when making this adjustment.
11 Tighten the striker pin nut.

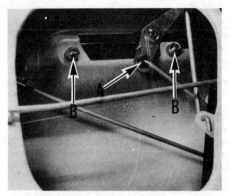

22.3 Door handle operating rod bush (A) and handle retaining nuts (B)

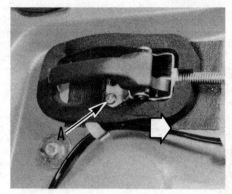

22.5 Undo the release lever retaining screw (A) and slide the lever rearwards to release it

22.7 Door lock retaining screws

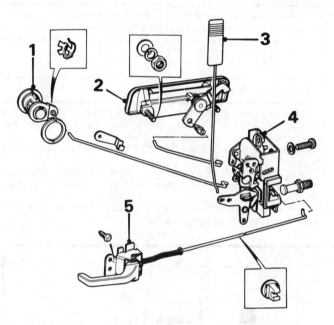

Fig. 12.5 Front door lock components (Secs 22 and 23)

1   Private lock
2   External handle
3   Lock button
4   Lock assembly
5   Internal release lever

23 Front door private lock – removal and refitting

Note: *On vehicles equipped with central locking, further information will be found in Chapter 10.*
1   Remove the door interior trim panel, as described in Section 20, and wind the window fully up.
2   Release the retaining clip and withdraw the operating rod from the private lock lever.
3   Prise out the retaining ring and withdraw the lock and seating washer from the door.
4   To remove the lock barrel, extract the circlip and remove the spring washer and lock lever. Withdraw the lock barrel from the lock body.
5   Refitting is the reverse sequence to removal.

24 Front door glass and regulator – removal and refitting

Note: *On vehicles equipped with electrically-operated windows, further information will be found in Chapter 10.*
1   Refer to Section 20 and remove the inner trim panel and, where fitted, the trim capping.

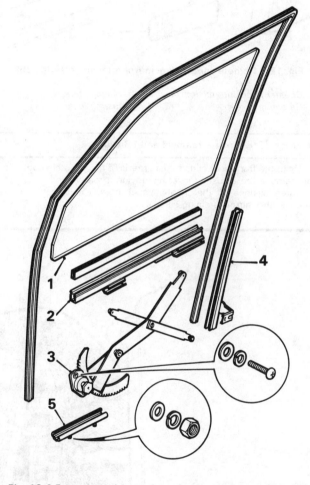

Fig. 12.6 Front door glass and regulator components (Sec 24)

1   Door window
2   Lifting channel
3   Window regulator
4   Rear window channel
5   Auxiliary slide

2   Refer to Section 25 and remove the door mirror.
3   Lower the window fully and insert a small block of wood in the base of the door to support the window after removal of the regulator.
4   Undo the three bolts securing the regulator to the door panel. Release the three nylon wheels of the regulator arms from their runners and manipulate the regulator out of the door aperture.
5   Tip the window forward and remove it upwards from the door.
6   Refitting is the reverse sequence to removal.

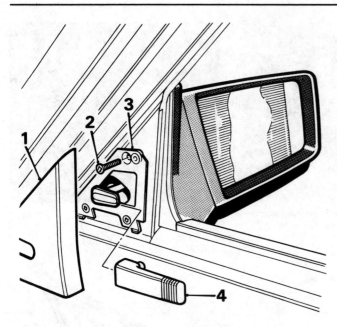

Fig. 12.7 Remote control door mirror attachments (Sec 25)

1   Cheater panel internal finisher   3   Retaining plate
2   Retaining screw   4   Adjusting lever

25 Front door mirror – removal and refitting

1   Open the front door and, if the mirror is of the remotely-controlled type, remove the adjusting lever by moving it rearwards.
2   Carefully prise off the cheater panel internal finisher to provide access to the mirror retaining screws.

3   Undo and remove the three screws, lift off the retaining plate and withdraw the mirror from the door.
4   Refitting is the reverse sequence to removal.
5   The design of the mirror is such that it will give slightly under impact, against spring tension, thus minimising the risk of damage or injury. On remotely-controlled mirrors this will cause the adjusting lever to disengage from the internal mechanism and become inoperative. The lever can be refitted, without removing the mirror, as follows.
6   Open the door and lower the window.
7   Gently move the mirror head toward the front of the car to give a working clearance between the mirror head and the base.
8   With the mirror held in this position, grasp the ball plate with one hand and the ball-head of the adjusting lever with the other. Firmly push the ball-head into the socket on the ball plate until it fully engages. Release the mirror head and check the operation.

26 Rear door inner trim panel – removal and refitting

1   Removal and refitting of the rear inner trim panel is basically the same as that for the front doors, as described in Section 20. The only difference is the retaining screw locations and this will be obvious after inspection.

27 Rear door – removal and refitting

1   If central locking is fitted, remove the inner trim panel, as described in Section 26, and disconnect the wiring harness connectors. Remove the harness from the door.
2   Carefully mark the outline of the door hinges on the pillar using a soft pencil.
3   With the help of an assistant to support the door, undo and remove the bolts securing the upper and lower hinges to the door. Carefully lift away the door.
4   Refitting is the reverse sequence to removal. Align the hinges with the pencil marks made during removal.

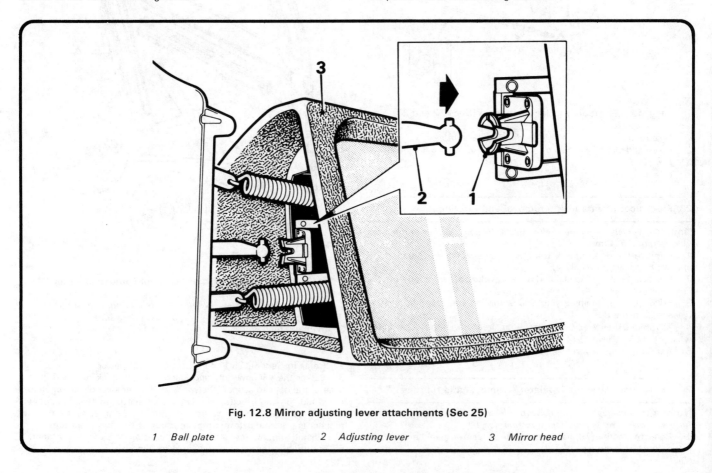

Fig. 12.8 Mirror adjusting lever attachments (Sec 25)

1   Ball plate   2   Adjusting lever   3   Mirror head

28 Rear door lock – removal, refitting and adjustment

Note: *On vehicles equipped with central locking, further information will be found in Chapter 10.*

1   Remove the door inner trim panel and capping, as described in Section 26.

2   Temporarily refit the window regulator handle and wind the window fully up. Unscrew the interior lock button.

3   Prise the operating rod out of its retaining bush on the door exterior handle operating lever.

4   Undo and remove the screw securing the internal release lever to the door. Release the operating rod from its steady clip and slide the lever rearwards to release it from its location.

5   Using a small screwdriver or punch, press out the locking pin from the centre of the bellcrank lever (photo), and withdraw the bellcrank and pivot (photo). Detach the lock operating rod from the bellcrank and from its steady clip.

6   Undo and remove the three screws securing the lock to the door and withdraw the unit, complete with operating rods, from the door aperture. Disconnect the courtesy lamp switch plug.

7   Refitting is the reverse sequence to removal, but adjust the lock striker pin as follows.

8   Check that the latch disc is in the open position, then loosen the striker pin nut and position the pin so that the door can be closed

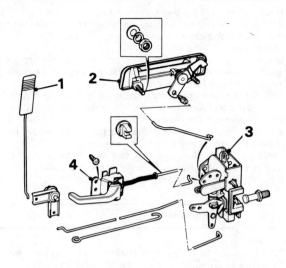

Fig. 12.9 Rear door lock components (Sec 28)

1 Lock button	3 Lock assembly
2 Door exterior handle	4 Internal release lever

easily, and is held firmly. Close the door gently, but firmly, when making this adjustment.

9   Tighten the striker pin nut.

29 Rear door glass and regulator – removal and refitting

1   Wind the window fully up, and then remove the inner trim panel and capping, as described in Section 26.

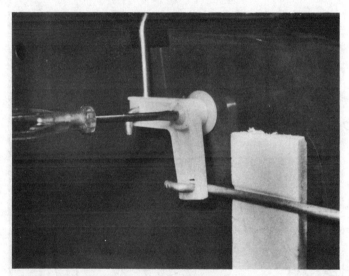

28.5A Press out the locking pin from the bellcrank lever ...

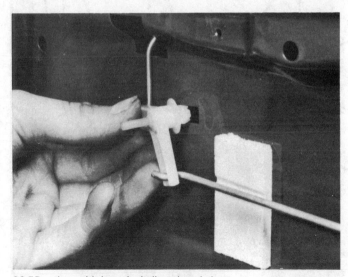

28.5B ... then withdraw the bellcrank and pivot

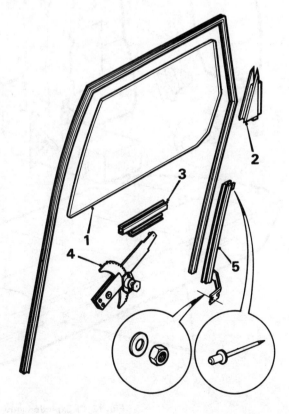

Fig. 12.10 Rear door glass and regulator components (Sec 29)

1 Door window	4 Window regulator
2 Cheater panel	5 Rear guide channel
3 Lifting channel	

2   Undo and remove the two bolts securing the window regulator to the door panel. Support the glass, disengage the regulator arm from the door glass lifting channel, and withdraw the regulator through the door aperture.
3   Lower the glass to the bottom of the door and then remove the rear guide channel retaining bolt. Release the channel from the glass and remove the felt from the channel.
4   Drill out the rivets securing the cheater panel to the door frame and lift off the panel. Drill an access hole in the bottom of the cheater panel and shake out the rivet debris.
5   Drill out the rivets securing the rear guide channel to the door, withdraw the channel and remove the rivet studs.
6   Carefully lift the glass upwards and out of the door. The lifting channel can now be removed from the glass, if necessary.
7   Refitting is the reverse sequence to removal.

30 Seats – removal and refitting

Front seat
1   Adjust the seat for access to the seat rail retaining bolts and undo the Torx type retaining bolts. Withdraw the seat from the car.
2   Refitting is the reverse sequence to removal; apply thread locking fluid to the bolt threads.

Rear seat
3   Push down on the base of the rear seat cushion to release the two retaining clips and remove the cushion.
4   Release the rear seat squab latches from inside the luggage compartment and fold down the squabs.
5   Undo the squab centre hinge bracket retaining screws and lift out the squabs.
6   Undo the bolt securing the seat squab extension lower edge to the wheel arch.
7   Slide the squab extension up and remove it from the car.
8   Refitting the squab extensions, seat guards and cushion is the reverse sequence to removal.

31 Seat belts – removal and refitting

Caution: *If the vehicle has been involved in an accident in which structural damage was sustained, all the seat belt components must be renewed.*

1   Move the front seats fully forward, ease the door seal from around the central door pillar, and remove the interior trim covering the seat belt mechanism.
2   Prise off the trim capping from the top mounting. Undo and

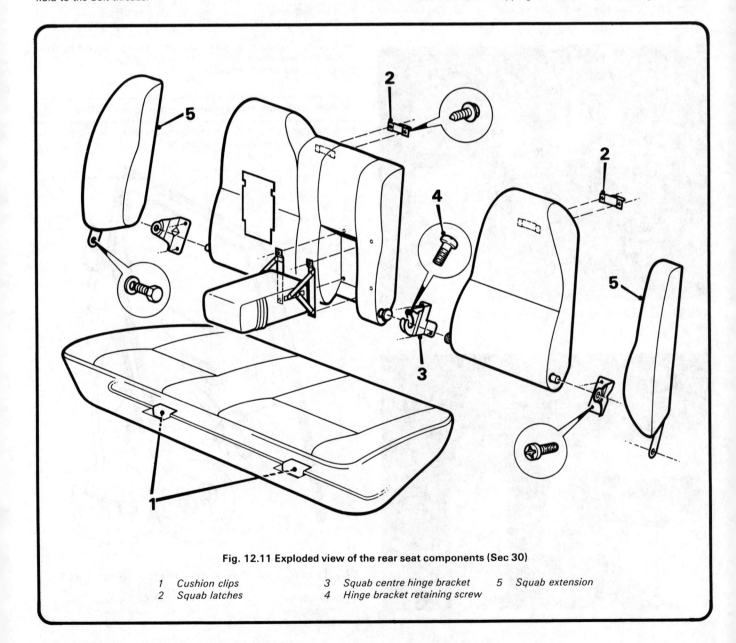

Fig. 12.11 Exploded view of the rear seat components (Sec 30)

1 Cushion clips	3 Squab centre hinge bracket	5 Squab extension
2 Squab latches	4 Hinge bracket retaining screw	

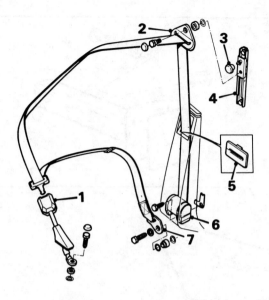

Fig. 12.12 Front seat belts and mountings (Sec 31)

1 Flexible stalk
2 Upper mounting
3 Adjustment slide knob
4 Adjustment slide

5 Belt guide
6 Inertia reel
7 Bottom mounting

remove the top, bottom and inertia reel mounting bolts, noting the position of the spacers and washers, and remove the assembly from the car.
3 Press out the belt guide from the interior trim and remove the belt from the trim.

4 Pull off the trim cover then undo and remove the bolt securing the flexible stalk to the floor. Note the position of the washers and spacer, and remove the stalk.
5 Refitting is the reverse sequence to removal, *but ensure that the mounting bolts are tightened to the specified torque.*

32 Bumpers – removal and refitting

Front bumper
1 Refer to Chapter 10 and remove the battery.
2 Undo the two bolts each side securing the bumper front support brackets to the body.
3 Undo the nuts and remove the washers and spacers securing the bumper and brackets to the front wings.
4 Detach the wheel arch liners from the bumpers and carefully withdraw the bumper from the front of the car.
5 Refitting is the reverse sequence to removal.

Rear bumper
6 Disconnect the battery negative terminal then remove the number plate lamps and detach the bulb holders. If necessary refer to Chapter 10.
7 Undo the two bolts each side, accessible from inside the luggage compartment, securing the bumper support brackets to the body.
8 Undo the two screws each side and remove the rear wheel arch liners.
9 Undo the nut securing each bumper front bracket to the rear valance then carefully withdraw the bumper from the car.
10 Refitting is the reverse sequence to removal.

33 Sunroof – removal and refitting

Sunroof panel
1 Open the trailing edge of the sunroof panel, ease each end of the

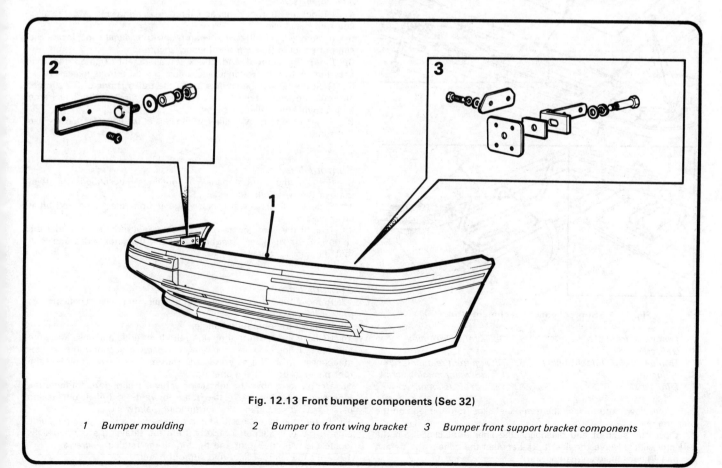

Fig. 12.13 Front bumper components (Sec 32)

1 Bumper moulding
2 Bumper to front wing bracket
3 Bumper front support bracket components

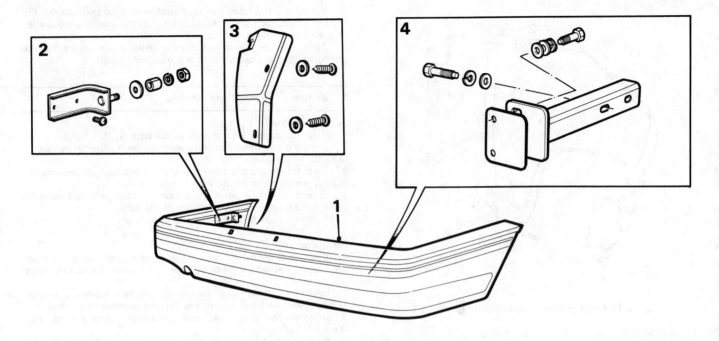

Fig. 12.14 Rear bumper components (Sec 32)

1   Bumper moulding   2   Bumper-to-rear valance   3   Wheel arch liner   4   Bumper support bracket
                                 support bracket

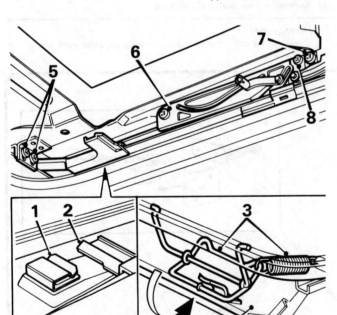

Fig. 12.15 Sunroof panel attachments (Sec 33)

1   Trim panel front clip   5   Deflector slide retaining
2   Deflector slide   screws
3   Trim panel clip and tensioning   6   Tilt arm front retaining screw
    spring   7   Tilt arm rear retaining screw
4   Trim panel   8   Tilt arm adjustment screw

panel liner down and release the trim panel clips from the legs on the
liner frame.
2   Close the sunroof and disengage the liner front clips from the
deflector slides. Move the liner back between the panels and detach
the panel liner tensioning springs.

3   Undo and remove the four screws on each side securing the
sunroof panel to the slide and tilt mechanism. *Do not remove the tilt
adjustment screws.*
4   Lift out the sunroof panel and, if required, remove the weatherstrip
and trim panel clips.
5   If removed, refit the weatherstrip and trim panel clips, locate the
sunroof panel in position and loosely secure.
6   Close the sunroof and check its alignment. Tighten the four
securing screws in sequence and adjust the tilt arm as necessary.
7   Refit the tensioning springs, open the trailing edge of the sunroof
and pull the liner forward, ensuring that the clips on the liner frame
engage with the deflector slides.
8   Engage the trim panel clips on the legs of the liner frame and close
the sunroof.

Sunroof liner
9   Remove the sunroof panel, as previously described, and then
remove the two deflector slides.
10 Slide the liner forward, disengage it from the frame and lift it
upwards and out.
11 To refit the liner, locate it on the frame with the slides under the
spigots. Slide the liner rearwards, refit the deflector slides and then
refit the sunroof panel.

Sunroof frame
12 Remove the sunroof panel, sunroof liner and the wind deflector, as
previously described.
13 Remove the operating handle.
14 Release the aperture trim along the front edge and ease away the
headliner. Undo and remove the two screws securing the winder
mechanism to the frame and front bracket. Take caare to protect the
roof panel in front of the aperture.
15 Undo and remove the remaining screws securing the frame to the
mounting assembly, move the frame forwards and lift it out, taking
care to avoid scratching the surrounding paintwork.
16 To refit the frame, position it in the aperture and engage the guide
pins in the rear support brackets. Refit the retaining screws loosely,
locate the wind deflector and tighten all the retaining screws.
17 Refit the sunroof liner and sunroof panel.

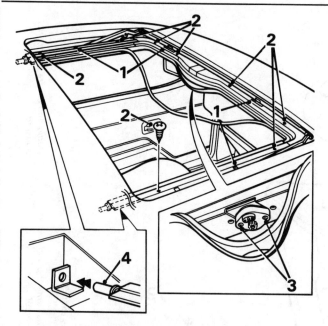

Fig. 12.16 Sunroof frame attachments (Sec 33)

1 Wind deflector retaining screws
2 Frame retaining screws
3 Winder-to-front bracket retaining screws
4 Guide pin and rear support bracket

34 Centre console – removal and refitting

1 Move both front seats fully forwards and undo the two console rear fixing screws.
2 Move the seats fully rearwards, prise out the seat belt stalk access panel and then unbolt and remove the stalks.
3 Remove both the front seat backrest control handles.
4 Remove the oddments tray from the front of the console to expose the console front fixing screws. Extract the screws.
5 Prise out the panel, forward of the handbrake lever.
6 On manual transmission models, lift the console upwards – the gear lever and gaiter will remain secured by a retaining plate and screws.
7 On automatic transmission models, move the console to the rear and then lift it upwards over the handbrake and the selector lever. The selector lever, index plate and bulbholder will remain attached to the floor.
8 Refitting is a reversal of removal, tighten the seat belt stalk bolts to the specified torque.

34.4 Centre console front retaining screws (arrowed)

35 Facia panel – removal and refitting

Note: *Removal of the facia is considerably involved, entailing extensive dismantling and the disconnection of many wiring harness plugs and connectors. Read through the entire Section before starting, and familiarize yourself with the procedure by referring to the illustration. Identify all electrical connections with a label before removal and, if necessary, make notes during dismantling.*
1 Disconnect the battery negative terminal.
2 Remove the steering column, as described in Chapter 11.
3 Remove the instrument panel, radio and facia switches, as described in Chapter 10.
4 From within the glovebox, release the two turnbuckles using a coin and lower the control unit panel from the glovebox roof. Disconnect the wiring multi-plugs and remove the panel.
5 Remove the courtesy lamp switch and disconnect its wiring plug.
6 Prise off the switch blanking plate on the right-hand side of the heater control knobs.
7 Undo the screw located behind the blanking plate.
8 Centralize the heater control knobs and ease the control panel, complete with knobs, off the facia. Disconnect the wiring multi-plugs and remove the control panel.
9 Release the cigarette lighter panel from the facia, disconnect the cigarette lighter wiring plug and, where fitted, the balance control wiring plug. Remove the panel.
10 Remove the fusebox cover from the right-hand side of the facia.
11 Prise off the trim caps from the two retaining screws at each end of the facia. Undo the four screws.
12 Undo the facia centre retaining screw situated at the base of the facia panel.
13 Pull the facia rearwards and disconnect any remaining wiring multi-plugs or connectors. Remove the facia from the car.
14 Refitting is a straightforward reverse of the removal sequence.

36 Heater – adjustments

1 The only adjustment possible is that of the heater flap linkage and is carried out as follows, with reference to Fig. 12.18.
2 Disconnect the battery negative terminal then remove the instrument panel, as described in Chapter 10.
3 Slacken the flap link rod trunnion screw situated behind the heater blend control lever.
4 Position the blend control in the fully raised position and move the heater flap lever as far as it will go towards the vehicle interior. Hold the linkage and control in this position and tighten the trunnion screw.
5 Check that the blend control lever moves freely from the fully raised to fully lowered position.
6 Refit the instrument panel and reconnect the battery.

37 Heater – removal and refitting

1 Drain the cooling system, as described in Chapter 2, and then remove the facia, as described in Section 35 of this Chapter.
2 From within the engine compartment slacken the clips and disconnect the two heater hoses at the outlets adjacent to the engine compartment bulkhead.
3 Disconnect the heater vacuum supply hose from the T-piece connector in the engine main vacuum line.
4 Undo the two screws securing the ventilator duct at the base of the heater.
5 Detach the windscreen demister ducts and the side vent hoses from the heater casing.
6 Disconnect the heater blower motor wiring multi-plug.
7 Undo the bolt securing the base of the heater to the support bracket and the two bolts and clamp plate securing the top of the heater to the facia rail.
8 Detach the heater drain hose from the engine compartment bulkhead, withdraw the heater and remove it from inside the car.
9 Refitting is the reverse sequence to removal. Refill the cooling system, as described in Chapter 2, after installation.

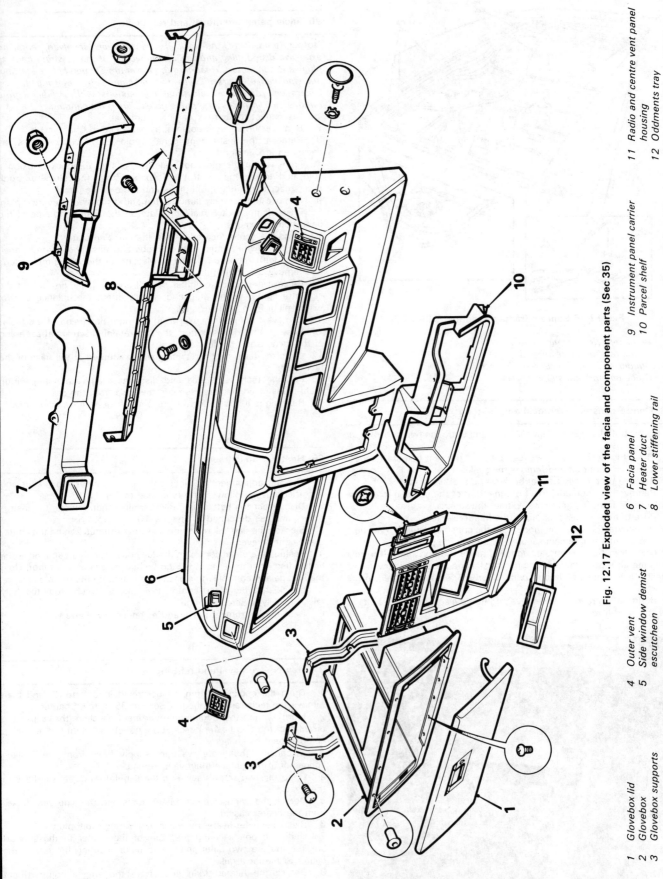

Fig. 12.17 Exploded view of the facia and component parts (Sec 35)

1 Glovebox lid
2 Glovebox
3 Glovebox supports

4 Outer vent
5 Side window demist
   escutcheon

6 Facia panel
7 Heater duct
8 Lower stiffening rail

9 Instrument panel carrier
10 Parcel shelf

11 Radio and centre vent panel
   housing
12 Oddments tray

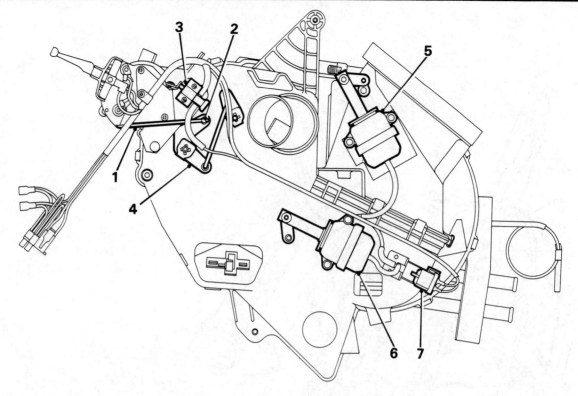

Fig. 12.18 Heater control and adjustment details (Sec 36)

1	Flap link rod	3	Vacuum switch	5	Upper vacuum actuator	7	Vacuum switch solenoid
2	Flap link rod trunnion screw	4	Heater flap lever	6	Lower vacuum actuator		

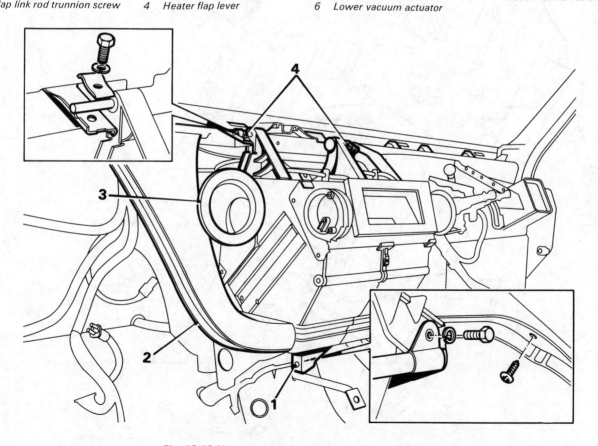

Fig. 12.19 Heater assembly attachment details (Sec 37)

1	Heater base retaining bolt	2	Demister duct	3	Side vent hose	4	Upper retaining bolts and clamp plate

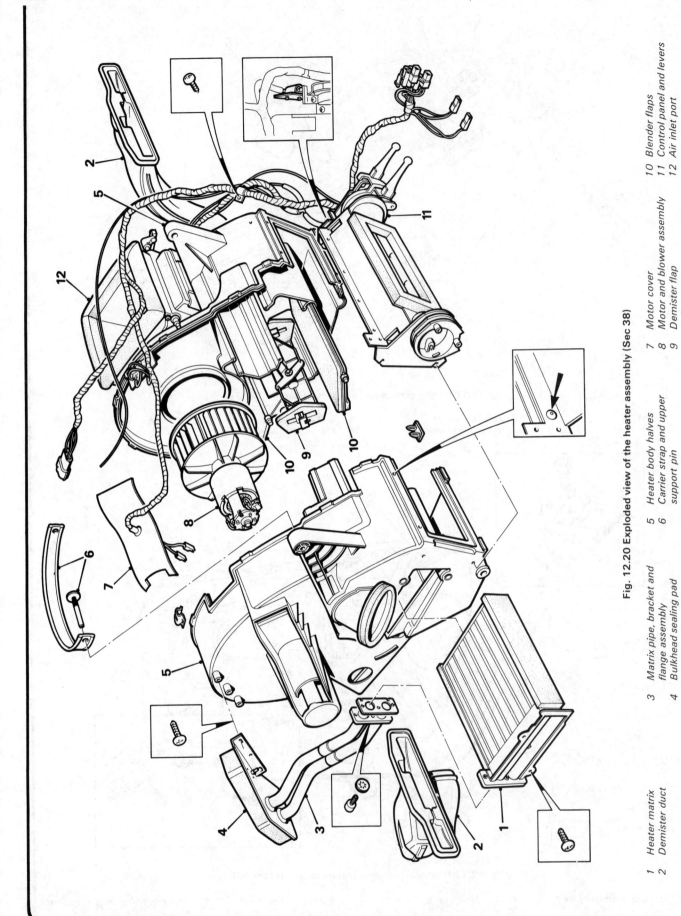

Fig. 12.20 Exploded view of the heater assembly (Sec 38)

1 Heater matrix	3 Matrix pipe, bracket and flange assembly	5 Heater body halves	7 Motor cover	10 Blender flaps
2 Demister duct	4 Bulkhead sealing pad	6 Carrier strap and upper support pin	8 Motor and blower assembly	11 Control panel and levers
			9 Demister flap	12 Air inlet port

38 Heater – dismantling and reassembly

1   With the heater removed from the car, remove the drain hose then release the retaining spring clips and remove the heater upper support pins and carrier strap.

2   Remove the side vent elbows.

3   Remove the Rokut rivets securing the heater control panel, remove the flap link rod from the trunnion then move the control levers and panel to one side.

4   Undo the screws securing the matrix pipe bracket, pipe support clip and matrix flange. Lift out the matrix and pipes. Undo the retaining screws and remove the pipes and gasket from the matrix.

5   Undo the eight screws and two clips securing the heater motor terminal cover. Lift off the cover and disconnect the motor wiring harness plugs. Remove the harness clips.

6   Undo the two screws and washers securing the demist flaps.

7   Drill out the pop rivet adjacent to the heater lower mounting. Remove the sealing tape from inside the heater control panel aperture.

8   Release the spring retaining clips and separate the two halves of the heater body. Withdraw the resistor unit and place the wiring harness to one side. Remove the two plastic hinge bushes from the flaps.

9   Using a flat screwdriver inserted between the heater body and the rear of the blower motor casing, carefully ease out the motor.

10 With the heater dismantled, inspect the matrix for signs of leaks and, if any are apparent, renew the matrix. If the matrix appears serviceable, brush off any accumulation of dirt or debris from the fins and then reverse flush the core. Inspect the remaining heater components for any signs of damage or distortion and renew as necessary.

11 Reassembly is the reverse sequence to dismantling, but adjust the heater flap linkage, as described in Section 36, on completion.

Chapter 13 Supplement:
Revisions and information on later models

Contents

1 Introduction

Since its introduction in 1984, the Montego has undergone a number of mechanical and cosmetic changes. These include the availability of an Estate version in October 1984, and the introduction of a new five-speed gearbox in October 1988.

This Supplement deals with the servicing and overhaul procedures which differ from those described in the first twelve Chapters of this manual. The procedures in this Chapter are either applicable to later models only, or they have been revised by the vehicle manufacturer for all models.

In order to use this Supplement to its best advantage, it is suggested that it is referred to before the main Chapters of the manual, in order to ensure that any relevant information is understood and absorbed into the procedures described in Chapters 1 to 12.

2 Specifications

The Specifications given below are a revision of, or are additional to, those given at the beginning of the first twelve Chapters.

Engine – 1.3 litre
Cylinder block (1988-on)

Note: *From 1988, the main bearing bores are colour-coded (see text)*

Main bearing bores:

Red	2.1466 to 2.1470 in (54.52 to 54.53 mm)
Green	2.1462 to 2.1466 in (54.51 to 54.52 mm)
Yellow	2.1458 to 2.1462 in (54.50 to 54.51 mm)

Crankshaft

Note: *From 1988, the main journal diameters are colour-coded (see text)*

Main journal diameter:	
With no colour code	2.0012 to 2.0017 in (50.83 to 50.84 mm)
Red	2.0005 to 2.0009 in (50.81 to 50.82 mm)
Green	2.0009 to 2.0013 in (50.82 to 50.83 mm)
Yellow	2.0013 to 2.0017 in (50.83 to 50.84 mm)
Main bearing running clearance	0.0007 to 0.0023 in (0.018 to 0.058 mm)
Crankshaft endfloat	0.002 to 0.0023 in (0.018 to 0.058 mm)
Crankshaft endfloat	0.002 to 0.003 in (0.051 to 0.076 mm)
Thrustwasher oversize	0.003 and 0.030 in (0.076 and 0.76 mm)

Valves (1988-on)

Valve seat angle	45° 30′
Head diameter (leaded fuel engines):	
Inlet	1.389 to 1.409 in (35.28 to 35.79 mm)
Exhaust	1.150 to 1.160 in (29.21 to 29.46 mm)
Head diameter (unleaded fuel engines):	
Inlet	1.307 to 1.312 in (33.20 to 33.32 mm)
Exhaust	1.151 to 1.156 in (29.24 to 29.36 mm)
Stem diameter:	
Inlet	0.2790 to 0.2797 in (7.087 to 7.104 mm)
Exhaust	0.2788 to 0.2793 in (7.082 to 7.094 mm)
Clearance in guide:	
Inlet	0.0015 to 0.0027 in (0.038 to 0.069 mm)
Exhaust	0.002 to 0.003 in (0.051 to 0.076 mm)
Valve clearances	0.012 in (0.30 mm)

Lubrication system (1988-on)

Oil pump:	
Outer rotor endfloat	0.005 in (0.13 mm)
Inner rotor endfloat	0.005 in (0.13 mm)
Outer rotor-to-body clearance	0.010 in (0.25 mm)

Engine – 1.6 litre

Compression ratio

1989-on, engine number prefix 16HE33 and 34	9.4 : 1
1989-on, engine number prefix 16HE60, 61, 62, 63, 64, 65 and 66	9.2 : 1
1989-on, engine number prefix 16HD86 and 16HD91	9.6 : 1

Crankshaft (1989-on)

Crankpin journal diameter	1.8748 to 1.8756 in (47.62 to 47.64 mm)
Endfloat	0.004 to 0.007 in (0.10 to 0.18 mm)
Thrustwashers available	0.089 to 0.092 in (2.26 to 2.34 mm) in increments of 0.001 in (0.025 mm)

Cylinder head (1989-on)

Maximum face distortion	0.008 in (0.20 mm)

Cylinder block (1989-on)

Maximum face distortion	0.004 in (0.01 mm)

Torque wrench settings

	lbf ft	Nm
Engine mountings:		
$\frac{5}{16}$ UNF	18	24
$\frac{3}{8}$ UNF	30	41

Cooling system (all models)

Test pressure	11 to 15 lbf/in² (0.8 to 1.0 bar)

Fuel system – 1.3 litre, 1984-on

Carburettor

Fast idle speed	1000 to 1200 rpm

Fuel octane rating

12H type engine	97 RON leaded

Fuel system – 1.6 litre, 1984-on

Carburettor

Engine number prefix 16HE33 and 16HE34, and 16HE60 to 16HE66 inclusive (as for 16H except for the following):

Type	SU HIF 44E with electronic idle speed control and cold drive enrichment

Carburettor (continued)

Engine number prefix 16HE33 and 16HE34, and 16HE60 to 16HE66 inclusive (as for 16H except for the following):

Identification:	
Manual gearbox	FZX 1517
Automatic transmission	FZX 1518
Base idle speed	550 to 650 rpm
Throttle position speed	725 to 775 rpm
ECU-controlled speed	750 rpm

Engine number prefix 16HD86 and 16HD91 (as for 16HE33 except for the following):

Base idle speed	650 rpm
Throttle position speed	750 rpm
ECU-controlled speed	750 rpm

Fuel octane rating

1.6 litre without catalytic converter:	
16H and 16D	97 RON leaded
16HE33, 34, 60, 61, 62, 63, 64, 65, 66	97 RON leaded, or 95 RON unleaded (minimum)
1.6 litre with catalytic converter:	
16HE60, 61, 64 and 65	95 RON unleaded only (minimum)

Ignition system – 1.3 litre, 1984-on

Ignition coil

Type	GCL 138 or 143
Primary resistance at 20°C (68°F)	0.71 to 0.81 ohms

Distributor

Make/type	Lucas 59DM4 or 65DM4

Ignition system – 1.6 litre, 1984-on

Ignition coil

Type	GCL 139 or 143
Primary resistance at 20°C (68°F)	0.71 to 0.81 ohms

Programmed ignition system

Electronic control unit type:	
Engine number prefix:	
16H	Lucas AB17-84185 or 84740
16HE33 and 16HE44	AUU 1644
16HE60 to 66 inclusive	AUU 1644 or MEQ 100003
16HD 86 and 16HD91	AUU 1526
Crankshaft sensor	Lucas 84229

Ignition timing

1989-on 16HE33, 16HE34, 16HD86 and 16HD91 (vacuum pipe connected)	13° to 27° BTDC @ 1500 rpm max with throttle switch closed
1989-on 16HE60, 16HE61, 16HE62, 16HE63, 16HE64, 16HE65 and 16HE66:	
Vacuum pipe connected	13° to 27° BTDC @ 1000 rpm max with throttle switch closed
Vacuum pipe disconnected	4° to 18° BTDC @ 1000 rpm max with throttle switch closed
1990-on 16HE60, 16HE61, 16HE64 and 16HE65:	
Vacuum pipe connected	20° BTDC @ 1000 rpm max with throttle switch closed
Vacuum pipe disconnected	11° BTDC @ 1000 rpm max with throttle switch closed

Ignition timing (at 1000 rpm)

Vacuum hose connected	13 to 27° BTDC
Vacuum hose disconnected	4 to 18° BTDC

Clutch – 1.6 litre, five-speed, 1989-on

General

Type	Single dry plate operated by self-adjusting cable
Clutch disc diameter	8.46 in (215 mm)

Manual gearbox – 1.6 litre, five-speed, 1989-on

General

Type	Five forward speeds, all synchromesh, and reverse. Final drive integral with main gearbox
Type code	T5 AR

Lubrication

Lubricant type/specification *	Multigrade engine oil, viscosity SAE 10W/40 (Duckhams QXR, Hypergrade, or 10W/40 Motor Oil)
Lubricant capacity	3.9 pints (2.2 litres)

*Note: Austin Rover specify a 10W/40 oil to meet warranty requirements for models produced after August 1983. Duckhams QXR and 10W/40 Motor Oil are available to meet these requirements

Gearbox ratios

1st	3.250 : 1
2nd	1.894 : 1
3rd	1.222 : 1
4th	0.906 : 1
5th	0.714 : 1
Reverse	3.000 : 1
Final drive ratio	4.062 : 1

Gearbox overhaul data

Mainshaft:
- Endfloat .... 0.005 to 0.008 in (0.13 to 0.20 mm)
- Adjustment .... Selective circlips

Countershaft:
- 1st gear endfloat .... 0.001 to 0.003 in (0.03 to 0.08 mm)
- Adjustment .... Selective thrust washer
- 2nd gear endfloat .... 0.001 to 0.003 in (0.03 to 0.08 mm)
- Adjustment .... Selective collar
- Minimum baulk ring-to-gear clearance .... 0.016 in (0.41 mm)
- Maximum reverse idler gear-to-shaft clearance ..... 0.005 in (0.13 mm
- Maximum selector fork-to-synchro sleeve clearance .... 0.04 in (1.00 mm)
- Maximum reverse gear fork-to-gear clearance .... 0.028 in (0.71 mm)
- Minimum end cover seal diameter .... 1.97 in (50.0 mm)

Differential:
- Endfloat .... 0.006 in (0.15 mm)
- Adjustment .... Selective circlips

Torque wrench settings

	lbf ft	Nm
Bellhousing to engine adaptor plate	66	90
Gearbox steady rod bolt	6	8
Gear lever to gearchange rod	18	25
Remote control housing to underbody	18	25
Oil filter plug	33	45
Oil drain plug	30	40
Countershaft access plug	52	70
Reversing lamp switch	18	25
Reverse idler shaft retaining bolt	49	67
Gearcase to bellhousing bolts	20	27
Speedometer pinion retaining bolt	8	11
Clutch release fork to operating lever bolt	21	29
Countershaft retaining nut	81	110
Reverse gear fork bracket bolts	10	14
Gearchange holder and interlock retaining bolts:		
M8 bolts	21	29
M6 bolts	9	13
Gearchange shaft detent plug	16	22

Driveshafts – 1.6 litre, 1989-on

Type
Unequal length solid steel, splined to inner and outer constant velocity joints

Lubrication

Overhaul only – see text

Lubricant type:
- Outer constant velocity joint .... Molycote grease VN2461/C, or equivalent
- Inner constant velocity joint .... Mobil 525 grease, or equivalent
- Hub bearing water shield .... Duckhams Admax, or equivalent

Quantity:
- Outer constant velocity joint .... 90 cc
- Inner constant velocity joint .... 190 to 205 cc

Torque wrench settings

	lbf ft	Nm
Driveshaft retaining nut:		
Castellated nut	140	190
Staked nut	150	203
Roadwheel nuts	53	72
Swivel hub to strut nuts	66	90
Tie-rod outer balljoint to steering arm	22	30

Electrical system

Starter motor

Type and application:
- Lucas M79 .... 1.3 models and 1.6 models with manual transmission
- Lucas M78R .... 1.6 models with automatic transmission
- Minimum brush length .... 0.15 in (3.81 mm)
- Commutator minimum diameter .... 1.134 in (28.8 mm)

Suspension and steering
Steering geometry
Camber angle:
    1986 to 1988.. 0°8' positive to 0°34' negative
    1988-on:
        Strut type A*.. 0°6' positive to 0°36' negative
        Strut type B*.. 0°9' positive to 0°51' positive
Castor angle (1988-on)... 0°7' positive 1°07' positive
Steering axis inclination (1989-on):
    Strut type A*... 12 to 13°
    Strut type B*... 12°03' to 13°03'
Toe setting (1988-on).. Parallel ± 0°8'
Steering angles (at kerb weight):
    Inner wheel... 36°59'
    Outer wheel.. 34°05'
For A and B, see text

Power steering
Steering wheel turns lock to lock 2.94
Pump drivebelt deflection.. 0.25 to 0.50 in (7 to 12 mm)
Fluid type/specification .... Dexron IID type ATF (Duckhams Uni-Matic or D-matic)
Power steering fluid reservoir capacity......................... 0.9 pints (0.5 litres)

Roadwheels – 1.6 litre, 1989-on
Wheel size:
    Base model.. 5J x 13
    Other models .... 5½J x 14

Tyres – 1.6 litre, 1989-on
Tyre size:
    Base model.. 165 SR 13
    Other models .... 185/65 TR 14

Torque wrench settings
Power steering

	lbf ft	Nm
Flexible hose to pump union	20	27
Flexible hose to rack unions	11	15
Reservoir-to-pump bracket bolts	18	24
Pipe to pipe adaptor	18	24

Weights (kerb) – 1989-on
Saloons
1.6... 2190 lb (995 kg)
1.6L... 2270 lb (1030 kg)
1.6SL... 2310 lb (1050 kg)
1.6L automatic... 2310 lb (1050 kg)
1.6SL automatic... 2355 lb (1070 kg)

Estates
1.6... 2345 lb (1065 kg)
1.6L... 2400 lb (1090 kg)
1.6SL... 2445 lb (1110 kg)
1.6L automatic... 2445 lb (1110 kg)
1.6SL automatic... 2485 lb (1130 kg)

3 Routine maintenance

Revisions – all models
1 The routine maintenance procedures given at the beginning of this manual have been retrospectively revised, with the following changes.
2 The air cleaner element now need only be renewed at 24 000 mile intervals.
3 On later 1.6 litre models fitted with the fuel evaporative emission control (EVAP) system, inspect the system components and connections for condition and security every 12 000 miles (20 000 km) or 12 months, whichever occurs first

4 Engine

Engine and gearbox (1.6 litre, 1989-on) – removal and refitting
1 From 1989, 1.6 litre models are fitted with a Honda-derived five-speed gearbox.

2 The removal and refitting of the engine and gearbox assembly is basically as described in Chapter 1, noting the different mountings described in this Section.
3 The procedure for removal of the gearbox on its own is given in Section 9 of this Supplement.

Oil filter (1.6 engine)
4 In order to prevent cold start rattle, caused by the engine oil draining back to the sump when the vehicle is standing, it is imperative that the later type of oil filter, incorporating an anti-drain valve, is always fitted at time of renewal. The correct oil filter to use is either Unipart GFE 180 or the Champion equivalent (see Chapter 1 Specifications).

Oil filler pipe emulsification
5 If oil emulsification ('salad cream' type consistency) is evident within the engine oil filler pipe, particularly in cold weather, it can be eliminated in the following way.
6 Remove the air cleaner element and then drill a 0.37 in (9.5 mm) hole through the casing, as shown at 'A' in Fig. 13.1, to accept a grommet.
7 Remove the oil filler pipe. Drill a 0.20 in (5.0 mm) hole and braze a

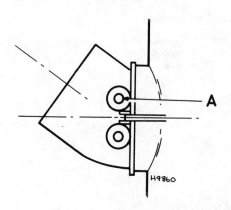

Fig. 13.1 Air cleaner drilling position (A) (Sec 4)

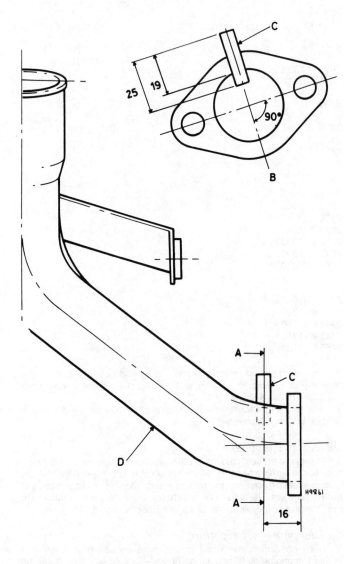

Fig. 13.2 Oil filler drilling position – dimensions in mm (Sec 4)

A	Pipe stub position	C	Tubing
B	Section A - A	D	Oil filler pipe

1.0 inch (25.0 mm) length of brake hydraulic pipe into the filler tube, as shown in Fig. 13.2.

8   Connect a 16.5 in (420.0 mm) length of flexible tubing between the newly fitted grommet and the oil filler pipe brazed stub. The flexible tubing must protrude into the air cleaner by 5.0 mm and slope downward throughout its length.

9   Start the engine, bring it to normal operating temperature, and check and adjust the idle speed and mixture settings.

10   If slight emulsification still occurs, the oil filler pipe and separator should be insulated. A special lagging kit is available.

Engine modifications for operation on unleaded fuel

11   Fuel octane ratings are given in the Specifications section of this Chapter. Certain 1.6 litre models without catalytic converters may be run on either unleaded or leaded fuel, but models fitted with a catalytic converter **must only** use unleaded fuel, otherwise the converter will be damaged.

12   Note that some engine types are capable of conversion to operate on unleaded fuel (95 RON octane minimum). This conversion must be entrusted to a Rover dealer and where this has been made, the letter 'U' will be stamped between the end of the engine number prefix and the serial number. Where doubt exists regarding the fuel type to use, consult a Rover dealer.

13   The valve seats on engines capable of using unleaded fuel have different angles and are specially hardened. Because of this, they cannot be re-worked using conventional machine tools.

Piston rings (1.6 models) – gap setting

14   It should be noted that, on later models, the second compression ring is stepped.

15   When fitting the oil control ring, fit the lower rail with its gap towards the flywheel, and the upper rail with its gap towards the timing belt. The expander should have its gap away from the thrust side of the piston (away from the dipstick side).

Flywheel retaining bolts – renewal

16   If the flywheel is removed during overhaul, always use new bolts at reassembly. Before fitting the bolts, clean away all trace of old thread-locking material from the threaded holes in the crankshaft. A suitable tap is recommended for this job.

Crankshaft main bearings (1.3 litre, 1988-on) – colour codes

17   As from 1988, all 1.3 litre engines are fitted with selective main bearings, the size being identified by a colour code – red, green or yellow.

18   The main bearing bores, crankshaft main bearing journals, main bearing caps and replacement bearings are all either marked with appropriate coloured dye, or stamped with the letters R, G or Y.

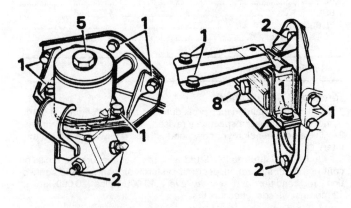

Fig. 13.3 Engine mounting nuts and bolts torque wrench settings – 1.3 models (Sec 4)

1	18 lbf ft (24 Nm)	5	66 lbf ft (89 Nm)
2	30 lbf ft (41 Nm)	6	60 lbf ft (81 Nm)
3	33 lbf ft (45 Nm)	7	45 lbf ft (61 Nm)
4	37 lbf ft (50 Nm)	8	90 lbf ft (122 Nm)

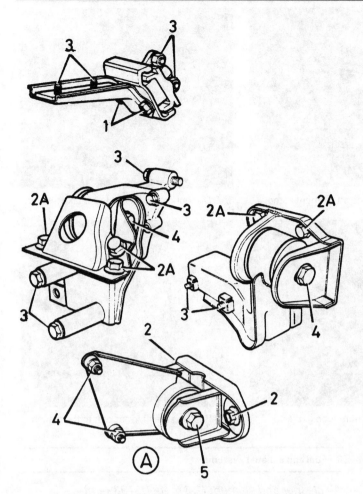
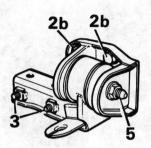
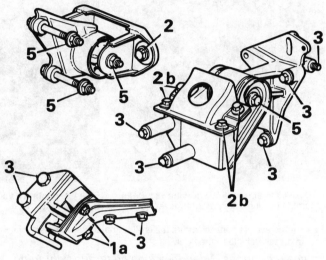

Fig. 13.4 Engine mounting nuts and bolts torque wrench settings – 1.6 litre models, 1985-on (Sec 4)

A Front mounting
1 18 lbf ft (25 Nm)
2 30 lbf ft (41 Nm)
2A 33 lbf ft (45 Nm)
3 33 lbf ft (45 Nm)
4 60 lbf ft (81 Nm)

Fig. 13.5 Engine mountings nuts and bolts torque wrench settings – 1.6 litre models, 1989-on (Sec 4)

1a 25 lbf ft (33 Nm)
2 30 lbf ft (41 Nm)
2b 33 lbf ft (45 Nm)
3 33 lbf ft (45 Nm)
5 60 lbf ft (81 Nm)

Engine overhaul with power-assisted steering

19   On vehicles fitted with optional power-assisted steering, the steering pump drivebelt must be removed before carrying out any of the following operations.

Renewal of the camshaft front oil seal
Tappet adjustment
Timing belt renewal

20   The steering pump must be unbolted and moved to one side before any of the following operations can be carried out.

Cylinder head – removal and refitting
Engine/transmission – removal and refitting
Engine right-hand mounting – renewal

21   The steering hydraulic pipelines must be unclipped before the engine sump pan can be removed.
22   Refer to Section 13 for details of power-assisted steering system.

Engine mountings (1.3 models) – revised removal procedure and dismantling

23   The following procedure should be substituted for that described for the right-hand mounting in Chapter 1, Section 29. See also Fig. 1.7.
24   Unbolt the coolant expansion tank and move it to one side.
25   Raise the right-hand front roadwheel and support the car securely on stands.
26   Remove the access panel from the right-hand wing valance.
27   Take the weight of the engine/transmission using a hoist or a jack and block of wood as an insulator.

28   Slacken the right-hand mounting through-bolt and remove the bolts which hold the mounting bracket to the engine bracket. Take off the restraint bracket.
29   Unscrew the bolts which hold the support strap to the wing valance, lower the engine slightly and remove the mounting assembly. Take off the valance bracket plate.
30   The mounting can be dismantled once the through-bolt is unscrewed. This bolt can be very tight, and may require the application of heat and the use of a long knuckle bar.
31   Apply thread-locking fluid to clean threads of the bolts which hold the mounting bracket to the engine bracket, and the support strap to the body bracket.
32   Locate the mounting so that the spacer is offset towards the top, then place the reinforced buffer plate on one side and engage the top locating hole on the mounting. Place the support strap over the mounting, and enter a new through-bolt from the buffer plate side.
33   Attach the assembly to the combined mounting bracket and fixed buffer plate, but do not tighten at this stage.
34   Place the mounting assembly on the valance bracket and fit the support strap, but do not tighten the bolts.
35   Raise the engine/transmission slightly and locate the engine bracket, mounting bracket and restraint bracket. Tighten the mounting bracket bolts to the specified torque (Fig. 13.3).
36   Tighten the mounting through-bolt to the specified torque.
37   Move the support strap and align the mounting rubber as shown in Fig. 1.7 (inset).
38   Tighten the support strap bolts to torque and check that the mounting is parallel to the buffer plates.

6.1A Removing the air cleaner top cover

6.1B Cold air intake tube attachment at the front body panel

6.1C Disconnecting the cold air intake tube at the air cleaner

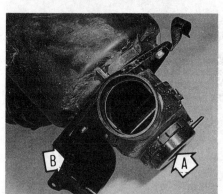

6.1D Air cleaner removed showing vacuum assembly (A) and attachment bracket (B)

6.4A Connecting a vacuum tube to the thermac unit

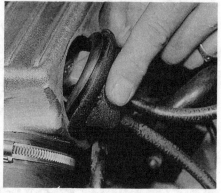

6.4B Refitting the thermac unit to the plenum chamber

39   Fit the access panel, remove the engine/transmission hoist or jack and lower the vehicle to the floor.

Engine mountings – identification and torque tightening fixings

40   In order to clarify the mounting bolt and nut torque wrench settings, refer to Figs. 13.3 and 13.4.
41   All other mounting nuts and bolts should be tightened as given in the Specifications for this Supplement.

Engine front mounting (1.6 models) – modified type

42   As from 1985 the front mounting (snubber) is of different design and the removal and refitting procedure is as described in the following paragraphs. Refer to 'A', Fig. 13.4.
43   Disconnect the battery.
44   Slacken the front mounting through-bolt.
45   Unscrew the bolts, nuts and washers which secure the mounting bracket to the body.
46   Unscrew and remove the nuts from the starter motor mounting bolts, detach the wiring harness bracket and remove the mounting bracket.
47   Remove the mounting through-bolt.
48   Refitting is a reversal of removal, tighten all nuts and bolts to the specified torque.

Engine mountings (1.6 models, 1989-on) – removal and refitting

49   The engine mountings on 1.6 litre models from 1989-on have been modified, to accept the Honda-derived five-speed gearbox.
50   Removal and refitting of the mountings is basically as described in Chapter 1, and Section 9 of this Supplement. Use the torque wrench settings given in this Supplement when tightening the mountings.

---

5 Cooling system

Cooling system – pressure testing

1   When using pressure test equipment to check the cooling system for leaks, it is important that the recommended test pressure is not exceeded (see Specifications).

---

6 Fuel and exhaust systems

Air cleaner and element (1.6 models, 1989-on) – removal and refitting

1   The procedure is similar to that described in Chapter 3, Section 4, but the assembly is slightly different and is shown in Fig. 13.6 (photos).
2   To remove the air cleaner from the engine the HT lead and clutch cable must first be released from the cleaner housing.
3   Note also that one of the housing retaining bolts also secures the thermostat housing. Before removing this bolt, remove the cap from the expansion tank to release any pressure in the cooling system, and refit the bolt after the air cleaner housing is removed to minimise coolant loss.
4   Removal and testing of the thermac unit is as described in Chapter 3 (photos).
5   When refitting the thermac unit hoses to the connector on the plenum chamber, the connections are as follows:

    Large diameter hose – thermac switch to thermac unit
    Small diameter hose – thermac switch to manifold vacuum hose T-piece connector

Fuel pump spacer block (1.3 litre, all models)

6   It is imperative when fitting a replacement fuel pump to all A-series engines that the correct spacer block is fitted between the pump and cylinder block.
7   Early, single-turret type pumps have a spacer with a large cut-out; later, twin-turret, types have a smaller cut-out (see Fig. 13.7).
8   The spacer block with the large cut-out must not be fitted to a later twin-turret pump because the block will not retain the lever pivot pin, which may come adrift and fall into the engine.

SU carburettor – adjustments

9   Certain later models are fitted with a carburettor which incorporates a progressive throttle cam (Fig. 13.8). When carrying out the adjustments described in Chapter 3, Section 12, the lost motion clearance should be set by bending the tag (8).

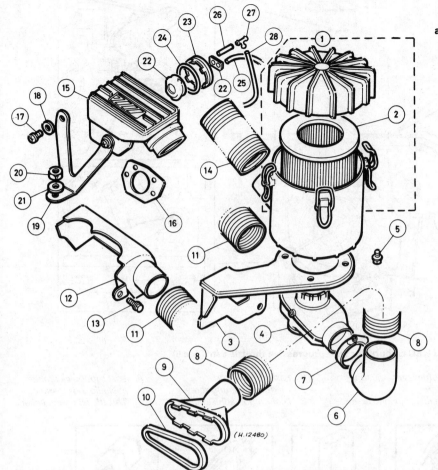

Fig. 13.6 Exploded view of the air cleaner assembly and associated components (Sec 6)

1 Air cleaner assembly
2 Filter element
3 Mounting bracket
4 Thermac unit
5 Bolt
6 Intake tube elbow
7 Retaining clip
8 Cold air intake tube
9 Intake tube adaptor
10 Rubber seal
11 Hot air intake tube
12 Hot air box
13 Bolt
14 Plenum chamber intake tube
15 Plenum chamber
16 Gasket
17 Screw
18 Washer
19 Support bracket
20 Nut
21 Spring washer
22 Thermac switch
23 Grommet
24 Seal
25 Large diameter vacuum hose
26 Small diameter vacuum hose
27 T-piece connector
28 Vacuum hose – inlet manifold to T-piece

(H.12480)

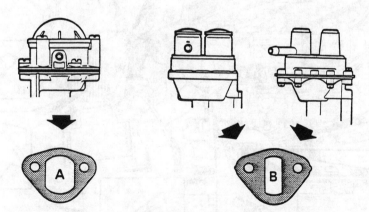

Fig. 13.7 Showing single and twin-turret fuel pumps and respective spacer blocks (Sec 6)

A – Spacer block with large cut-out
B – Spacer block with small cut-out

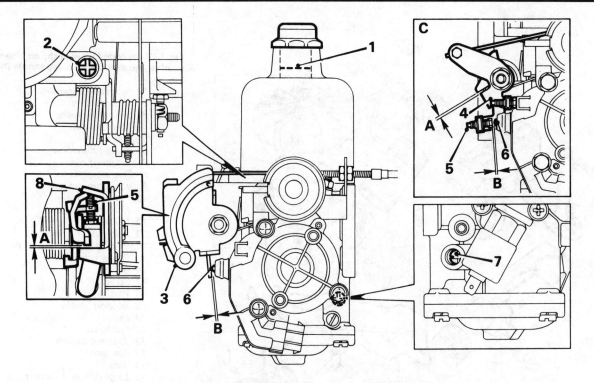

Fig. 13.8 SU carburettor with progressive throttle cam (Sec 6)

1	Piston damper oil level	4	Throttle lever adjusting screw
2	Idle speed adjusting screw	5	Fast idle adjusting screw
3	Progressive throttle cam		

6	Fast idle pushrod	A	Lost motion clearance
7	Mixture adjusting screw	B	Fast idle rod clearance
8	Lost motion adjusting tag	C	Throttle lever and linkage

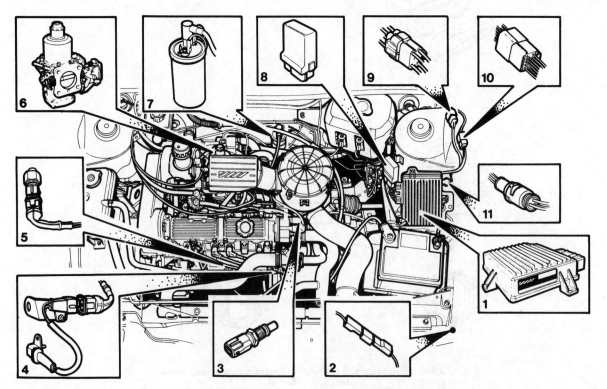

Fig. 13.9 ERIC component location diagram – 1.6 models, 1989-on (Sec 6)

1	ECU	3	Coolant temperature sensor
2	Ambient air temperture sensor – located behind horns	4	Crankshaft sensor
		5	Knock sensor
		6	Carburettor

7	Ignition coil	10	13-way harness connector
8	Engine multi-function unit	11	Engine tuning and diagnostic connector
9	4-way harness connector		

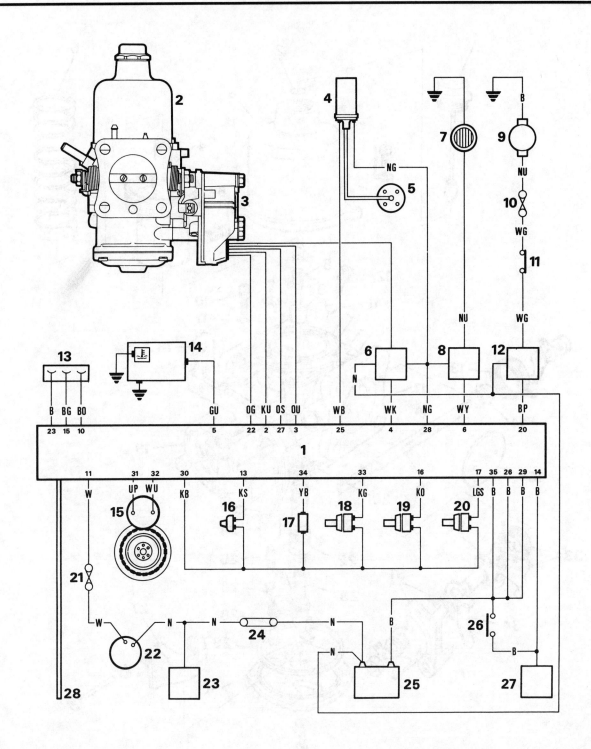

Fig. 13.10 Electronically Regulated Ignition and Carburation system (ERIC) wiring diagram – 1.6 models, 1989-on (Sec 6)

1 ECU	9 Electric fuel pump (not on 1.6)	16 Throttle pedal switch
2 Carburettor	10 Fuse No 17 in fusebox	17 Ambient air temperture
3 Stepper motor	11 Fuel cut-off switch (inertia)	sensor
4 Ignition coil	12 Fuel pump relay (not on 1.6)	18 Coolant thermistor
5 Distributor cap	13 Engine tuning and diagnostic	19 Inlet air temperature sensor
6 Main relay	socket	(not on 1.6)
7 Manifold heater	14 Temperature gauge	20 Knock sensor
8 Manifold heater relay	15 Camshaft sensor	21 Fuse No 1 in fusebox

22 Ignition switch
23 Ignition relay
24 Fusible link
25 Battery
26 Automatic transmission
inhibitor switch
27 Starter relay
28 Manifold vacuum tube

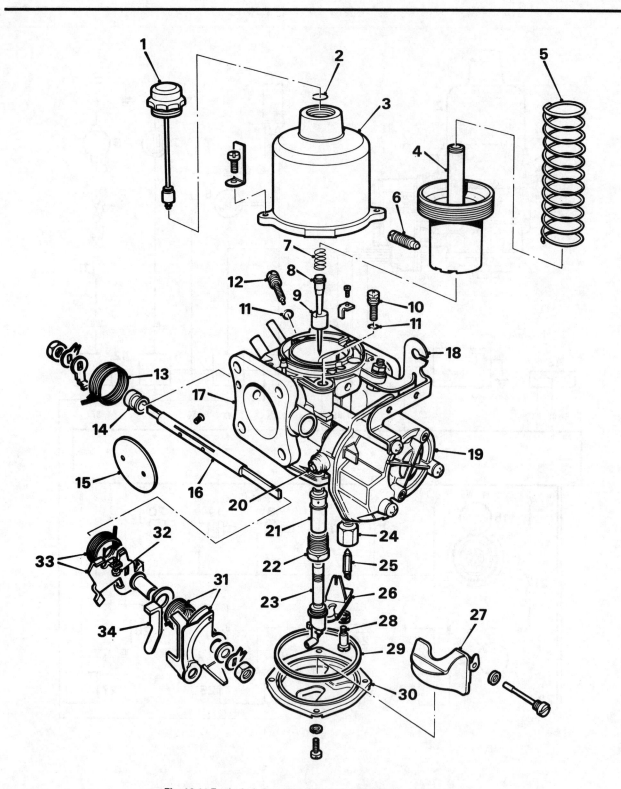

Fig. 13.11 Exploded view of the SU HIF 44E carburettor (Sec 6)

1 Piston damper
2 Spring ring – piston
3 Suction chamber
4 Piston
5 Piston spring
6 Needle retaining screw
7 Needle bias spring
8 Jet needle
9 Needle guide
10 Idle speed adjustment screw
11 Adjusting screw seal
12 Mixture adjustment screw
13 Throttle return screw
14 Throttle spindle seal
15 Throttle disc
16 Throttle spindle
17 Carburettor body
18 Mounting bracket
19 Mixture control stepper motor
20 Fast idle push-rod
21 Jet bearing
22 Jet bearing nut
23 Jet assembler
24 Float needle seat
25 Float needle
26 Bi-metal jet lever
27 Float
28 Jet retaining screw
29 Float chamber cover seal
30 Float chamber cover
31 Throttle cam and return spring
32 Throttle position screw
33 Lost motion link and return spring
34 Fast idle lever

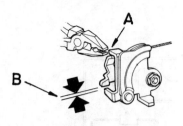

Fig. 13.12 Throttle position speed adjustment on the SU HIF 44E carburettor (Sec 6)

Fig. 13.13 Lost motion clearance adjustment on the SU HIF 44E carburettor (Sec 6)

A – adjustment tag
B – clearance = 0.140 to 0.160 in (3.5 to 4.0 mm)

Electronically Regulated Ignition and Carburation system (ERIC), (1.6 models, 1989-on) – general description

10   The Electronically Regulated Ignition and Carburation system (ERIC) is a complete engine management system, which meets a very high standard of emission control. Using a single Electronic Control Unit (ECU), which effectively combines the fuel ECU and Ignition ECU modules used in the programmed systems of earlier models, ERIC controls and synchronises all fuel and ignition requirements of the engine. The ECU receives its information from the same source as the earlier systems, and acts on the carburettor and ignition system in the same manner. For the purposes of this Supplement, the fuelling aspects of the system are dealt with in this Section and the ignition aspects in Section 7.

ERIC system – component location, removal and refitting

11   The layout of the ERIC system components and their location is shown in Fig. 13.9.
12   The removal and refitting of the components is as described in Chapters 3 and 4.

ERIC system – tuning, adjustment and fault diagnosis

13   The accurate tuning, adjustment and fault diagnosis of the ERIC system requires the use of special test equipment which is likely to be only available to Austin Rover dealers.
14   For this reason there is no alternative but to take the vehicle to an Austin Rover dealer for tuning, adjustment and fault diagnosis, and subsequent rectification.
15   The ERIC system is equipped with a diagnostic and tuning socket connector into which the special test equipment can be connected for speedy testing and fault diagnosis. The ECU is equipped with a memory interface in which are stored intermittent faults which can be read off on the test equipment, so simplifying the test procedure and pin-pointing faults.

Carburettor (SU HIF 44E) – overhaul, tuning and adjustment

16   The HF 44E carburettor fitted to 1.6 litre models with the ERIC system is basically the same as the HIF 44 carburettor described in Chapter 3, and can be overhauled using the same procedure, bearing in mind the following differences.
17   There is no fuel shut-off valve.
18   The throttle linkage is slightly different, and the lost motion clearance is adjusted by bending the tag on the throttle lever as described later.
19   Without the special tuning and diagnostic equipment mentioned earlier the carburettor cannot be adjusted accurately. Approximate settings can be obtained following the procedure given in Chapter 3, Section 12, noting the following.
20   Base idle is set to the specified speed using the idle speed adjustment screw, and the CO content is set using the mixture screw (both these screws are as shown in Fig. 3.5 in Chapter 3).
21   Throttle position speed is adjusted on the fast idle adjustment screw.

22   Lost motion clearance is adjusted by bending the tag on the fast idle lever, and must be set to the specified clearance shown in Fig. 13.13.

Fuel octane rating

23   Refer to the Specifications section and Section 4, paragraphs 11 to 13, of this Chapter for details.

Catalytic converter – general

24   Some models are fitted with a catalytic converter in the exhaust system, the function of which is to minimise the pollutants passing through the exhaust system. An 'open-loop' converter system is used, thus there is no information feedback to the fuel and ignition systems via a sensor in the exhaust as on 'closed-loop' systems.
25   The catalytic converter is connected to the exhaust downpipe at its front end and to the exhaust intermediate section at the rear. It has an integral ceramic and precious metals filter to reduce the harmful exhaust emissions and operates at very high temperatures. The catalytic converter is a fragile device which can be easily damaged. With this in mind, it is imperative that the following precautionary observations be noted and adhered to whenever working on, or in the vicinity of, the catalytic converter in order to ensure its efficiency and maximise service life.

Catalytic converter – precautions

26   The catalytic converter is a reliable and simple device which needs no maintenance in itself, but there are some facts of which an owner should be aware if the converter is to function properly for its full service life.

(a)   DO NOT use leaded petrol in a car equipped with a catalytic converter – the lead will coat the precious metals, reducing their converting efficiency and will eventually destroy the converter.

(b)   Always keep the ignition and fuel systems well-maintained in accordance with the manufacturer's schedule – particularly, ensure that the air cleaner filter element and the spark plugs are renewed at the correct interval – if the intake air/fuel mixture is allowed to become too rich due to neglect, the unburned surplus will enter and burn in the catalytic converter, overheating the element and eventually destroying the converter.

(c)   If the engine develops a misfire, do not drive the car at all (or at least as little as possible) until the fault is cured – the misfire will allow unburned fuel to enter the converter, which will result in its overheating, as noted above.

(d)   DO NOT push- or tow-start the car – this will soak the catalytic converter in unburned fuel, causing it to oveheat when the engine does start – see (b) above.

(e)   DO NOT switch off the ignition at high engine speeds – if the ignition is switched off at anything above idle speed, unburned fuel will enter the (very hot) catalytic converter, with the possible risk of its igniting on the element and damaging the converter.

(f)   DO NOT use fuel or engine oil additives – these may contain substances harmful to the catalytic converter.

(g)   DO NOT continue to use the car if the engine burns oil to the extent of leaving a visible trail of blue smoke – the unburned carbon deposits will clog the converter passages and reduce its efficiency; in severe cases the element will overheat.

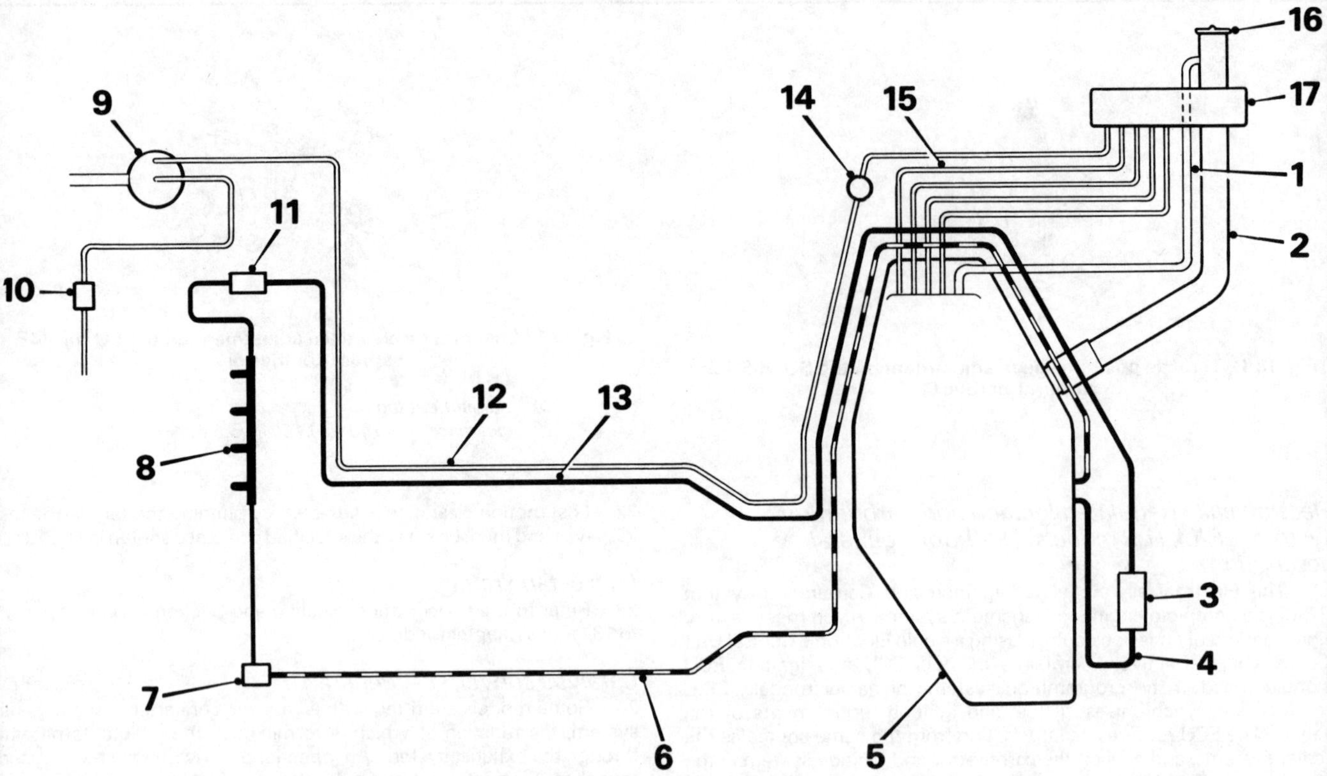

Fig. 13.14 Fuel evaporative loss system (EVAP) layout (Sec 6)

1	Fuel tank breather	6	Fuel return pipe
2	Fuel filler pipe	7	Pressure regulator
3	Fuel pump	8	Injectors
4	Fuel feed pipe	9	Charcoal canister
5	Fuel tank		

10	Purge valve	14	2-way valve
11	Fuel filter	15	Fuel tank to vapour
12	Fuel tank vent pipe		separation tank pipes
13	Fuel feed pipe	16	Fuel filler cap
		17	Separation tank

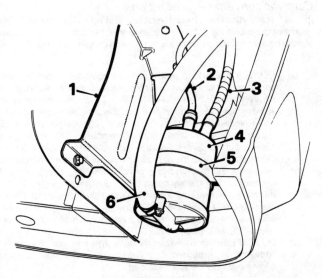

Fig. 13.15 Charcoal canister and connections (Sec 6)

1	Mounting bracket	4	Charcoal canister
2	Vapour pipe	5	Clamp
3	Purge pipe	6	Drain pipe

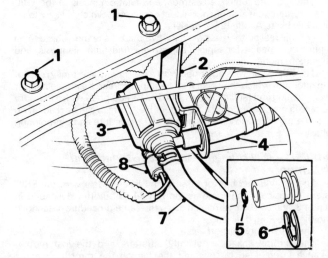

Fig. 13.16 Purge valve connections (Sec 6)

1	Mounting bracket	5	O-ring
	securing bolts	6	Circlip
2	Mounting bracket	7	Vacuum pipe
3	Purge control valve	8	Wiring multi-plug
4	Hose connector		

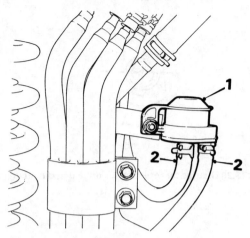

Fig. 13.17 2-way valve and connections (Sec 6)

   1   *2-way valve*
   2   *Hoses (note fitted positions before disconnecting)*

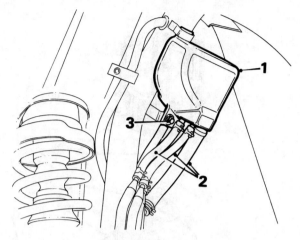

Fig. 13.18 Vapour separation tank and connections (Sec 6)

   1   *Separation tank*
   2   *Hoses*
   3   *Retaining screw*

(h)   *Remember that the catalytic converter operates at very high temperatures – hence the heat shields on the car's underbody – and the casing will become hot enough to ignite combustible materials which brush against it. DO NOT, therefore, park the car in dry undergrowth, over long grass or piles of dead leaves.*

(i)   *Remember that the catalytic converter is FRAGILE – do not strike it with tools during servicing work, take great care when working on the exhaust system, ensure that the converter is well clear of any jacks or other lifting gear used to raise the car and do not drive the car over rough ground, road humps, etc., in such a way as to 'ground' the exhaust system.*

(j)   *In some cases, particularly when the car is new and/or is used for stop/start driving, a sulphurous smell (like that of rotten eggs) may be noticed from the exhaust. This is common to many catalyitic converter-equipped cars and seems to be due to the small amount of sulphur found in some petrols reacting with hydrogen in the exhaust to produce hydrogen sulphide (H_2S) gas; while this gas is toxic, it is not produced in sufficient amounts to be a problem. Once the car has covered a few thousand miles the problem should disappear – in the meanwhile a change of driving style or of the brand of petrol used may effect a solution.*

(k)   *The catalytic converter, used on a well-maintained and well-driven car, should last for between 50 000 and 100 000 miles – from this point on, careful checks should be made at all specified service intervals of the CO level to ensure that the converter is still operating efficiently – if the converter is no longer effective it must be renewed.*

(l)   *Do not overfill the engine oil level. Excess oil may enter the exhaust system, cause the catalytic converter to be clogged and excess HC emissions will result as it is being burnt off.*

(m)   *Do not apply an exhaust sealant to any of the system joints forward of the catalytic converter. Any excess sealant compressed into the system when tightening the joints may be transferred into the converter.*

Fuel evaporative loss control system (EVAP) – general

27   A fuel evaporative loss system is fitted to 1.6 litre models from November 1988 on. The function of the system is to prevent the fuel vapours that are produced in the fuel tank when the engine is turned off escaping to the atmosphere. This is achieved by transferring the vaporised fuel to a charcoal canister. When the engine is restarted, the stored fuel vapour is transferred from the canister to the induction system and burnt off in the normal combustion process. A schematic diagram of the system is shown in Fig. 13.14.

28   The charcoal canister (located under the right-hand front wing), contains charcoal granules which absorb and store the fuel system vapours. A purge valve, mounted in the engine compartment (at the rear of the right-hand headlamp), prevents purging between the charcoal canister and the throttle body under certain operating conditions which

would otherwise effect the operation of the engine and/or catalytic converter (where applicable).

29   The vapours pass from the fuel tank to a separation tank, mounted under the right-hand rear wing which collects fuel vapour, separates the fuel from it and passes the remaining vapour to the charcoal canister via a 2-way valve located under the separation tank. The valve opens under pressure (approx 0.4 lbf/in) to enable the fuel vapour to pass from the separation tank to the charcoal canister. At the same time, the valve also opens the fuel tank to atmosphere to relieve the depression in the fuel tank. The liquid fuel collected in the separation tank is passed back into the main fuel tank.

Fuel evaporative loss control system (EVAP) components – removal and refitting

Charcoal canister

30   Loosen off (but do not remove at this stage) the right-hand front roadwheel, then raise and support the front of the vehicle on axle stands. With the vehicle securely supported, remove the roadwheel.

31   Where applicable, undo the retaining screws and remove the front spoiler.

32   Release the retaining studs and detach the wheel arch liner from the leading edge on the underside of the right-hand front wing.

33   Undo the mounting bracket retaining screw and nuts, then lower the canister plate. Loosen off the securing clip and detach the drain pipe from the canister. Compress the vapour and purge pipe hose clips, lower the canister and detach the hoses from their connectors at the top.

34   Refit in the reverse order of removal. Ensure that the hoses are in good condition and are securely connected.

Purge valve

35   Undo the bolts securing the valve mounting bracket, then move the bracket to enable the multi-plug to be detached.

36   Disconnect the vacuum pipe from the purge valve, then detach the hose connector circlip and remove the connector. The purge valve can now be slid free from its bracket and removed.

37   Refit in the reverse order of removal. Renew the O-ring when fitting it to the hose connector; smear it with clean engine oil to ease assembly. Ensure that all connections are cleanly and securely made.

2-way valve

38   Loosen off (but do not remove at this stage) the right-hand rear roadwheel, then raise and support the rear of the vehicle on axle stands. With the vehicle securely supported, remove the roadwheel.

39   Unscrew and remove the 2-way valve retaining screw. Note the fitted positions of the hoses, compress their retaining clips and detach the hoses from the valve. Remove the valve.

40   Refit in the reverse order of removal. Ensure that the hoses are correctly secured to the valve prior to refitting it into position.

Vapour separation tank

41   Loosen off (but do not remove at this stage) the right-hand rear roadwheel, then raise and support the rear of the vehicle on axle stands.

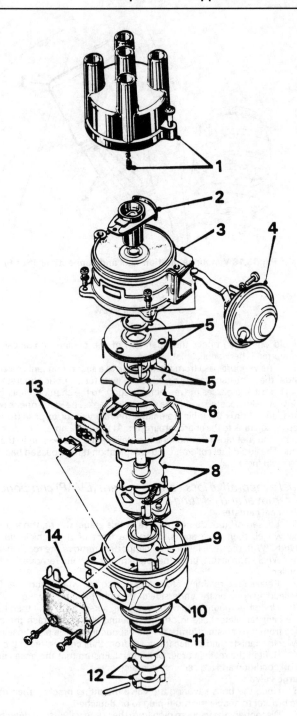

Fig. 13.19 Exploded view of the 65 DM4 distributor (Sec 7)

1 Distributor cap, brush and spring	8 Reluctor, centrifugal advance and shaft
2 Rotor arm	9 Internal thrust washer
3 Upper body	10 Lower body
4 Vacuum unit	11 O-ring seal
5 Stator pack, thrust washers and circlip	12 External thrust washer and drive dog
6 Pick-up winding assembly	13 Connector and gasket
7 Clamp ring	14 Amplifier module

With the vehicle securely supported, remove the roadwheel.
42 Unscrew and remove the separation tank retaining screws. Note the fitted positions of the hoses, compress their retaining clips and detach them from the tank. Remove the tank.

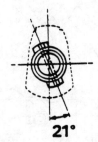

Fig. 13.20 Distributor rotor arm-to-drive dog pin offset (Sec 7)

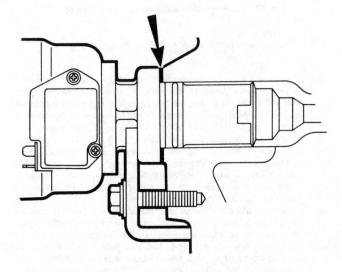

Fig. 13.21 Fitted position of spacer (arrowed) on 65 DM4 distributor (Sec 7)

43 Refit in the reverse order of removal. Ensure that the hoses are correctly and securely connected to the separation tank prior to refitting it into position.

---

7   Ignition system

Distributor (1.3 models) – removal, overhaul and refitting
  Warning: *The amplifier unit in the distributor contains beryllia and must not be opened.*
1 The type 65DM4 distributor fitted to later models is removed as described for the type 59DM4 in Chapter 4, except that a spacer is fitted between the distributor mounting pedestal and the engine.
2 Commence overhaul by removing the rotor arm.
3 Extract the fixing screws and remove the amplifier module, its gasket and connector block. Do not attempt to open the amplifier.
4 Extract the retaining screws and separate the upper and lower bodies.
5 Lift the clamp ring and the pick-up winding from the upper body.
6 Extract the vacuum unit fixing screw.
7 Extract the circlip, remove the thrustwasher and withdraw the stator pack from the vacuum link and then remove the vacuum unit.
8 Support the drive dog and drive out the pin. Remove the dog and thrustwasher.
9 Pull the driveshaft from the lower body and remove the spacer. Do not attempt to remove the reluctor from the shaft.
10 Clean and examine all components and renew as necessary. Reassembly is a reversal of dismantling.

11   As reassembly progresses, lightly oil the bushes, pivots and bearing surfaces.

12   When fitting the amplifier, smear its mounting face with heat conductive silicone paste available from your dealer or auto-electrical engineer.

13   If a new mainshaft is being fitted, fit the rotor arm to the original shaft and use it as a guide for drilling the 0.13 in (3.20 mm) hole for the drive dog (see Fig. 13.20).

14   Fit a new O-ring and, if fitted, the spacer, and apply a drop of oil to the shaft centre bearing lubrication hole.

15   To refit the distributor, first check that No 1 piston is still at TDC.

16   Hold the distributor over its mounting hole so that the vacuum unit is pointing towards No 1 spark plug and the amplifier is at the 4 o'clock position.

17   Insert the distributor and turn the rotor arm slightly to engage the drive dog torque with the coupling.

18   Align the distributor body-to-engine reference marks and clamp the distributor.

19   Refit the cap and connect all leads, the multi-plug and vacuum hose.

20   Check the ignition timing dynamically using a stroboscope, as described in Chapter 4, Section 5.

Rotor arm securing screw seizure – programmed ignition

21   Seizure of the rotor arm securing screw is due to thread-locking compound remaining in the threads of the hole in the camshaft after the screw is removed.

22   Whenever the rotor arm is removed, the threads in the camshaft must be cleaned out using an M6 tap and die.

23   Apply thread-locking compound in the threads of the securing screw on refitting.

Spark plugs

24   It is recommended that when renewing the spark plugs, the copper-cored type is used, as given in the Specifications.

Ignition warning lamp circuitry

25   The ignition warning lamp live feed is connected to the instrument supply line through a fused circuit controlled by the ignition switch.

26   Should the instrument supply fuse blow, with consequent loss of instrument readings, it should be realised that the ignition (charge) warning lamp will also be extinguished and there will not be any alternator output at low engine speeds. Failure of the ignition warning lamp bulb will also affect alternator output.

27   On later models, a resistor is connected in parallel with the ignition warning lamp to prevent lack of charge.

Ignition switch/steering column lock (1986-on)

28   The ignition switch used on later models incorporates a safety feature which prevents the starter being energised while the engine is running.

29   The starter cannot be energised twice consecutively unless the key has been turned to position 1 between starting attempts.

Electronically Regulated Ignition and Carburation system (ERIC) – general description

30   As described in Section 6 of this Supplement, the Electronically Regulated Ignition and Carburation system (ERIC) combines the functions of the fuel and ignition systems into one system.

31   Reference should be made to Section 6 for details of the system and to Chapter 4, Part 2 for details of the ignition components of the system, which are the same as used on the earlier programmed ignition system.

8 Clutch

Clutch cable adjustment (all models) – checking

1   Whenever the clutch cable is disturbed for any reason the following adjustment check must be carried out.

2   Using moderate hand pressure, push the clutch lever on the bellhousing downward (in the opposite direction to its normal travel) through its full stroke. When released, the lever should return to its original position.

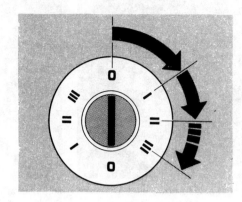

Fig. 13.22 Ignition switch positions (Sec 7)

I   Steering unlocked, radio, blower and windscreen wiper can be operated
II   Ignition on, all electrical circuits operational
III Starter motor operates

3   Press the adjuster body into its abutment on the engine bulkhead with one hand, and with the other, pull the outer cable away from the bulkhead until resistance is felt, holding the outer cable at a point just behind the spring.

4   Release both parts of the cable.

5   This should ensure the self-adjusting mechanism is correctly set. If the adjuster fails to operate correctly, refer to the fault diagnosis (Section 8) of Chapter 5.

Clutch pedal freeplay (all models)

6   The ratchet and spring self-adjusting mechanism which compensates for clutch friction wear operates in steps. The pedal freeplay will therefore diminish between steps and increase again when the next step on the ratchet mechanism is reached.

7   The pedal freeplay can be considered normal if at all times it is between a minimum of 0.47 in (12.0 mm) and a maximum of 1.1 in (28.0 mm).

Clutch (1.6 five-speed, 1989-on) – general description

8   The clutch fitted to 1989-on 1.6 litre models with five-speed gearbox is of conventional diaphragm spring type, operated by self-adjusting cable.

The clutch components comprise a steel cover assembly, clutch disc or driven plate, release bearing and release mechanism. The cover assembly which is bolted and dowelled to the rear face of the flywheel contains the pressure plate and diaphragm spring.

The clutch disc is free to slide along the gearbox input shaft splines and is held in position between the flywheel and pressure plate by the pressure of the diaphragm spring.

Friction material is riveted to the clutch disc, which has a spring-cushioned hub to absorb transmission shocks and to help ensure a smooth take-up of the drive.

Depressing the clutch pedal moves the clutch release lever on the gearbox by means of the clutch cable. This movement is transmitted to the release bearing which moves inwards against the fingers of the diaphragm spring. The spring is sandwiched between two annular rings which act as fulcrum points. As the release bearing pushes the spring fingers in, the outer circumference pivots out, so moving the pressure plate away from the flywheel and releasing its grip on the clutch disc.

When the pedal is released, the diaphragm spring forces the pressure plate into contact with the friction linings of the clutch disc. This disc is now firmly sandwiched between the pressure plate and the flywheel, this transmitting engine power to the gearbox.

Adjustment of the clutch to compensate for wear of the clutch disc friction linings is automatically taken up by the self-adjusting mechanism incorporated in the cable.

Clutch cable (1.6 five-speed, 1989-on) – removal and refitting

9   The procedure is as described in Chapter 5, Section 2.

8.18 Refitting the clutch disc and cover

8.20 Centralise the disc so that the clutch disc hub is in line with the hole in the crankshaft and the circle formed by the diaphragm spring fingers

8.24 Slip off the retaining wire ends and withdraw the release bearing

Clutch assembly (1.6 five-speed, 1989-on) – removal, inspection and refitting

10   Remove the gearbox as described in Section 9.

11   In a diagonal sequence, half a turn at a time, slacken the bolts securing the clutch cover assembly to the flywheel.

12   When all the bolts are slack, remove them and then ease the cover assembly off the locating dowels. Collect the clutch disc which will drop out when the cover assembly is removed.

13   With the clutch assembly removed, clean off all traces of asbestos dust using a dry cloth. This is best done outside or in a well ventilated area; *asbestos dust is harmful, and must not be inhaled.*

14   Examine the linings of the clutch disc for wear and loose rivets, and the disc for rim distortion, cracks, broken torsion springs and worn splines. The surface of the friction linings may be highly glazed, but, as long as the friction material pattern can be clearly seen, this is satisfactory. If there is any sign of oil contamination, indicated by a continuous,

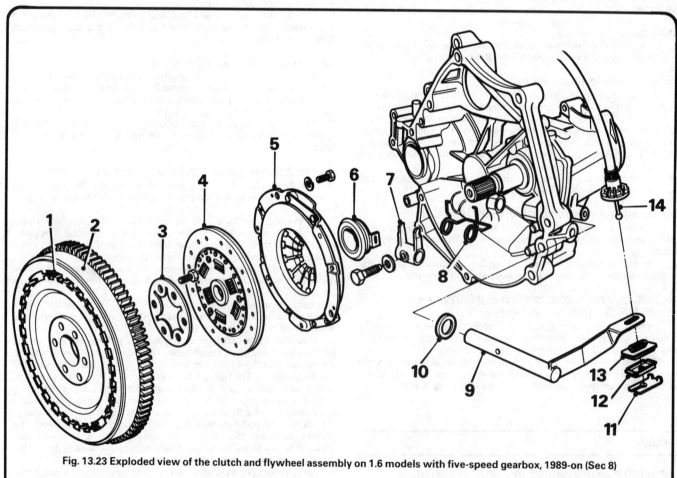

Fig. 13.23 Exploded view of the clutch and flywheel assembly on 1.6 models with five-speed gearbox, 1989-on (Sec 8)

1 Reluctor ring teeth	6 Release bearing	9 Operating lever	12 Cable seating plate
2 Flywheel	7 Release fork	10 Operating lever seal	13 Rubber pad
3 Locking plate	8 Release bearing retaining	11 Cable retaining clip	14 Clutch cable
4 Clutch disc	spring wire		
5 Clutch cover assembly			

or patchy, shiny black discoloration, the disc must be renewed and the source of the contamination traced and rectified. This will be either a leaking crankshaft oil seal or gearbox mainshaft oil seal – or both. Renewal procedures are given in Chapter 1, Section 27 for the engine, whilst for the gearbox refer to Chapter 6, Section 17 (or section 9 of this Chapter for later models fitted with the Honda gearbox).The disc must also be renewed if the lining thickness has worn down to, or just above, the level of the rivet heads.

15   Check the machined faces of the flywheel and pressure plate. If either is grooved, or heavily scored, renewal is necessary. The pressure plate must also be renewed if any cracks are apparent, or if the diaphragm spring is damaged or its pressure suspect.

16   With the gearbox removed it is advisable to check the condition of the release bearing, as described in the following sub-section.

17   To refit the clutch assembly, plate the clutch disc in position with the raised portion of the spring housing facing away from the flywheel. The words FLYWHEEL SIDE will also usually be found on the other side of the disc that faces the flywheel.

18   Hold the disc in place and refit the cover assembly loosely on the dowels (photo). Refit the retaining bolts and tighten them finger tight so that the clutch disc is gripped, but can still be moved.

19   The clutch disc must now be centralised so that when the engine and gearbox are mated, the gearbox input shaft splines will pass through the splines in the centre of the hub.

20   Centralisation can be carried out quite easily by inserting a round bar or long screwdriver through the hole in the centre of the clutch disc, so that the end of the bar rests in the hole in the centre of the crankshaft. Moving the bar sideways or up and down will move the clutch disc in whichever direction is necessary to achieve centralisation. With the bar removed, view the clutch disc hub in relation to the hole in the end of the crankshaft and the circle created by the ends of the diaphragm spring fingers (photo). When the hub appears exactly in the centre, all is correct. Alternatively, if a clutch aligning tool can be obtained this will eliminate all the guesswork, obviating the need for visual alignment.

21   Tighten the cover retaining bolts gradually, in a diagonal sequence to the specified torque wrench setting.

22   The gearbox can now be refitted as described in Section 9.

Clutch release bearing (1.6 five-speed, 1989-on) – removal, inspection and refitting

23   Remove the gearbox as described in Section 9.

24   Slip the retaining spring wires ends out of the slots on the bearing body then slide the bearing off the gearbox input shaft (photo).

25   Check the bearing for smoothness of operation and renew it if there is any roughness or harshness as the bearing is spun.

26   Refitting is the reverse sequence of removal, but ensure that the retaining spring wire ends are located behind the release fork.

9   Manual gearbox

Gearbox – general

1   1.3 models and early 1.6 models are fitted with Volkswagen four- and five-speed transmission units.

2   From 1989, a Honda-designed, Austin Rover-built unit is fitted to all 1.6 models with five-speed transmission.

Gearchange linkage (four-speed and pre-1989 five-speed) – lubrication

3   It is important that the linkage is regularly lubricated in the following way to maintain smooth and easy gear selection.

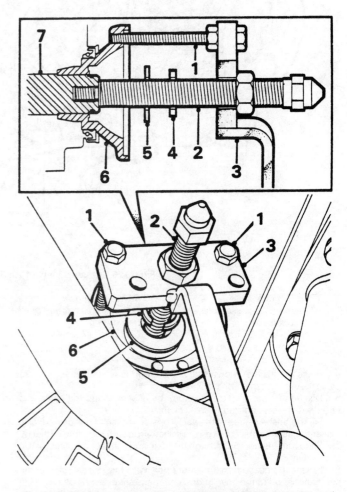

Fig. 13.25 Tool for compressing drive flange spring on four-speed and pre-1989 five-speed models (Sec 9)

1	Bolts (75 x 8.0 mm)	5	Dished washer
2	Centre screw	6	Drive flange
3	Tool bridge piece	7	Pinion shaft
4	Circlip		

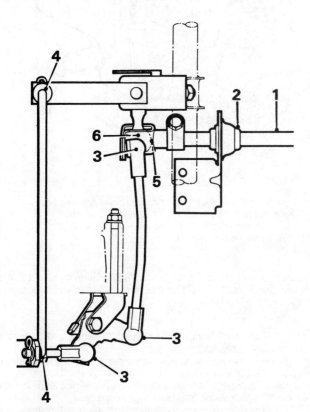

Fig. 13.24 Gearchange linkage lubrication points (Sec 9)

1	Selector rod	4	Pivot
2	Bush	5	Selector rod lever
3	Ball	6	Ball-stud

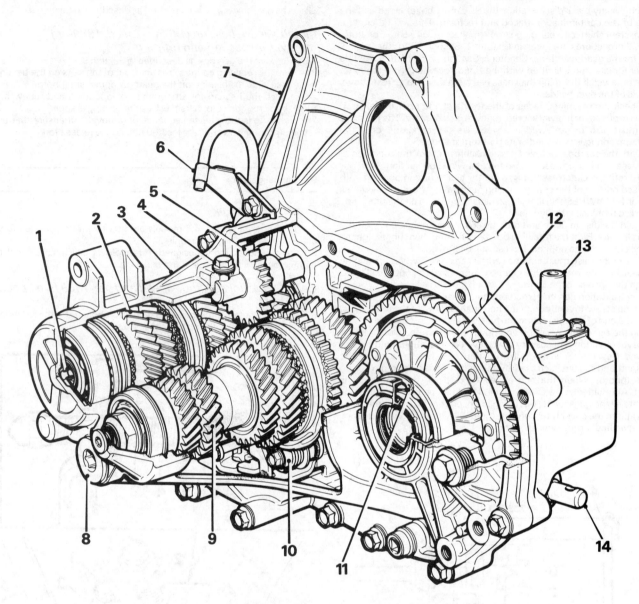

Fig. 13.26 Cut-away view of 1989-on five-speed gearbox (Sec 9)

1	Oil guide plate	5	Reverse idler gear
2	Mainshaft assembly	6	Gearbox breather and bracket
3	Gearcase	7	Bellhousing
4	Reverse idler shaft retaining bolt	8	Countershaft access plug

9	Countershaft assembly	12	Final drive differential
10	Gearchange holder and interlock assembly	13	Speedometer pinion
11	Differential endfloat circlip shim	14	Gearchange shaft

4   Raise the front of the vehicle and support it securely on stands, or place it over an inspection pit.

5   Disconnect the selector rod from the gearlever and rod lever, and withdraw the selector rod from its bush.

6   Fill the bush cavities with special grease (consult your dealer); also smear the bush contact area on the selector rod with the same lubricant.

7   Slide the selector rod through the bush and connect the selector rod to the lever.

8   Smear the gearlever shouldered bolt with the special grease and connect the rod to the gear lever. Tighten the shouldered bolt nut.

9   Apply the special grease to all other pivot points, except the balljoints which are self-lubricating.

10   Peel back the flexible boot from the gearbox selector shaft, smear the shaft with general purpose grease and refit the boot.

11   Check the gearlever adjustment (described in Chapter 6, Section 24).

12   It may also be found that on early models the spring clips which

secure each end of the rod at the pivot points (item 4 in Fig. 13.24) are broken. Modified clips are available from Austin Rover dealers which overcome this problem.

Differential drive flange oil seal (four-speed and pre-1989 five-speed) – renewal

13   A leaking oil seal can be renewed without the need to remove the transmission from the vehicle.

14   Remove the driveshaft from the appropriate side as described in Section 10, and prise the plug from the drive flange.

15   A compressor will now be required to compress the spring at the back of the drive flange, so that the flange dished washer and retaining circlip can be extracted.

16   Although a special tool (18G 1389) is available for the purpose, a simple alternative tool can be made, similar to the one shown in Fig. 13.25. This is bolted to the drive flange using two 75 x 8.0 mm bolts.

9.22A Gearbox filler plug location (arrowed)

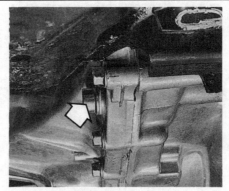

9.22B Gearbox drain plug location (arrowed)

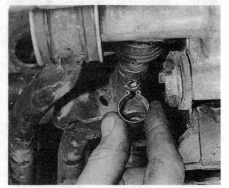

9.44A Remove the clip from the selector rod roll pin

9.44B Tap out the pin

9.44C Disconnect the selector rod

9.45 Remove the bolt from the gearbox steady rod

17 Once the drive flange has been removed, prise out the defective oil seal.
18 Apply grease to the lips of the new seal and drive it squarely into place.
19 Push (do not tap) the drive flange into position, and use the special tool to compress the centre spring. Fit the dished washer and the circlip, remove the tool and fit the plug to the drive flange.
20 Refit the driveshaft and check for oil leaks.

Gearbox (1.6 five-speed, 1989-on) – general description
21 The gearbox is of Honda design and is equipped with five forward and one reverse gear. Synchromesh gear engagement is used on all forward gears.
The mainshaft and countershaft carry the constant mesh gear cluster assemblies and are supported on ball and roller bearings. The short input end of the mainshaft eliminates the need for additional support from a crankshaft spigot bearing.
The synchromesh gear engagement is by spring rings which act against baulk rings under the movement of the synchroniser sleeves. Gear selection is by means of a floor mounted lever connected by a remote control housing and gear change rod to the gear change shaft in the gearbox. Gear change shaft movement is transmitted to the selector forks via the gear change holder and interlock assembly.
The final drive (differential) unit is integral with the main gearbox and is located between the bellhousing and gearcase. The gearbox and final drive components both share the same lubricating oil.

Gearbox (1.6 five-speed, 1989-on) – maintenance and inspection
22 Maintenance and inspection of the gearbox is as described in Chapter 6, Section 2 (photos).

Gearbox (1.6 five-speed, 1989-on) – overhaul (general)
23 Dismantling, overhaul and reassembly of the gearbox is reasonably straightforward and can be carried out without recourse to the manufacturer's special tools. It should be noted however that any repair or overhaul work on the final drive differential must be limited to the renewal of the carrier support bearings. Owing to the complicated nature of this unit and the costs involved, the advice of an Austin Rover dealer should be sought if further repair is necessary.

24 Before starting any work on the gearbox, clean the exterior of the casings using paraffin or a suitable solvent. Dry the unit with a lint-free rag. Make sure than an uncluttered working area is available with some small containers and trays handy to store the various parts. Label everything as it is removed.
25 Before starting reassembly, all the components must be spotlessly clean and should be lubricated with the recommended grade of gear oil during reassembly.

Gearbox (1.6 five-speed, 1989-on) – removal and refitting
26 Remove the bonnet as described in Chapter 12, Section 8.
27 Disconnect the battery negative terminal (Chapter 10, Section 4).
28 Remove the air cleaner element as described in Section 6 of this Supplement.
29 Remove the starter motor with reference to Chapter 10, Section 10, noting that the starter retaining bolts also secure the manifold support tube and fuel overflow pipe bracket.
30 Disconnect the speedometer cable as described in Chapter 10, Section 40, noting that the cable is secured by a clip, not a bolt as on previous models.
31 Disconnect the reversing light switch at the connector in the lead, not at the switch terminals.
32 Disconnect the crankshaft TDC sensor as described in Chapter 4, Section 13.
33 Disconnect the clutch cable (Chapter 5, Section 2).
34 Raise the vehicle onto axle stands, sufficiently high to enable the gearbox to be manoeuvred out from underneath it.
35 Remove both front wheels.
36 Remove the access plate from the left-hand wheel well.
37 Position a suitable container underneath the gearbox, unscrew the drain plug and drain the oil from the gearbox (see photo 9.22B). When the oil is drained, refit the drain plug.
38 Remove the crankshaft sensor noting the spacer under the flange.
39 Remove the top suspension strut-to-hub bolt and loosen the lower bolt (refer to Chapter 11, Section 3).
40 Refer to Section 10 of this Supplement and lever the inboard end of the left-hand driveshaft from the differential.
41 Being extremely careful not to strain the brake hose, tilt the hub downwards at the same time pulling it outward, and manoeuvre the inboard end of the driveshaft upwards, securing it out of the way.

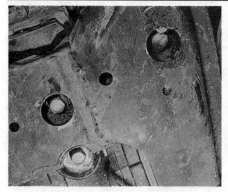

9.46A Rear engine mounting bracket bolts seen through the holes in the subframe

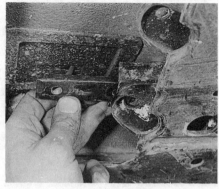

9.46B Removing the threaded plate from on top of the subframe

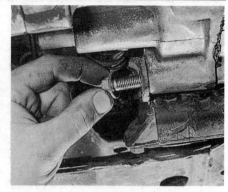

9.46C Removing the mounting bolts

9.47 Engine front mounting

9.48A Lifting gear arrangement seen from above

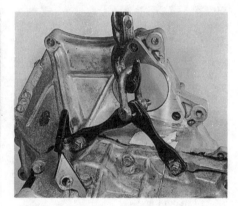

9.48B Close-up of the lifting gear

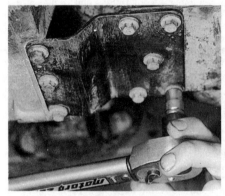

9.49 Removing the engine reinforcing plate bolts

9.51 Lowering the gearbox to the ground

9.54 Unscrew the reversing lamp switch

9.56 Gearbox breather, bracket and case retaining bolt

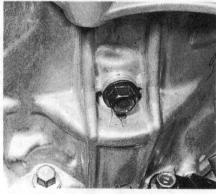

9.57 Reverse idler shaft retaining bolt

9.58 Countershaft access plug (arrowed)

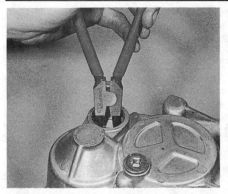

9.59A Release the countershaft bearing retaining circlip ...

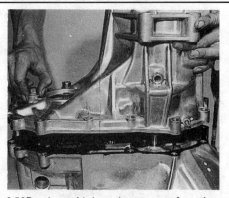

9.59B ... then withdraw the gearcase from the bellhousing and gear clusters

9.63 Lift the mainshaft and countershaft, then remove the selector forks and shafts

9.64 Withdraw the mainshaft and countershaft together from the bellhousing

9.65 Remove the differential from the bellhousing

9.74A Slide the 3rd gear needle-roller bearing onto the mainshaft ...

9.74B ... followed by 3rd gear

9.75A Place the baulk ring and spring ring on 3rd gear ...

9.75B ... then fit the 3rd/4th synchro hub and sleeve assembly

9.76A Locate the spring ring and baulk ring on the synchro unit ...

9.76B ... then fit 4th gear with its needle-roller bearing

9.77 Fit the distance collar with its shoulder towards 4th gear

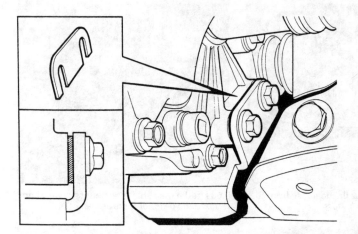

Fig. 13.27 Position of rear engine mounting shim on 1.6 five-speed models (Sec 9)

42   Repeat this operation on the right-hand driveshaft.
43   Support the engine on a jack, inserting a block of wood between the jack head and the engine, and being careful not to obstruct the gearbox-to-engine reinforcing plate.
44   Remove the clip from the selector rod roll pin, tap out the pin and disconnect the selector rod (photos).
45   Remove the bolt from the gearbox steady rod, noting which way the dished washer fits, and disconnect the steady rod (photo).
46   Remove the rear engine mounting bracket. Access to the bracket-to-sub-frame bolts is gained through the holes in the sub-frame. Note the threaded plate on top of the sub-frame (photos).
47   Remove the engine front mounting (photo).
48   Arrange a suitable sling and lifting gear over the engine bay, attach the sling to the gearbox and just begin to take the weight. The arrangement we used is shown in the photographs (photos).
49   Remove all the gearbox-to-adaptor plate bolts. The gearbox-to-engine reinforcing plate will have to be removed to reach the lower bolts (photo).
50   Remove the left-hand engine mounting, noting the earth lead under one of the bolts. On some models the lead may be under an end-case bolt.
51   Gently taking some weight on the sling, separate the gearbox from the engine, pulling the clutch shaft clear, then carefully lower the gearbox to the ground, twisting it as necessary to clear obstructions (photo).
52   Release the lifting gear from the gearbox and manoeuvre it from under the vehicle.
53   Refitting the gearbox is a reversal of the removal procedure, bearing in mind the following points.

(a)   Tighten all bolts to the specified torque, but do not fully tighten the engine mounting bolts until all the mountings are in place and have been centralised.

(b)   Refill the gearbox with the specified lubricant.
(c)   Lubricate the selector gear rods with a lithium-based grease before refitting.
(d)   Carry out the clutch cable adjustment check detailed in Section 8 after refitting the clutch cable.

Note: If oil leakage from the bell-housing-to-gearcase joint in the area of the rear engine mounting is evident, the probable cause is the rear mounting bolts pulling the casing apart. Shims are available from Austin Rover dealers to insert between the mounting and the gearcase to prevent this happening, but no more than three shims should be used. If more shims than this are required, the mounting bracket must be renewed.

Gearbox (1.6 five-speed, 1989-on) – dismantling into assemblies

54   Stand the gearbox on its bellhousing face on the bench, and begin dismantling by removing the reversing lamp switch (photo).
55   Undo the retaining bolt and plate and lift out the speedometer pinion assembly if this was not done during gearbox removal.
56   Undo all the gearcase-to-bellhousing retaining bolts, noting the location of the breather pipe and bracket, which are also retained by one of the case bolts. Remove the breather pipe and bracket (photo).
57   Undo the reverse idler shaft retaining bolt located on the side of the gearcase (photo).
58   Using a large Allen key, hexagonal bar, or suitable bolt with two nuts locked together, undo the countershaft access plug on the end of the gearcase (photo).
59   Using circlip pliers inserted through the access plug aperture, spread the countershaft retaining circlip while at the same time lifting upwards on the gearcase. Tap the case with a soft mallet if necessary. When the circlip is clear of its groove, lift the case up and off the bellhousing and gear clusters (photos).
60   Undo the two retaining bolts and remove the reverse gear fork and bracket.
61   Lift out the reverse gear idler shaft with reverse gear.
62   Undo the three bolts and remove the gearchange holder and interlock assembly. Note that the holder locates in a slot in the 1st/2nd selector shaft.
63   With the help of an assistant, lift up the mainshaft and countershaft as an assembly approximately 0.5 in (12.0 mm) and withdraw the selector shafts and forks from the bellhousing and gear clusters (photo).
64   Lift the mainshaft and countershaft out of their respective bearings in the bellhousing (photo).
65   Finally, remove the differential from the bellhousing (photo).

Mainshaft (1.6 five-speed, 1989-on) – dismantling and reassembly

66   Remove the mainshaft bearing, using a two or three-legged puller if necessary, unless the bearing remained in the gearcase during removal.
67   Withdraw the 5th gear synchroniser hub and sleeve assembly from the mainshaft using a two or three-legged puller if it is tight. Recover the 5th gear baulk ring from the cone face of 5th gear and place it, together with the spring ring, on the synchroniser unit.

9.78A Position the 5th gear needle-roller bearing over the collar ...

9.78B ... then slide on 5th gear

9.79 Fit the 5th gear synchro unit complete with spring ring and baulk ring

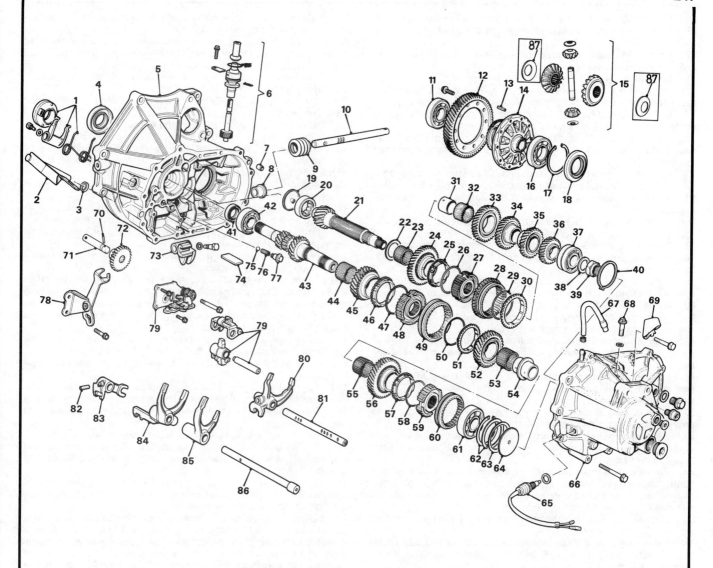

Fig. 13.28 Exploded view of 1989-on five-speed gearbox (Sec 9)

1 Clutch release bearing, fork and retaining spring wire	22 Thrustwasher	46 Baulk ring	69 Gearbox breather bracket
2 Clutch operating lever	23 Needle roller bearing	47 Spring ring	70 Roll pin
3 Clutch operating lever oil seal	24 1st gear	48 3rd/4th synchro hub	71 Reverse idler shaft
4 Differential oil seal	25 Baulk ring	49 3rd/4th synchro sleeve	72 Reverse idler gear
5 Bellhousing	26 Spring ring	50 Spring ring	73 Gearchange arm
6 Speedometer pinion asembly	27 1st/2nd synchro hub	51 Baulk ring	74 Magnet
7 Locating dowel	28 1st/2nd synchro sleeve	52 4th gear	75 Detent ball
8 Gearchange shaft oil seal	29 Spring ring	53 Needle roller bearing	76 Detent spring
9 Rubber boot	30 Baulk ring	54 Distance collar	77 Detent plug
10 Gearchange shaft	31 Distance collar	55 Needle roller bearing	78 Reverse gear fork and bracket
11 Final drive support bearing	32 Needle roller bearing	56 5th gear	79 Gearchange holder and interlock assembly
12 Final drivegear	33 2nd gear	57 Baulk ring	80 1st/2nd gear selector fork
13 Roll pin	34 3rd gear	58 Spring ring	81 1st/2nd gear selector shaft
14 Final drive casing	35 4th gear	59 5th gear synchro hub	82 Roll pin
15 Differential sun and planet gear components	36 5th gear	60 5th gear synchro sleeve	83 5th/reverse gear selector
16 Final drive support bearing	37 Countershaft ball bearing	61 Mainshaft ball bearing	84 5th/reverse gear selector shaft
17 Differential endfloat circlip shim	38 Tongued washer	62 Selective circlips	85 3rd/4th gear selector fork
18 Differential oil seal	39 Retaining nut	63 Belleville washer	86 5th gear selector fork
19 Oil guide plate	40 Circlip	64 Oil guide plate	87 Thrust washers
20 Countershaft roller bearing	41 Mainshaft oil seal	65 Reversing lamp switch	
21 Countershaft	42 Mainshaft ball bearing	66 Gearcase	
	43 Mainshaft	67 Gearbox breather	
	44 Needle roller bearing	68 Reverse idler shaft retaining bolt	
	45 3rd gear		

68   Slide off 5th gear followed by the 5th gear needle roller bearing.
69   Withdraw the distance collar followed by 4th gear and the needle roller bearing.
70   Remove the 3rd/4th synchro hub and sleeve assembly complete with baulk rings and spring rings.
71   Remove 3rd gear and its needle roller bearing.
72   Carry out a careful inspection of the mainshaft components as described later in this Section and obtain any new parts as necessary.
73   During reassembly, lightly lubricate all the parts with the specified grade of gear oil as the work proceeds.
74   Slide the 3rd gear needle roller bearing onto the mainshaft, followed by 3rd gear with its flat face towards the other gears on the shaft (photos).
75   Place the 3rd gear baulk ring and spring ring on the cone face of 3rd gear, then fit the 3rd/4th synchro hub and sleeve assembly. Ensure that the lugs on the baulk ring engage with the slots in the synchro hub (photos).
76   Locate the 4th gear spring ring and baulk ring in the synchro unit, then slide on 4th gear with its needle roller bearing (photos).
77   Fit the distance collar with its shoulder towards 4th gear (photo).
78   Place the 5th gear needle roller bearing over the collar then slide 5th gear onto the bearing (photos).
79   Locate the 5th gear baulk ring and spring ring in the 5th gear synchro unit then fit this assembly to the mainshaft (photo).
80   If the mainshaft or any of its components, the gearcase or bellhousing have been renewed, then the mainshaft endfloat must be checked, and if necessary adjusted, as follows. To do this it will be necessary to remove the mainshaft ball bearing in the gearcase, remove the selective circlips, Belleville washer and oil guide, then refit the bearing.
81   Position the assembled mainshaft in the bellhousing, fit the gearcase and temporarily secure it with several evenly spaced bolts. Tighten the bolts securely.
82   Support the bellhousing face of the gearbox on blocks, so as to provide access to the protruding mainshaft.
83   Place a straightedge across the bellhousing face, in line with the mainshaft, then accurately measure and record the distance from straightedge to mainshaft.
84   Turn the gearbox over so that the bellhousing is uppermost and gently tap the mainshaft back into the gearcase using a soft-faced mallet. Take a second measurement of the mainshaft to straightedge distance.
85   Subtract the first measurement from the second measurement and identify this as dimension A.
86   Measure the thickness of the Belleville washer and add an allowance of 0.006 in (0.17 mm) which is the nominal mainshaft endfloat. Identify this as dimension B.
87   Subtract dimension B from dimension A and the value obtained is the thickness of selected circlip(s) required to give the specified mainshaft endfloat.
88   Remove the gearcase and the mainshaft. Remove the bearing from the gearcase, refit the oil guide, Belleville washer and circlips of the required thickness, then refit the bearing.

Countershaft (1.6 five-speed, 1989-on) – dismantling and reassembly

89   Support the pinion gear on the countershaft in a vice between two

blocks of wood. Tighten the vice just sufficiently to prevent the countershaft turning as the nut is undone.
90   Using a small punch, release the staking on the countershaft nut then undo and remove the nut. Note that the nut has a left-hand thread and must be turned clockwise to unscrew it.
91   Remove the tongued washer then draw off the countershaft bearing using a two or three-legged puller.
92   Slide 5th, 4th, 3rd and 2nd gears off the countershaft noting their fitted directions.
93   Remove the 2nd gear baulk ring and spring ring.
94   Slide off the 2nd gear needle roller bearing followed by the distance collar. Use two screwdrivers to lever off the collar if it is tight.
95   Remove the 1st/2nd synchro hub and sleeve assembly followed by the 1st gear baulk ring and spring ring.
96   Slide off 1st gear followed by the needle roller bearing and thrustwasher.
97   Carry out a careful inspection of the countershaft components as described later in this Section and obtain any new parts as necessary.
98   During reassembly, lightly lubricate all the parts with the specified grade of gear oil as the work proceeds.
99   Fit the thrustwasher to the countershaft followed by the needle roller bearing and 1st gear (photos).
100   Fit the baulk ring and spring ring to the cone face of 1st gear then slide on the 1st/2nd synchro unit. The synchro unit must be fitted with the selector fork groove in the synchro sleeve away from 1st gear. As the unit is fitted ensure that the lugs on the baulk ring engage with the slots in the synchro hub (photos).
101   Warm the distance collar in boiling water then slide in onto the countershaft with the oil hole offset towards 1st gear (photo).
102   Fit the 2nd gear needle roller bearing to the distance collar (photo).
103   Locate the 2nd gear baulk ring and spring ring on the synchro unit then slide 2nd gear into place over the needle roller bearing (photos).
104   Fit 3rd gear to the countershaft with its longer boss away from 2nd gear (photo).
105   Fit 4th gear with its boss towards the 3rd gear boss (photo).
106   Fit 5th gear with its flat face towards 4th gear then tap the countershaft bearing into position using a hammer and suitable tube (photos).
107   Fit the tongued washer followed by a new countershaft nut (photos). Hold the pinion between blocks of wood in the vice as before and tighten the nut to the specified torque.
108   Using feeler gauges, measure the clearance between the rear face of the pinion and 1st gear, and between the 2nd and 3rd gear faces (photos). Compare the measurements with the endfloat dimension given in the specifications. If the recorded endfloat is outside the tolerance range, dismantle the countershaft again and fit an alternative thrustwasher or distance collar.
109   With the countershaft assembled and the endfloat correctly set, recheck the torque of the countershaft nut then peen its edge into the countershaft groove using a small punch.

Gearcase (1.6 five-speed, 1989-on) – inspection and overhaul

110   Check the gearcase for cracks or any damage to its bellhousing mating face. Renew the case if damaged.

9.99A Fit the thrustwasher to the countershaft ...

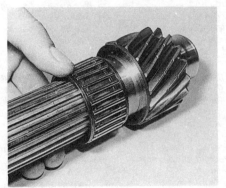

9.99B ... followed by the needle-roller bearing ...

9.99C ... and 1st gear

9.100A Fit the baulk ring and spring ring to 1st gear ...

9.100B ... then slide on the 1st/2nd synchro unit

9.101 Fit the distance collar with its oil hole (arrowed) offset towards 1st gear

9.102 Locate the 2nd gear needle-roller bearing over the distance collar

9.103A Fit the baulk ring and spring ring ...

9.103B ... followed by 2nd gear

9.104 Fit 3rd gear with its boss away from 2nd gear

9.105 Fit 4th gear with its boss towards 3rd gear

9.106A Fit 5th gear ...

111   Check the condition of the mainshaft bearing in the gearcase and ensure that it spins smoothly with no trace of roughness or harshness. The bearing must be removed if it is worn, if the gearcase is to be renewed, or if it is necessary to gain access to the mainshaft endfloat selective circlips located behind it.

112   Removal of the bearing entails the use of a slide hammer with adaptor consisting of internally expanding flange or legs, to locate behind the inner race. An Austin Rover special tool is available for this purpose, but it may be possible to make up a suitable alternative with readily available tools. Whichever option is chosen it is quite likely that the oil guide plate will be damaged or broken in the process. If so a new one must be obtained.

113   If any of the mainshaft components are being renewed during the course of overhaul, do not refit the bearing, circlips, Belleville washer or oil guide plate until after the mainshaft endfloat has been checked and adjusted.

114   When the bearing is fitted this can be done by tapping it squarely into place using a hammer and tube of suitable diameter in contact with the bearing outer race.

115   If there is any sign of leakage, the differential oil seal in the gearcase should be renewed. Drive or hook out the old seal and install the new one with its open side facing inward ie towards the differential (photo). Tap the seal squarely into place using a suitable tube or the old seal. Smear a little grease around the sealing lip to aid refitting of the driveshaft. **Note:** *If the differential or differential bearings have been renewed or disturbed from their original position, do not fit the oil seal until the gearbox has been completely reassembled.* The differential bearing clearances are checked through the oil seal aperture, and cannot be done with the seal in place.

Bellhousing (1.6 five-speed, 1989-on) – inspection and overhaul

116   With the mainshaft and countershaft removed, lift out the magnet from its location in the bellhousing edge (photo).

9.106B ... followed by the countershaft bearing ...

9.106C ... then drive the bearing fully onto the shaft

9.107A Engage the tongued washer with the countershaft groove ...

9.107B ... and screw a new nut onto the shaft

9.108A Check the clearance between the pinion and 1st gear ...

9.108B ... and between 2nd and 3rd gear

9.115 Fit a new differential oil seal with its open side facing inward

9.116 Remove the magnet from the bellhousing

9.118 Undo and remove the clutch release fork peg bolt

9.120A Undo the gearchange shaft detent plug ...

9.120B ... then lift out the detent spring ...

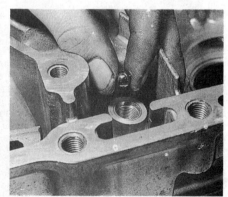

9.120C ... and detent ball

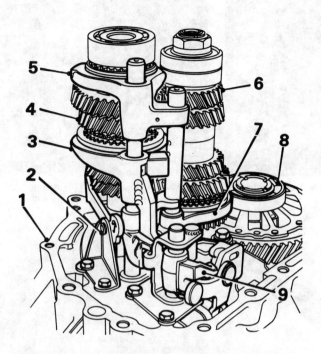

Fig. 13.29 Geartrain and gear selector components on 1989-on five-speed gearbox (Sec 9)

1 Bellhousing	6 Countershaft assembly
2 Reverse gear fork and bracket	7 1st/2nd gear selector fork
3 3rd/4th gear selector fork	8 Final drive differential
4 Mainshaft assembly	9 Gearchange holder and
5 5th gear selector fork	interlock assembly

117   Remove the release bearing as described in Section 8.

118   Undo the retaining peg bolt securing the release fork to the clutch operating lever (photo).

119   Withdraw the lever from the fork and lift out the fork noting the fitted position of the release bearing retaining spring wire.

120   Undo the gearchange shaft detent plug and lift out the detent spring and ball (photos).

121   Undo the bolt securing the gearchange arm to the shaft and slide the arm off the shaft (photos).

122   Withdraw the gearchange shaft from the bellhousing and recover the rubber boot.

123   Examine the bellhousing for cracks or any damage to its gearcase mating face. Renew the case if damaged.

124   Check the condition of the ball and roller bearings in the bellhousing ensuring that they spin smoothly with no trace of roughness or harshness.

125   Renewal of the bearings entails the use of a slide hammer with adaptor consisting of an internally expanding flange or legs, to locate through the centre of the bearing. An Austin Rover special tool is available for this purpose, but it may be possible to make up a suitable substitute with readily available tools. Another alternative would be to take the bellhousing along to your dealer and have him renew the bearings for you. Whichever option is chosen it is quite likely that the oil guide plate behind the countershaft roller bearing will be damaged or broken in the process (assuming this bearing is to be renewed) and if so a new guide plate must be obtained (photo).

126   Refit the bearings by tapping them squarely into place using a hammer and tube of suitable diameter. Ensure that the oil hole in the countershaft bearing faces the gearbox interior.

127   Carefully inspect all the oil seals in the bellhousing and renew any that show signs of leakage. The old oil seals can be driven out with a tube or punch and the new seals tapped squarely into place using a block of wood or the old seal. Ensure that in all cases the open side of the seal faces inward. In the case of the mainshaft oil seal it will be necessary to remove the mainshaft bearing to enable a new seal to be fitted.

128   Inspect the gearchange shaft for distortion or wear across the detent grooves and check the gearchange arm for wear of the forks. Renew these components if wear is evident.

129   With the new bearings and seals in position and any other new parts obtained as necessary, refit the gearchange shaft and rubber boot with the detent grooves facing outward, ie towards the gear clusters.

130   Slide on the gearchange arm so that its forked side is facing away from the bellhousing starter motor aperture. Refit the retaining bolt and washer and tighten to the specified torque.

131   Refit the detent ball followed by the spring and plug bolt. Tighten the bolt to the specified torque.

132   Slide the clutch operating lever into the bellhousing and engage the release fork. Ensure that the joined end of the bearing retaining spring wire is positioned behind the release fork arms. Refit and tighten the retaining bolt.

133   Refit the magnet with its forked end down, then refit the clutch release bearing as described in Section 8.

Mainshaft and countershaft components and synchro units (1.6 five-speed, 1989-on) – inspection and overhaul

134   With the mainshaft and countershaft dismantled, examine the shafts and gears for signs of pitting, scoring, wear ridges or chipped teeth. Check the fit of the gears on the mainshaft and countershaft splines and ensure that there is no lateral free play.

135   Check the smoothness of the bearings and check for any signs of scoring on the needle roller bearing tracks and distance collars.

136   Check the mainshaft and countershaft for straightness, check for damaged threads or splines and ensure that the lubrication holes are clear (photos).

137   Mark one side of each synchro hub and sleeve before separating the two parts, so that they may be refitted in the same position.

138   Withdraw the hub from the sleeve and examine the internal gear teeth for wear or ridges. Ensure that the hub and sleeve are a snug sliding fit with the minimum of lateral movement.

139   Check the fit of the selector forks in their respective synchro sleeve grooves. If the clearance exceeds the figure given in the Specifications, check for wear ridges to give an indication of whether it is the fork or the sleeve groove that has worn. As a general rule the selector fork usually wears first, but not always. If in doubt compare with new parts.

140   Place each baulk ring on the cone face of its respective gear and measure the distance between the baulk ring and the gear face. If the clearance is less than specified renew the baulk ring. Renew them also if there is excessive wear or rounding off of the dog teeth around their periphery, if they are cracked, or if they are in any way damaged. If the gearbox is in reasonable condition and is to be rebuilt it is advisable to renew all the baulk rings as a matter of course. The improvement in the synchromesh action when changing gear will justify the expense.

141   When reassembling the synchro units make sure that the two oversize teeth in the synchro sleeve engage with the two oversize grooves in the hub (photo).

142   If any of the gears on the mainshaft are to be renewed then the corresponding gear on the countershaft must also be renewed and vice versa. This applies to the countershaft and differential final drivegear as well.

Selector forks, shafts and gearchange mechanism (1.6 five-speed, 1989-on) – inspection and overhaul

143   Visually inspect the selector forks for obvious signs of wear ridges, cracks or deformation.

144   Slide the selector forks off the shafts, noting their fitted position and check the detent action as the fork is removed. Note that the detent balls and springs are located in the selector forks and cannot be removed. If the detent action is weak or if there is evidence of a broken spring or damaged ball, the fork must be renewed.

145   Check the fit of the selector forks in their respective synchro sleeves. If the clearance exceeds the figure given in the Specifications check for wear ridges to give an indication of whether it is the fork or the sleeve groove that is worn. If in doubt compare them with new parts and renew any that are worn.

146   Examine the selector shafts for wear ridges around the detent grooves and for any obvious signs of distortion. Renew any suspect shafts.

9.121A Undo the gearchange arm retaining bolt ...

9.121B ... slide off the arm and remove the shaft

9.125 Mainshaft ball bearing (A), countershaft roller bearing (B), oil guide plate (C) and bearing oil holes (D)

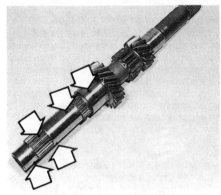

9.136A Gearbox mainshaft showing oil holes (arrowed) ...

9.136B ... and gearbox countershaft

9.141 Oversized teeth in the synchro sleeve engaged with corresponding grooves in the hub

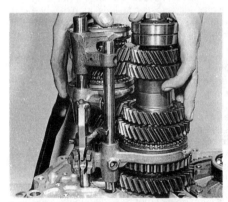

9.158 Lift up the mainshaft and countershaft to allow fitment of the selector forks and shafts

9.159 Fit the gearchange holder and interlock assembly ...

9.160 ... and secure with the three securing bolts (arrowed)

9.161 Fit the reverse idler shaft and gear

9.162 Engage the reverse gear fork with reverse gear and with the 5th/reverse selector peg (arrowed)

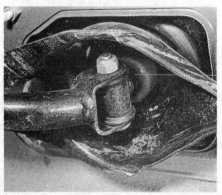

9.171 Gear lever-to-gearchange remote control rod retaining bolt

147 Examine the gearchange holder and interlock assembly for any visible sign of wear or damage. It is advisable not to dismantle this mechanism unless it is obviously worn and is in need of renewal. If this is the case it can be separated into three main units and the worn part renewed.

148 Having obtained any new parts as required, reassemble the selector forks back onto the shafts and reassemble the gearchange holder and interlock components if these were dismantled for renewal. **Note:** *On later gearboxes, the 5th/reverse gear selector shaft assembly is slightly different from earlier types; the selector forks are retained on the shaft by a machined collar and not a circlip as on the earlier type.*

Final drive differential (1.6 five-speed, 1989-on) – inspection and overhaul

149 The only parts which can be renewed as a practical proposition are the two main support bearings on the final drive casing. The differential unit should be examined for any signs of wear or damage, but if any is found it is recommended that you seek the advice of your dealer. Differential parts are supplied in sets, ie final drive gear and matching countershaft; sun gears and matching planet gears etc, and consequently are extremely expensive. The cost of individual parts may even equal the price of a complete exchange box.

150 Check that the bearings spin freely with no sign of harshness or roughness. If renewal is necessary, remove the bearings by levering them off the differential using two screwdrivers or use a small puller.

151 Fit the new bearings by tapping them into place using a hammer and tube in contact with the bearing inner race.

Gearbox (1.6 five-speed, 1989-on) – reassembly

152 Position the differential in its location in the bellhousing and tap it down gently, using a soft-faced mallet to ensure that the bearing is fully seated.

153 Fit the gearcase to the bellhousing and secure it temporarily with several bolts tightened to the specified torque.

154 Using feeler gauges inserted through the oil seal aperture in the gearcase, measure the clearance between the bearing and the circlip type shim in the bearing recess. If the clearance is not equal to the differential endfloat dimension given in the Specifications, slacken the case retaining bolts, extract the circlip through the oil seal aperture and substitute a thicker or thinner circlip as required from the range available. Repeat this procedure until the correct endfloat is obtained then remove the gearcase. Gearbox reassembly can now proceed as follows.

155 Insert the magnet into its location in the edge of the bellhousing, forked end first.

156 Refit the gearchange shaft and arm as described in Section 9 if this has not already been done.

157 With the bearings in place in the bellhousing, hold the assembled mainshaft and countershaft together and insert them into their locations.

158 With the help of an assistant, lift up the mainshaft and countershaft assemblies together approximately 0.5 in (12.0 mm). Engage the selector forks with their respective synchro sleeves and locate the selector shafts in the bellhousing (photo). Return the mainshaft and countershaft to their original positions ensuring that the selector shafts engage fully with their holes in the bellhousing.

159 Refit the gearchange holder and interlock assembly, noting that the holder locates in a slot in the 1st/2nd selector shaft (photo).

160 Refit the three gearchange holder and interlock retaining bolts and tighten them to the specified torque (photo).

161 Refit the reverse gear idler shaft with reverse gear, noting that the long boss on the gear must face the bellhousing, and the hole in the top of the shaft faces away from the gear clusters (photo).

162 Engage the reverse gear fork over the reverse gear teeth and over the peg on the 5th/reverse selector. Secure the reverse gear fork bracket with the two retaining bolts tightened to the specified torque (photo).

163 Apply a continuous bead of RTV sealant to the gearcase mating face. Lower the gearcase over the gear clusters and engage the shafts and bearings in their locations. Using circlip pliers inserted through the countershaft access plug aperture, spread the circlip and tap the gearcase fully into position using a soft-faced mallet. Release the circlip ensuring that it enters the groove on the countershaft bearing.

164 Refit the gearcase retaining bolts and breather bracket then tighten the bolts progressively to the specified torque.

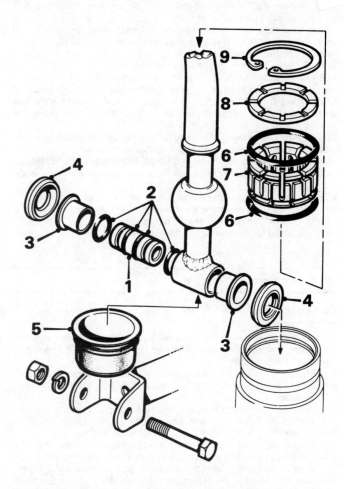

Fig. 13.30 Exploded view of gear lever on 1989-on five speed models (Sec 9)

1 Spacer	6 O-ring
2 O-rings	7 Gear lever seat
3 Bushes	8 Retaining ring
4 Sealing washers	9 Circlip
5 Dust cover	

165 Refit the reverse idler shaft retaining bolt and tighten it to the specified torque.

166 Apply RTV sealant to the thread of the countershaft access plug, refit the plug and tighten it to the specified torque.

167 Refit the speedometer drive pinion, if this was not removed when removing the gearbox from the car, and the reversing light switch.

168 Finally refit the final drive differential oil seal to the gearcase if not already done.

169 Check the operation of the gearchange mechanism ensuring that all gears can be engaged, then refit the gearbox to the car as described earlier.

Gear lever and remote control housing (1.6 five-speed, 1989-on) – removal and refitting

Gear lever

170 Jack up the front of the car and support it on axle stands.

171 From under the car undo the nut and withdraw the bolt securing the gear lever to the gearchange remote control rod (photo).

172 From inside the car unscrew the gear lever knob then remove the gaiter and rubber boot.

173 Remove the retaining circlip and withdraw the gear lever from the remote control housing.

174 Release the sealing washers, extract the bushes and slide out the spacer complete with O-ring seals located at the base of the gear lever. Examine the components for wear.

175   Inspect the O-rings and gear lever seat in the remote control housing for signs of wear or damage.
176   Renew any worn or damaged parts then reassemble and refit the gear lever using the reverse sequence to removal.

Remote control housing
177   Remove the gear lever as described above.
178   From under the front of the car, remove the spring clip to expose the gearchange rod to gearbox, gearchange shaft retaining roll pin.
179   Using a suitable punch, drive out the roll pin and slide the gearchange rod rearwards off the shaft. Remove the rod from under the car.
180   Undo and remove the small bolt in the centre of the gearbox steady rod, at the gearbox end of the remote control housing.
181   Remove the dished washer, slide off the steady rod and remove the flat washer. Note that the lip on the inner washer faces the steady rod rubber bush.
182   At the remote control housing end, undo the two bolts securing the mounting bracket to the vehicle floor and withdraw the remote control assembly from under the car (photo).
183   Refitting is the reverse sequence of removal.

---

10   Driveshafts

Driveshafts (1.3 and 1.6, pre-1989) – revised removal procedure

1   While the vehicle is standing on its wheels, firmly engage the handbrake and put the transmission in gear, or in the P position if automatic transmission is fitted.
2   Prise off the wheel trim and extract the driveshaft nut retaining split pin. Using a suitable socket and bar, slacken the nut, but do not remove it at this stage.
3   Slacken the wheel nuts, jack up the front of the car and support it on axle stands. Remove the roadwheel and return the transmission to neutral. Remove the driveshaft nut and washer.
4   From underneath the car, make an alignment mark between the inner constant velocity joint flange and the differential drive flange, as an aid to reassembly. On vehicles equipped with automatic transmission, the left-hand driveshaft must be withdrawn from the drive flange before the alignment marks can be made.
5   Remove the protective caps over the inner joint retaining bolts and using an Allen key of the appropriate size, unscrew and remove the bolts.
6   The procedure now varies slightly depending on whether the left-hand or right-hand driveshaft is being removed.

Left-hand driveshaft
7   On vehicles with manual transmission, ease the inner constant velocity joint away from the differential drive flange and lower the inner end of the shaft. Withdraw the outer constant velocity joint from the wheel hub, lower the driveshaft to the ground and remove it from under the car. Withdraw the bearing water shield from the outer joint.
8   On vehicles with automatic transmission, undo and remove the nuts and washers, then withdraw the two bolts securing the suspension strut to the upper part of the swivel hub. Separate the swivel hub from the strut, ease the inner constant velocity joint away from the differential drive flange and withdraw the outer constant velocity joint from the

wheel hub. Remove the driveshaft from under the car and slide the bearing water shield off the outer joint.

Right-hand driveshaft
9   On vehicles with manual transmission, extract the retaining clip and flat washer from the gearchange linkage pivot point directly above the inner constant velocity joint. Release the linkage from its pivot location and place it to one side.
10   Compress the inboard joint and ease it from the drive flange, then raise the inboard end of the shaft and rest it on the drive flange.
11   Withdraw the outer constant velocity joint from the wheel hub and remove the driveshaft from under the wheel arch. Withdraw the bearing water shield from the outer joint.
12   On vehicles with automatic transmission, the removal operations are as described for manual transmission vehicles, but of course ignore all reference to the gearchange control linkage.

Driveshafts (1.3 and 1.6, pre-1989) – revised refitting procedure

13   When refitting a driveshaft-to-hub securing nut, the following procedure should be used instead of that described in Chapter 8, Section 3, paragraph 13(c).
14   Tighten the inboard joint retaining bolts to the specified torque. On automatic transmission models, tighten the swivel hub retaining nuts on the left-hand driveshaft to the specified torque.
15   Where a castellated type nut is used, have an assistant apply the footbrake and tighten the nut to a torque of 140 lbf ft (190 Nm). If the split pin hole is aligned, fit a new split pin and bend over its ends.
16   If the hole is not aligned, slowly tighten the nut as necessary to a maximum torque of 160 lbf ft (215 Nm) until alignment is achieved.
17   If the split pin hole is still not in alignment, unscrew the nut and fit a thrustwasher of alternative thickness from the three sizes available.

   0.240 in (6.10 mm) thick – silver
   0.244 in (6.20 mm) thick – green
   0.248 in (6.30 mm) thick – dark grey

18   Again tighten to the specified torque and fit a new split pin. Where a staked type nut is fitted, always use a new nut and stake it into the shaft groove.

Driveshafts (manual gearbox from 1989) – general description

19   The driveshafts fitted to 1989-on 1.6 litre models with the Honda-derived five-speed gearbox are different to those fitted to earlier models, being of unequal length, solid steel construction, splined to the inner and outer constant velocity joints. Note that the driveshafts fitted to automatic transmission models remain as described in Chapter 8, and earlier in this Section.

Driveshafts (manual gearbox from 1989) – removal and refitting

20   While the vehicle is standing on its wheels, firmly apply the handbrake and put the transmission in gear.
21   Remove the wheel trim and extract the driveshaft retaining nut split pin. If a staked retaining nut is fitted, refer to paragraphs 41 to 43 for its removal and refitting details. Using a suitable socket and bar, undo and remove the driveshaft nut. Recover the thrustwasher (photo).
22   Slacken the wheel nuts, jack up the front of the car and support it

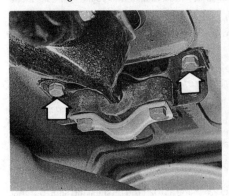

9.182 Remote control housing bracket retaining bolts (arrowed)

10.21 Remove the driveshaft retaining nut and recover the thrustwasher

10.25 Suspension strut-to-swivel hub retaining nuts

10.26 Remove the outer constant velocity joint from the hub

10.28 Withdraw the driveshaft assembly and recover the bearing water shield (A)

10.29 Enter the inner joint splines into the differential sun gear

on axle stands. Remove the roadwheel and return the gearlever to neutral.

23   Place a suitable container beneath the gearbox drain plug (see Section 9), undo the plug and allow the oil to drain. Refit the plug after draining.

24   Undo and remove the nut securing the tie-rod outer balljoint to the swivel hub steering arm. Release the balljoint from the steering arm using a balljoint separator tool.

25   Undo and remove the nuts and washers then withdraw the two bolts securing the suspension strut to the swivel hub (photo).

26   Pull the upper part of the swivel hub outwards as far as possible without placing undue strain on the flexible brake hose. Push the driveshaft inwards and manoeuvre the outer constant velocity joint from the hub (photo).

27   Using a suitable flat bar or large screwdriver, lever between the inner constant velocity joint and differential housing to release the joint from the differential sun gear.

28   Withdraw the inner joint fully from the differential then remove the

driveshaft assembly from under the wheel arch. Recover the plastic bearing water shield from the end of the outer constant velocity joint (photo).

29   To refit the driveshaft, place it in position under the car and enter the inner joint splines into the differential sun gear. Push the driveshaft firmly inward to engage the internal spring ring on the sun gear with the groove in the inner joint splines (photo).

30   Position the bearing water shield on the flange of the outer joint and fill the water shield groove with the specified grease.

31   Pull the swivel hub out at the top and enter the outer constant velocity joint into the hub. Refit the thrustwasher and driveshaft retaining nut and use the nut to draw the joint fully into place.

32   Refit the swivel hub to the suspension strut and secure with the two bolts, washers and nuts tightened to the specified torque.

33   Refit the steering tie-rod outer balljoint to the swivel hub arm. Screw on the nut and tighten it to the specified torque.

34   Refit the roadwheel then lower the car to the ground.

35   With the car standing on its wheels, tighten the wheel nuts to the

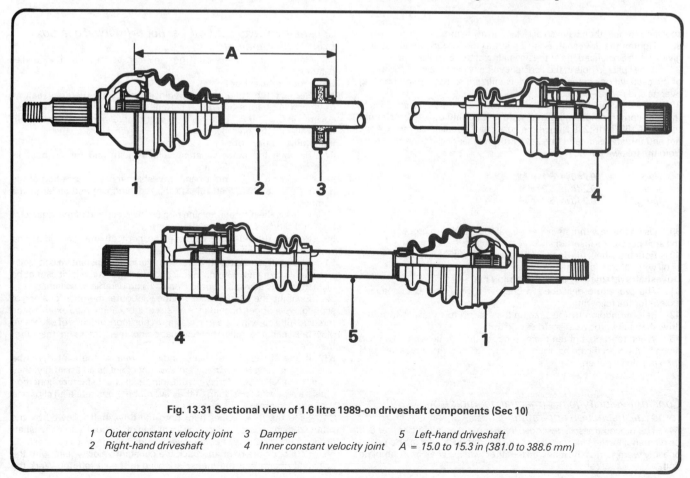

Fig. 13.31 Sectional view of 1.6 litre 1989-on driveshaft components (Sec 10)

1 Outer constant velocity joint	3 Damper	5 Left-hand driveshaft
2 Right-hand driveshaft	4 Inner constant velocity joint	A = 15.0 to 15.3 in (381.0 to 388.6 mm)

10.44A Constant velocity joint retaining circlip (arrowed) fitted to driveshaft groove

10.44B Fitting the outer constant velocity joint to the driveshaft

10.54A Fit the inner joint rubber boot ...

10.54B ... and secure its small end with the rubber retaining ring

10.55 Pack the joint with the special grease

specified torque, then have an assistant firmly depress the footbrake.

36   Tighten the driveshaft retaining nut to the torque wrench setting given in the Specifications at the beginning of this Supplement.

37   If the split pin hole in the constant velocity joint is aligned with one of the castellations in the nut, fit a new split pin and secure by bending over its ends.

38   If the split pin hole does not align with one of the castellations in the nut, continue tightening until it does, but only up to a maximum torque of 158 lbf ft (215 Nm). If the hole is still not aligned, remove the nut and thrustwasher and fit a new thrustwasher of different thickness from the following range

Silver	0.240 in (6.1 mm)
Green	0.244 in (6.2 mm)
Dark grey	0.248 in (6.3 mm)

39   Using the new thrustwasher, repeat the tightening procedure until the split pin can be inserted.

40   Refit the wheel trim, and refill the gearbox with the specified grade of oil with reference to the Specifications.

Driveshafts retained by staked retaining nut

41   The procedure described in Chapter 8, Section 3 is unchanged except for the following.

42   Before undoing the nut, tap up the staking to release it from the driveshaft joint groove. Discard the old nut.

43   When refitting, tighten the new nut to the torque wrench setting given in the Specifications, then tap its flange into the driveshaft joint groove using a punch.

Constant velocity joints (manual gearbox from 1989) – removal, inspection and refitting

44   The procedure is as described in Sections 4 and 5 of Chapter 8. If on dismantling, the inner constant velocity joint is found to be damaged or badly worn, it must be renewed together with the shaft as a complete unit.

Constant velocity joint rubber boots (manual gearbox from 1989) renewal

45   Remove the driveshaft from the car as described in earlier paragraphs.

Outer joint rubber boot

46   Remove the outer constant velocity joint from the driveshaft as described in Chapter 8, Section 5. Renewal of the rubber boot is also covered in Chapter 8, Section 5 as it is an integral part of the outer joint removal and refitting procedures.

Inner joint rubber boot

47   Remove the outer constant velocity joint and rubber boot as described in Chapter 8, Section 5.

48   If working on the right-hand driveshaft, mark the position of the damper on the shaft then release the retaining clip and slide off the damper.

49   Release the rubber retaining ring on the small end of the inner joint boot and slide the ring off the driveshaft.

50   Release the retaining clip, slip the boot off the inner joint and withdraw it from the driveshaft.

51   Clean out as much of the grease in the inner constant velocity joint as possible using a wooden spatula and old rags. As the joint cannot be dismantled it is advisable not to clean it using paraffin or solvents.

52   Examine the bearing tracks in the joint outer member for signs of scoring, wear ridges or evidence of lack of lubrication. Also examine the three bearing caps in the same way and check for evidence of excessive play between the roller bearing caps and their tracks in the outer member.

53   If any of the above checks indicate wear in the joint it will be necessary to renew the driveshaft and inner joint as an assembly; they are not available separately. If the joint is in a satisfactory condition obtain a repair kit consisting of the new rubber boot, retaining clips and the specified grease.

54   Slide the new rubber boot onto the driveshaft followed by the rubber retaining ring. Position the small end of the boot in the driveshaft groove then slip the rubber ring over it to secure it in place (photos).

55   Fold back the boot and pack the constant velocity joint with the specified quantity of the grease supplied in the kit (photo). Work the

10.57 Use a screwdriver to stretch the retaining clip slot over the tag

10.58 Tighten the clip fully by squeezing the raised portion

10.60 Secure the damper in place with the retaining clip

grease well into the joint while moving it from side to side. Fill the boot with any excess.

56   Position the large end of the boot over the joint outer member so that it locates squarely in the groove.

57   Position the retaining clip over the boot and engage one of the slots in the clip end over the small tag. Make sure the clip is as tight as possible and, if necessary, use a screwdriver to ease the slot over the tag (photo).

58   Fully tighten the clip by squeezing the raised portion with pincers (photo).

59   Refit the damper to the driveshaft (where applicable) and position it against the marks made on the shaft during removal. If a new driveshaft has been fitted or if the marks have been lost, use the setting dimension given in Fig. 13.31.

60   Secure the damper in position with the retaining clip as described in paragraphs 57 and 58 (photo).

61   Refit the outer constant velocity joint and rubber boot as described in Chapter 8, Section 5, then refit the driveshaft as described in Chapter 8, Section 3.

11   Braking system

Front brake disc checking – general

1   To avoid the possibility of brake judder being caused by a disc which is wearing unevenly, the front brake disc run-out should be checked whenever a disc is removed.

2   Before refitting the disc, check the thickness using a micrometer at four equidistant points around the disc, and at 0.4 in (10.0 mm) in from

the outer edge. If the thickness is outside the tolerance given in the Chapter 9 Specifications the disc must be renewed. Under no circumstances may the disc be refaced or machined in any way whilst removed from the vehicle. It may be possible to surface grind the disc in situ on the hub (not removed from it), but this is a task for a Rover dealer or suitably-equipped automotive engineer.

3   With the disc in place, use a dial test indicator to check the run-out with the probe positioned 0.25 in (6.3 mm) in from the outer edge. Slowly rotate the disc and check that the runout does not exceed the figure given in the Chapter 9 Specifications. If it does, remove the disc, turn it through 180°, refit and check the run-out again. Ensure that complete cleanliness of the disc and drive flange mating faces is maintained, and check that there are no burrs or irregularities on the surface which may affect the readings.

4   If the run-out still exceeds the specified figure, remove the disc and check the drive flange run-out. If this is excessive, renew the drive flange. If the flange run-out is satisfactory, renew the disc.

5   It should be noted that new replacement brake discs sometimes have a protective coating applied to combat corrosion during storage. Where this is the case, it is normal for the manufacturer to specify a suitable solvent with which to clean the lining contact faces of a replacement disc prior to fitting it.

6   Harsh braking during the running-in period of new discs will cause the coating to wear unevenly, leading to noise, judder and overheating of brake pads.

7   During the running-in period therefore, it is recommended that harsh braking is avoided, as far as possible.

Rear brake drum – inspection

8   Whenever a brake drum is removed, take the opportunity to inspect it for cracks or grooving caused by the shoes.

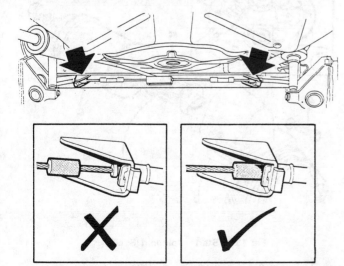

Fig. 13.32 Location and correct positioning of the handbrake cable seal (Sec 11)

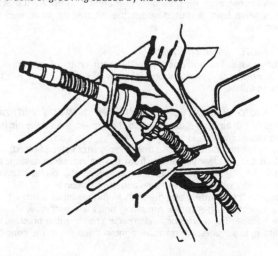

Fig. 13.33 Handbrake cable and body bracket (Sec 11)

1   Bracket hole

9   If these conditions are evident, renew the drum; do not have the drum ground or machined as the original diameter of the drum must not be increased.

Brake fluid level indicator – malfunction

10   Overfilling of the brake master cylinder reservoir or inversion of the filler cap during servicing can lead to contamination of the low level warning switch contacts giving rise to illumination of the warning light even when the reservoir is full.
11   If this problem should occur, thoroughly clean the outside of the cap and reservoir using lint-free cloth.
12   Disconnect the switch leads, unscrew the cap, keeping it the right way up and allowing any excess fluid to drain back into the reservoir.
13   Using a piece of cloth to prevent drips, transfer the cap to a clean bench.
14   Prise off the plastic cap and remove the screwed cap.
15   Invert the cap, allowing the float to drop and uncover the switch contacts, which should be cleaned using lint-free cloth.
16   Reassemble the cap, and refit to the reservoir in reverse order of removal.

Brake fluid – approved type

17   Brake hydraulic fluid to DOT 4 specification is now the specified brake system fluid. Austin Rover state that fluid to DOT 3 must no longer be used for topping up the fluid level.
18   DOT 4 has a higher boiling point than DOT 3, although it is completely compatible with, and can be used to top up systems containing, DOT 3 fluid.
19   Note that Duckhams Universal Brake and Clutch Fluid meets the requirements of a DOT 4 specification fluid.

Handbrake cable seal – displacement

20   The seal on each handbrake cable where the exposed portion of the inner cable enters the outer cable (see Fig. 13.32) is fitted to prevent dirt and water ingress.
21   During routine servicing operations and when a new cable is fitted, ensure that the seal is correctly fitted. The handbrake must be released (after chocking the wheels) while the check is carried out.
22   It is also possible for the cable to rattle against the sides of the hole in the bracket through which it passes. A rubber sleeve is available for fitting to prevent this problem.

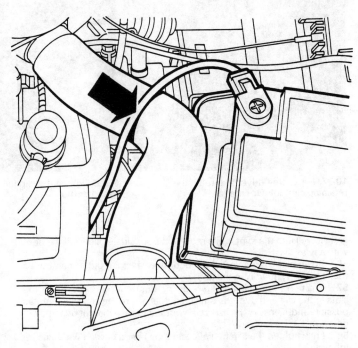

Fig. 13.34 Battery positive lead mounting on 1.6 models (Sec 12)

10   Disconnect the lead from the solenoid 'STA' terminal, extract the solenoid fixing screws and withdraw the solenoid. At the same time, disengage the solenoid plunger from the engaging lever.
11   Extract the screws and remove the nuts which hold the commutator end bracket. Withdraw the bracket from the yoke.
12   Remove the grommet from the yoke and ease the brush box off the commutator.
13   Remove the brush springs and unclip and remove the earth brushes.

12   Electrical system

Battery positive lead – routing

1   In order to eliminate the possibility of chafing, it is important that the battery positive lead is routed above the air cleaner hose, and not underneath it.

Starter motor (later models)

2   Later models are equipped with different types of pre-engaged starter motors (see Specifications). Removal, overhaul and refitting procedures follow.

Starter motor (Lucas type M78R) – removal and refitting

3   Remove the air intake trunking and disconnect the intake air temperature sensor multi-plug.
4   Disconnect the breather hose from the diverter valve (if fitted).
5   Unbolt the air cleaner bracket from the transmission casing, and move the air cleaner enough to gain access to the outlet trunking. Release the clip, remove the trunking and the air cleaner.
6   Disconnect the battery and the leads from the starter solenoid.
7   Unscrew the starter motor mounting bolts and withdraw the unit from the vehicle. The starter bolts also secure the snubber bracket.
8   Refitting is a reversal of removal; tighten the bolts to the specified torque.

Starter motor (Lucas type M78R) – overhaul

9   With the starter motor removed from the vehicle, clean away external dirt and grease.

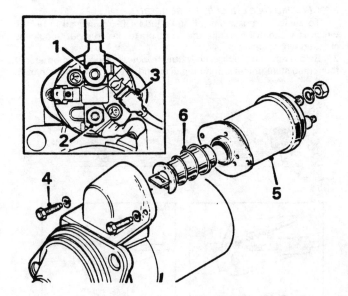

Fig. 13.35 Starter solenoid (Sec 12)

1 Battery (BAT) terminal	4 Retaining screws
2 Starter (STA) terminal	5 Solenoid
3 Solenoid (50) terminal	6 Solenoid plunger

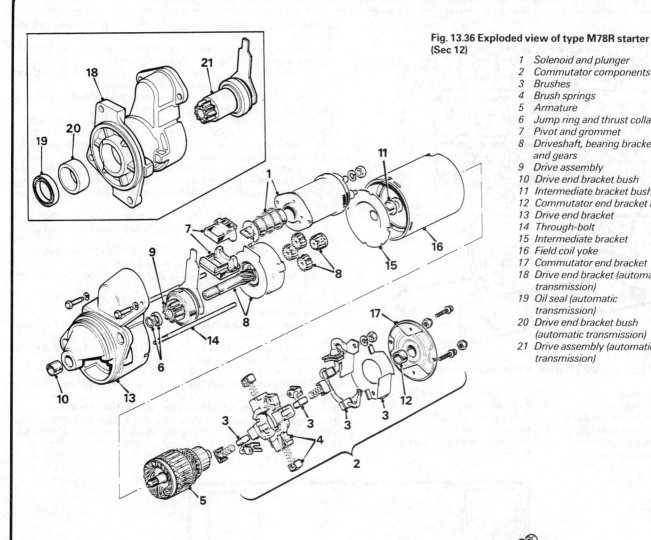

Fig. 13.36 Exploded view of type M78R starter (Sec 12)

1 Solenoid and plunger
2 Commutator components
3 Brushes
4 Brush springs
5 Armature
6 Jump ring and thrust collar
7 Pivot and grommet
8 Driveshaft, bearing bracket and gears
9 Drive assembly
10 Drive end bracket bush
11 Intermediate bracket bush
12 Commutator end bracket bush
13 Drive end bracket
14 Through-bolt
15 Intermediate bracket
16 Field coil yoke
17 Commutator end bracket
18 Drive end bracket (automatic transmission)
19 Oil seal (automatic transmission)
20 Drive end bracket bush (automatic transmission)
21 Drive assembly (automatic transmission)

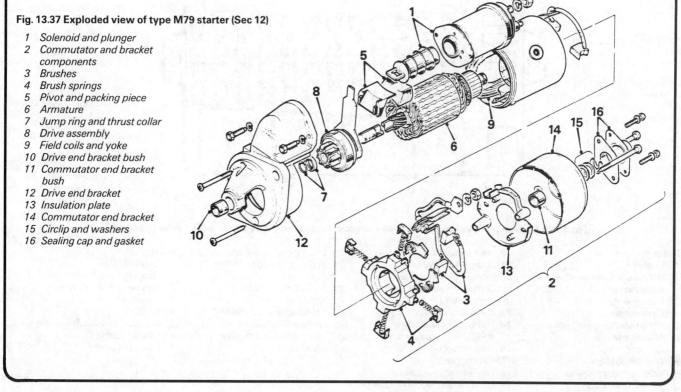

Fig. 13.37 Exploded view of type M79 starter (Sec 12)

1 Solenoid and plunger
2 Commutator and bracket components
3 Brushes
4 Brush springs
5 Pivot and packing piece
6 Armature
7 Jump ring and thrust collar
8 Drive assembly
9 Field coils and yoke
10 Drive end bracket bush
11 Commutator end bracket bush
12 Drive end bracket
13 Insulation plate
14 Commutator end bracket
15 Circlip and washers
16 Sealing cap and gasket

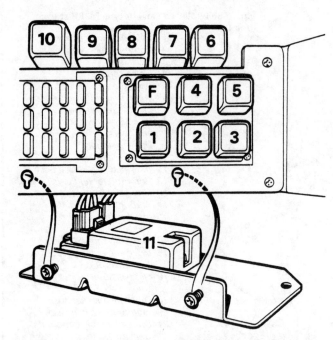

14   Remove the insulation plate and withdraw the brushes complete with bus bar.
15   If the brushes are worn below their specified length, renew them and make sure that they slide freely in their holders.
16   Remove the sun and planet gears and push out the drive and bearing bracket assembly from the drive end bracket.
17   Tap the thrust collar towards the pinion to expose the jump ring, which should be prised from its groove and removed.
18   Remove the collar and drive assembly from the armature shaft.
19   Clean the commutator with a fuel-moistened cloth. If it is badly discoloured polish it with fine glasspaper. Do not undercut the insulators between the segments.
20   This will normally be the limit of economical overhaul. Where the commutator is found to be deeply grooved or the armature shaft bushes are worn, then it is recommended that a new or factory-reconditioned starter is obtained.
21   Check the drive pinion assembly for wear or damage; also check that the pinion rotates in one direction only, independently of the clutch body.
22   Apply grease to the drive assembly pivot and lever, and fit the drive assembly and the thrust collar to the armature shaft.
23   Fit the jump ring into its groove and tap the thrust collar over the ring.
24   Insert the driveshaft into the drive end bracket bush.
25   Fit the planet and sun gears.
26   Locate the through-bolts, the intermediate bracket and yoke on the drive end bracket.
27   Fit the bus bar to the brush box then locate the brushes and the insulation plate. Insert the earth brushes and fit the clips.
28   Offer the assembly over the commutator and fit the brush springs.
29   Slide the armature and brush box inside the field coils and fit the yoke grommet.
30   Align the commutator end bracket and fit it into the yoke.
31   Engage the solenoid plunger with the engagement lever, fit the solenoid mounting screws then connect the lead to the 'STA' terminal.

Fig. 13.38 Location of relays on 1986 and later models (Sec 12)

1   Ignition
2   Headlamp
3   Auxiliary ignition 2
4   Starter solenoid
5   Auxiliary ignition 1
6   Heated rear screen
7   Carburettor manifold heater
8   Interior lamp
9   Windscreen wiper
10   Rear screen wiper
11   Windscreen wiper delay unit
F   Flasher unit

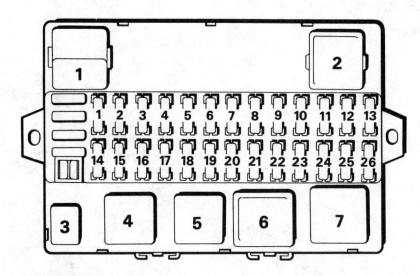

Fig. 13.39 Fuses and relays in fusebox, 1989-on models fuse rating is by colour code – refer to page 166

Fuses
1   Engine
2   RH side, tail and number plate lamps
3   Instruments
4   Cigar lighter
5   RH headlamp main beam
6   Electric mirrors
7   Sunroof
8   Dim-dip system
9   Stop-lamps, reversing lamps, indicators
10   Rear foglamps
11   Heated rear screen
12   Interior lamp, luggage compartment lamp, central door locking
13   Cooling fan
14   Heater blower motor
15   Windscreen wiper/washer
16   Tailgate wiper
17   Electric fuel pump
18   LH headlamp main beam
19   Horn, hazard warning lights
20   Electric front windows
21   Turbo carburettor cooling fan
22   Radio/cassette player
23   Spare
24   RH headlamp dip beam
25   LH headlamp dip beam
26   LH side, tail and number plate lamps

Relays
1   Courtesy lamp delay/lights-on alarm
2   Heated rear window timer
3   Headlamp
4   Ignition
5   Auxiliary cooling fan motor
6   Flasher relay
7   Dim-dip control

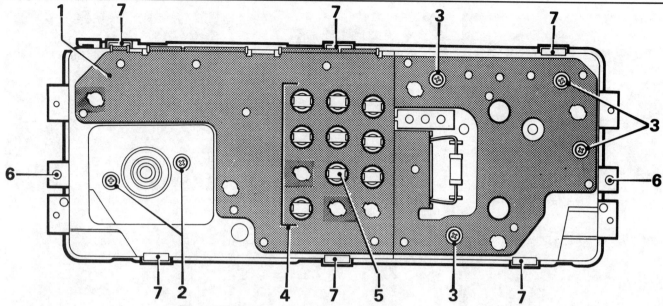

Fig. 13.40 Rear view of the three-gauge instrument panel (Sec 12)

1 Printed circuit
2 Speedometer securing screws
3 Fuel and temperature gauge securing screws
4 Warning light cluster
5 No-charge warning light bulb
6 Shroud retaining screws
7 Shroud retaining clips

Starter motor (Lucas type M79) – removal and refitting

32   The operations are similar to those described earlier for the type M78R, but refer to Chapter 3 for details of the air cleaner trunking and attachments used on 1.3 models. Note also that the engine mounting snubber bracket is held by the starter motor bolts only on 1.6 models.

Starter motor (Lucas type M79) – overhaul

33   Carry out the operations described earlier in paragraphs 9 and 10.
34   Remove the sealing cup and gasket from the commutator end bracket.
35   Extract the circlip and washers from the armature shaft.
36   Note the end bracket to yoke alignment marks, remove the retaining screws and carefully withdraw the end bracket from the yoke.
37   Remove the brush springs and withdraw the earth bushes, then lift the brush box from the commutator.
38   Remove the insulation plate and withdraw the positive brushes with bus bar.
39   If the brushes are worn below their specified length, renew them and make sure that they slide freely in their holders.
40   Remove the pivot and grommet from the drive end bracket.
41   Extract the screws and remove the yoke.
42   Withdraw the armature from the drive end bracket.
43   Tap the thrust collar towards the pinion to expose the jump ring, then prise the ring from its groove and remove it together with the collar and drive assembly.
44   Carry out the operations described in paragraphs 19 to 24 of this Section.
45   Locate the yoke on the drive end bracket and fit the retaining screws.
46   Place the bus bar on the brush box and fit the brushes and the insulation plate. Insert the earth brushes and fit the clips.
47   Offer the assembly over the commutator and fit the brush springs.
48   Slide the commutator end bracket over the brush springs and engage the grommet. Align the reference marks and fit the screws.
49   Fit the washers and circlip to the armature shaft, then fit the gasket and sealing cup.
50   Engage the solenoid plunger with the engagement lever and secure it with the screws. Make sure that the 'STA' terminal is adjacent to the starter motor. Fit the lead to the 'STA' terminal.

Fuses, relays and control units – later models

1986 to 1988 models

51   For these models, some of the relays and control units are relocated in a compartment below the steering column, the cover being secured by two turn buckles (Fig. 13.38).

52   In addition, the electronic control unit (ECU) for the fuel system is located behind the panel in the roof of the glovebox.
53   The ignition ECU is mounted on the left-hand valance within the engine compartment, and the fuel cut-off inertia switch, if fitted, is positioned behind and to the right of the radio aperture.

1989-on models

54   In addition to those relays mounted in the fusebox (see Fig. 13.39), the following relays and control units are also used.
55   The ERIC fuel and ignition ECU is mounted on the left-hand valance of the engine compartment.
56   The multi-function unit is also mounted on the left-hand valance of the engine compartment.
57   The electric window lift control unit is located behind a panel in the roof of the glovebox.
58   A delay unit, mounted on the fusebox in the glovebox, allows the electric windows, sunroof and radiator cooling fan to be operated for approximately 45 seconds after the ignition has been switched off.
Warning: *In the case of the cooling fan which is switched on by the thermal switch, it means that the fan can cut in and start running even when the ignition is switched off, so great care must be taken when working on a hot engine which has just been switched off.*
59   A timer unit, also mounted in the fuse box, controls the operation of the heated rear screen, limiting its use to ten minutes.

Heater fuse (all models, 1987-on) – higher rating

60   Internal modifications to the heater have resulted in the blower motor drawing more current and overloading the existing fuse, which can blow.
61   If this happens repeatedly, with no evidence of malfunction within the heater circuits, renew the existing 15 amp fuse with one of 20 amp rating. The fuse label should be amended accordingly for future reference.

Instrument panel switches (1989-on) – removal and refitting

62   Remove the screws from and tilt forward the instrument panel as described in later paragraphs.
63   Pull off the multi-plug from the switch (photo).
64   Depress the clips on the side of the switch and withdraw the switch from the instrument panel (photo).
65   The switch illumination bulb is housed in the multi-plug connector, and is a push fit in the holder.
66   To refit the switch, reconnect its wiring, then push the switch into the instrument panel until it is felt to clip securely into position.

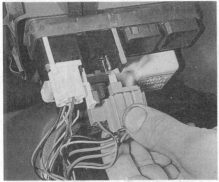

12.63 Pull the multi-plug from the switch

12.64 Depress the clips on the side of the switch and withdraw the switch

12.74 View of the rear of the digital clock with bezel withdrawn

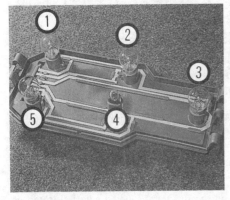

12.79 Bulb positions in rear lamp cluster
1   *Indicator light*
2   *Reversing light*
3   *Foglight*
4   *Side/parking light*
5   *Stop /tail light*

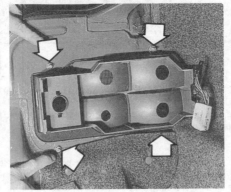

12.80 Rear lens unit securing nuts (arrowed)

12.82 Removing the rear stowage box

Instrument panel (1989-on) – removal and refitting

67   Although of different design to earlier models, the procedure for removing the instrument panel is as described in Chapter 10, Section 19.

68   There are two types of panel in use, one with three gauges and one with four. A view of the rear of each panel appears in Figs. 13.40 and 13.41.

69   Note also that progressively from 1988, instrument panels manufactured by Nippon Seiki are being used in place of those produced by Lucas.

70   The complete instrument panels are fully interchangeable, but individual instruments are not.

71   The manufacturer's name is clearly marked on the rear of the panels, and to identify them when they are in the fitted position Nippon Seiki types have seven-digit odometers and round trip reset buttons, while Lucas have six-digit odometers and rectangular reset buttons.

Digital clock (1989-on) – removal and refitting

72   Remove the instrument panel bezel as described in the previous sub-Section.

73   Disconnect the clock multi-plug.

74   Remove the clock retaining screws and withdraw the clock (photo).

75   Refit in reverse order.

Dim-dip headlamp system

76   In order to comply with current regulations, all post-October 1986 models are fitted with a headlamp dim-dip system.

77   The system used a relay-controlled resistor circuit, which is incor-

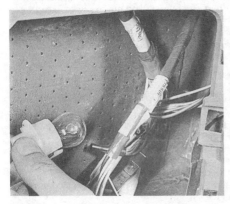

12.83 Removing a rear lamp cluster bulb

12.86A Rear number plate lamp

12.86B Rear number plate lamp bulb

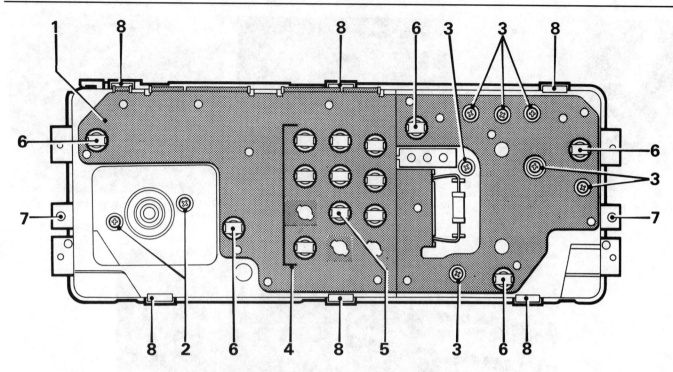

Fig. 13.41 Rear view of the four gauge instrument panel (Sec 12)

1   Printed circuit
2   Speedometer securing screws
3   Tachometer, fuel and temperature gauge securing screws
4   Warning light cluster
5   No-charge warning light bulb
6   Panel illumination light bulb
7   Shroud retaining screws
8   Shroud retaining clips

porated in a control unit mounted at the front of the vehicle, to the rear of the left-hand headlamp.

78   When the parking lamps are illuminated and the ignition is switched on, the headlamps come on automatically at one-sixth of their normal (dipped beam) power. This system prevents the vehicle being driven with only the parking lamps illuminated.

Rear lamp cluster (Saloon models, 1989-on)
Bulb renewal
79   The procedure is as described in Chapter 10, Section 25. The bulb positions are detailed in the accompanying photograph (photo).
Lens renewal
80   The lens is secured to the rear panel by four nuts, accessible after taking down the carpeting in the luggage compartment (photo).
81   Refit in reverse order, ensuring the foam seal is in good condition.

Rear lamp cluster bulbs (Estate models) – renewal
82   Remove the rear storage box cover, then undo the retaining screws and lift out the stowage box (photo).
83   Reach into the aperture and withdraw the relevant bulbholder by

turning it anti-clockwise (photo).
84   Press and turn the bulb to remove it from the holder.
85   Refitting is the reverse sequence to removal.

Number plate lamp bulb (Estate models) – renewal
86   Undo the two lens retaining screws and withdraw the lens and bulb (photos).
87   Withdraw the bulb from the contacts.
88   Refitting is the reverse sequence to removal.

Side repeater lamp – bulb renewal
89   Later models have side repeater lamps fitted to the front wings.
90   To renew a bulb, push the unit rearwards, then pull it out from the wing panel (photo).
91   Pull the lens unit from the bulbholder to gain access to the bulb, which is a push fit in the holder (photo).

Glovebox lamp – bulb renewal
92   Open the lid of the glovebox and gently prise the lamp from its

12.90 Removing the side repeater lamp from the wing

12.91 Pulling the lens unit from the bulb holder

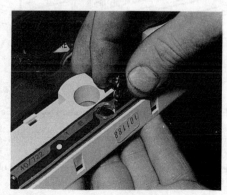

12.97 Removing a bulb from the heater control panel

12.102 Removing wiper blade from arm

12.104 Removing tailgate washer jet

12.111 Prising out a trim moulding retaining button

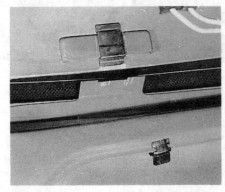

12.112 Tailgate trim moulding upper retaining tag

12.114 Tailgate wiper motor showing washer tube

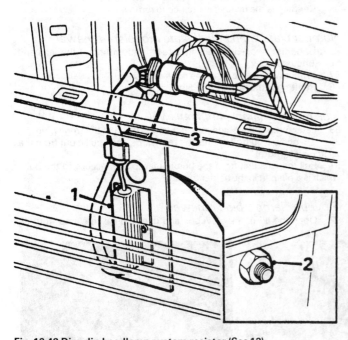

Fig. 13.42 Dim-dip headlamp system resistor (Sec 12)

1 Resistor   2 Mounting nut   3 Multi-plug

location. Use a small screwdriver to prise the bulb end cap out of its retaining clips.

Courtesy lamp switches and interior lamps

93   Courtesy lamp switches are incorporated in the door pillars and at

the tailgate body flange, or at the door locks, according to model type and year.
94   The lamp-mounted switch has three positions: on, off and door switch controlled.
95   A delay relay is fitted into the front door switch circuit to keep the interior light on for a short period after closing the door.

Heater control panel illumination bulb (1989-on) – renewal

96   Remove the heater control panel as described in Section 14.
97   The bulbs are a bayonet fix in the holder (photo).
98   Refit the panel in reverse order of removal.

Windscreen wiper arms – setting position

99   Operate the wipers and switch them off by means of the wiper switch, not the ignition key. Remove the wiper arms if they require adjustment.
100   When refitting the wiper arms to their spindles, position the arms so that they contact the intake moulding on the finisher below the windscreen. After tightening the retaining nut, lift the arm and locate it against the upper face of the stop peg.
101   Operate the wipers with a wet windscreen and check that they park with the arms contacting the stop pegs.

Tailgate wiper arm and blade (Estate models) – removal and refitting

102   To remove the wiper blade, lift the arm away from the window, release the spring retaining catch and separate the blade from the wiper arm (photo).
103   Insert the new blade into the arm, making sure that the spring retainer catch is correctly engaged.
104   To remove the wiper arm, spring back the hinged cover and withdraw the washer jet from the spindle (photo).
105   Unscrew the retaining nut, then carefully prise the arm off the spindle.
106   Before refitting the arm, switch on the ignition, then turn the

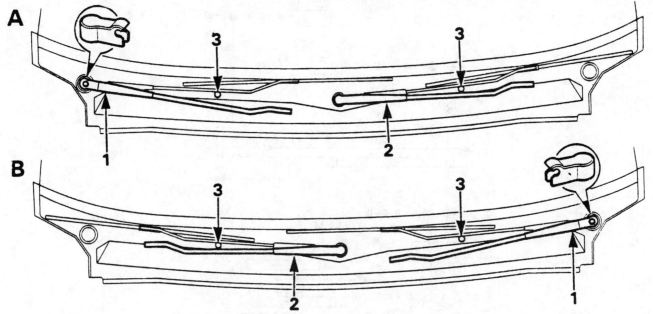

Fig. 13.43 Windscreen wiper arm setting diagram (Sec 12)

A Right-hand steering	1 Arm setting point –	2 Arm setting point –	3 Parked position against
B Left-hand steering	driver's side	passenger's side	stop pegs

wiper on and off, allowing the motor to return to the 'Park' position. Switch off the ignition.

107   Refit the arm to the spindle with the blade positioned just below the lower heater element in the window.

108   Refit and tighten the retaining nut, then refit the washer jet so that the jets are positioned on either side of the arm.

109   Close the hinged cover.

Tailgate wiper motor (Estate models) – removal and refitting

110   Remove the tailgate wiper arm and blade, as described previously.

111   Open the tailgate and release the interior trim moulding by prising out the retaining buttons (photo). Some of these will probably break during removal, so new buttons should be obtained before refitting.

112   Release the trim from the upper retaining tags (photo), disconnect the speaker leads (where fitted) and remove the trim.

113   Disconnect the wiring multi-plug and tailgate washer hose at the wiper motor.

114   Undo the wiper mounting bracket retaining bolts and the self-tapping screw from the support plate (photo). Withdraw the motor and mounting bracket from the tailgate.

115   With the assembly on the bench undo the bolts and separate the wiper motor from the mounting bracket.

116   Refitting is the reverse sequence to removal.

Automatic rear wipe in reverse gear (Estate models) – modification

117   Automatic rear wipe when reverse gear is selected can be obtained by fitting a standard 4-pin relay into the reversing light circuit, joining together the red/light green wires (see Fig. 13.45).

118   This will provide intermittent wipe when reverse is selected, irrespective of the position of the facia panel rear wipe switch.

119   By connecting the common terminal of the relay to the light green wire of the facia rear wipe switch (dotted line in Fig. 13.45), continuous rear wipe can be obtained whenever reverse is selected.

120   In both cases, the relay feed should be connected to the reversing light circuit.

Continuous rear wipe (Estate models) – modification

121   The provision of continuous rear wipe can be obtained by fitting an of/off switch to connect the two red/light green wires (pin numbers 1

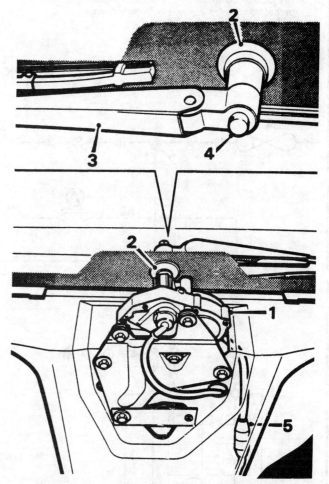

Fig. 13.44 Tailgate wiper – Estate models (Sec 12)

1 Wiper motor	4 Washer jet
2 Grommet	5 Wiring connector
3 Wiper arm	

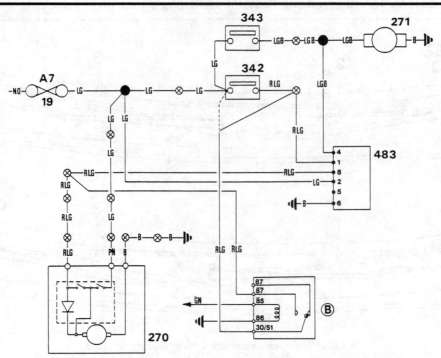

Fig. 13.45 Wiring diagram – automatic rear wipe modification (Sec 12)

B   Relay (30/51 – common
    terminal, 85 relay feed)
19 Fusebox

270 Wiper motor
271 Washer pump

342 Wiper switch
343 Washer switch
483 Programmed wash/wipe unit

For colour code, refer to the first page of
wiring diagrams at the end of the manual

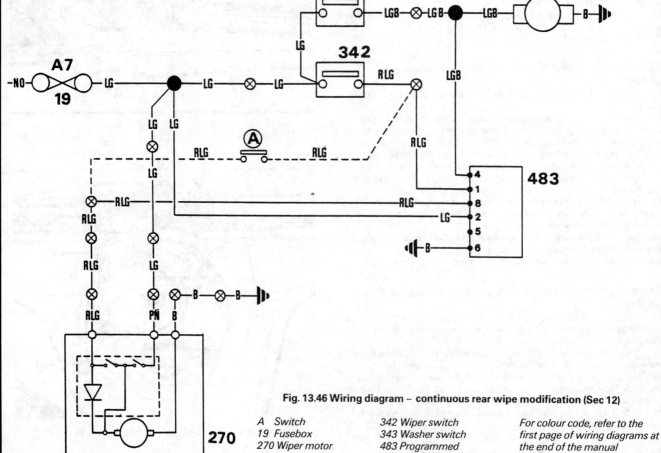

Fig. 13.46 Wiring diagram – continuous rear wipe modification (Sec 12)

A   Switch
19 Fusebox
270 Wiper motor
271 Washer pump

342 Wiper switch
343 Washer switch
483 Programmed
    wash/wipe unit

For colour code, refer to the
first page of wiring diagrams at
the end of the manual

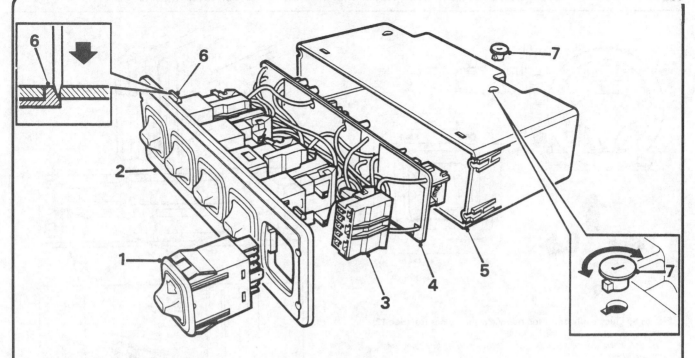

Fig. 13.47 Electric window lift control unit on 1986-88 models (Sec 12)

1 Switch
2 Escutcheon
3 Multi-plug
4 Printed circuit board
5 Housing
6 Escutcheon plate retainers
7 Printed circuit board retainers

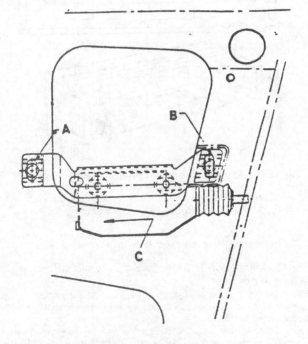

Fig. 13.48 Door lock solenoid (Sec 12)

A Front fixing screw
B Rear fixing screw
C Forward setting of unit

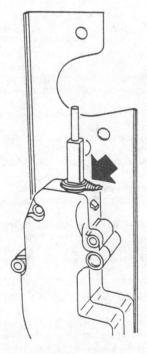

Fig. 13.49 Soft iron wire (arrowed) wrapped around groove of motor casing (Sec 12)

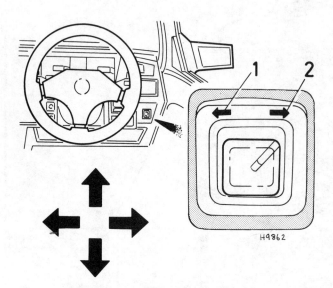

Fig. 13.50 Electrically operated rear view mirror control (Sec 12)

1 *Left-hand mirror selector position*   2 *Right-hand mirror selector position*

and 8) on the rear wiper programme unit.
122   The connections are shown by the dotted line in Fig. 13.46, the switch being marked 'A'.
123   With the switch in the open position, normal programmed wipe with delay is available.
124   With the switch closed, continuous wipe is available whenever the facia panel rear wipe switch is operated.

Horns
125   As from February 1986, twin horns are fitted as standard to all models.

Electric window lift control unit and switches
1986 to 1988 models
126   The control unit and switches are located in a housing on the facia panel.
127   Removal and refitting is as described for the electronic radio/cassette player fitted to some models, described later in this Section.
1989-on models
128   The control unit is located behind a panel in the roof of the glovebox.
129   The switches are located in the centre console.

Central door lock switch/solenoid – modified fitting procedure
130   More precise fitting instructions are required to ensure correct operation of the central door lock unit than are contained in Chapter 10, Section 46.
131   Locate the assembly in the door so that the front fixing screw is located in the upper half of its slot and the control unit pushed towards the front of the vehicle as far as it will go. Tighten the screw.
132   Tighten the rear fixing screw, again with it within the upper half of its slot.

Tailgate lock solenoid (Estate models) – removal and refitting
133   To remove the tailgate lock solenoid on models equipped with central locking, open the tailgate and remove the interior trim moulding with reference to previous paragraphs 111 and 112.
134   Disconnect the operating rod from the cylinder lock lever.
...   the nut securing the solenoid mounting plate to the lock
...   the wiring multi-plug and position the mounting plate
...   perture. Undo the retaining screws and withdraw the
...   ehind the mounting plate.

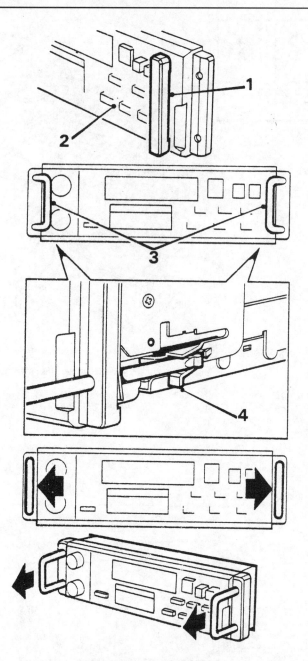

Fig. 13.51 Electronic tune radio/cassette player removal (Sec 12)

1 *Blanking plate*   3 *Removal tools*
2 *Radio/cassette player*   4 *Retaining clip*

137   Refitting is the reverse sequence to removal.

Boot/tailgate central door locking solenoid (all models) – failure to operate
138   Should the boot/tailgate central door locking solenoid fail to operate, this may be due to 'bounce-back' off the mechanism and can be corrected as follows:
139   Remove the boot/tailgate central locking solenoid and motor assembly as described in Chapter 10, Section 10, Section 47 for the boot, and in this Section for the tailgate.
140   Ensure the slots through which the nylon plunger, which carries the lock bolt, slides are free from manufacturing butts.
141   Using soft iron wire, wrap one turn around the groove of the motor casing as shown in Fig. 13.49, twisting the ends of the wire together using flat-nosed pliers, until **slight** resistance of the plunger is felt.

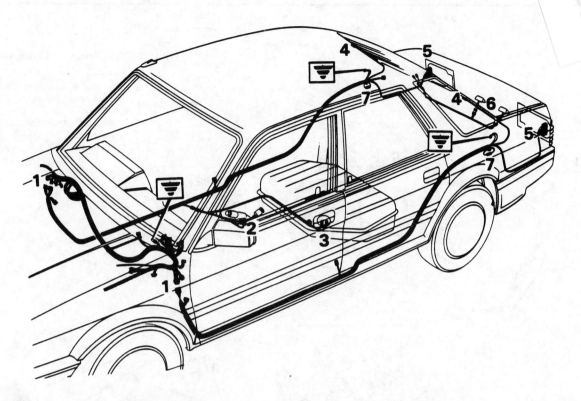

Fig. 13.52 Wiring harness – body, fuel contents unit and trailer (Sec 12)

1 Main harness connectors
2 Handbrake switch and rear cigar lighter
3 Fuel tank sender unit
4 Heated rear screen connectors
5 Rear lamps
6 Rear number plate lamps
7 Trailer connector

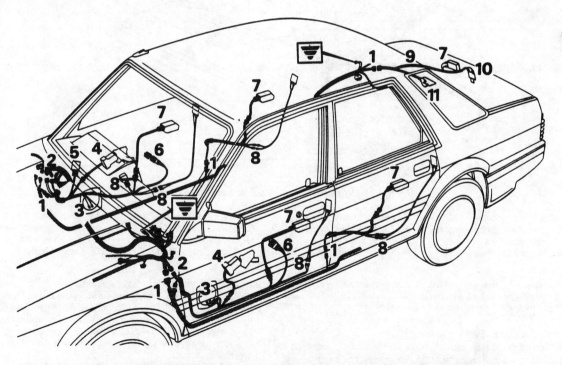

Fig. 13.53 Wiring harness – electric windows, and central door locking (Sec 12)

1 Connectors
2 Connectors
3 Front speakers
4 Window motors
5 'One touch' window lift unit
6 Electric window switch
7 Central door lock solenoid
8 Relay
9 Boot harness
10 Boot lamp switch
11 Boot lamp

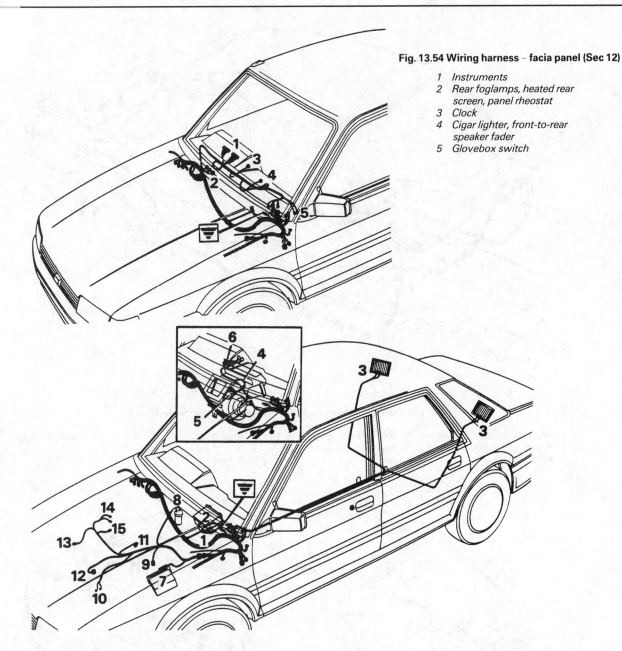

Fig. 13.54 Wiring harness – facia panel (Sec 12)

1   Instruments
2   Rear foglamps, heated rear
    screen, panel rheostat
3   Clock
4   Cigar lighter, front-to-rear
    speaker fader
5   Glovebox switch

Fig. 13.55 Wiring harness – engine, heater, radio (Sec 12)

1 Radio connector	5 Demister solenoid	9 Speed transducer (1.6)	13 Alternator
2 Main harness connectors	6 Heater blower switch	10 Starter motor	14 Stepper motor and fuel
3 Rear speakers	7 Ignition ECU	11 Crankshaft sensor (1.6)	cut-off valve
4 Heated harness	8 Ignition coil	12 Knock sensor (1.6)	15 Coolant sensor

142   Cut off any surplus wire, and ensure that the remaining wire end cannot foul any part of the lock mechanism.
143   Refit the lock solenoid and motor assembly.

Electrically operated rear view mirror
144   Later top of the range models have electrically operated and demisted exterior rear view mirrors. The heating elements are automatically energised as soon as the ignition is switched on.
145   Removal is similar to the procedure described in Section 14 for the manually operated type, except that the wiring harness must be disconnected as the mirror is withdrawn.
146   If the glass is to be removed, operate the control switch until the top of the glass is tilted as far as possible into the casing.
147   Insert a forked tool into the gap at the centre of the lower edge of

the glass and release the glass retainers.
148   Again operate the control switch until the bottom of the glass is tilted as far as possible into the casing. Release the glass upper retainers.
149   Disconnect the wiring connectors and withdraw the glass.
150   Before refitting the glass, centralise the control switch (ignition on) and connect the wiring. Check that the insulated connector is uppermost.
151   Align the pegs on the retaining ring, and then press the mirror glass firmly and evenly to engage the retaining clips. It is recommended that protective gloves are used during this work.

Radio/cassette player – removal and refitting
152   In addition to the plate retaining design described in Chapter 10, Section 41, some in-car entertainment units may be retained by clips.

Fig. 13.56 Wiring harness – engine, steering column, battery, fuses (Sec 12)

1 Main harness
2 Ignition/starter switch connector
3 Steering column switch harness
4 Fusebox and relays
5 Fuel ECU wiper control unit, interior lamp
6 Stop-lamp switch
7 Wiper motor connectors
8 Brake fluid level switch, manifold heater
9 Washer pump
10 Battery lead with fusible links
11 Horn connectors
12 Front lamp connectors
13 Coolant fan connectors
14 Air temperature sensor
15 Oil pressure switch
16 Courtesy lamp harness
17 Disc pad wear sensor

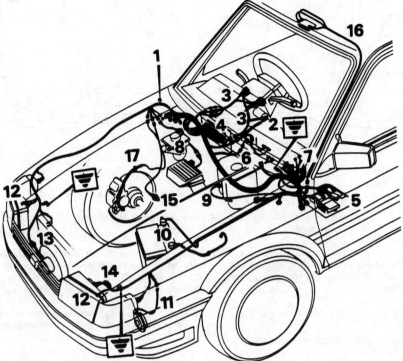

Fig. 13.57 Wiring harness – tailgate (Sec 12)

1 Main harness junction
2 Lamp header
3 Right-hand light cluster
4 Tailgate harness junction
5 Heated rear window
6 Right-hand number plate lamp
7 Central door locking switch
8 Left-hand number plate lamp
9 Central door locking motor
10 Wiper motor
11 Luggage compartment lamp
12 Tailgate speaker harness junction
13 Heated rear window
14 Left-hand light cluster
15 Rear earthing point
16 Left-hand speaker
17 Right-hand speaker

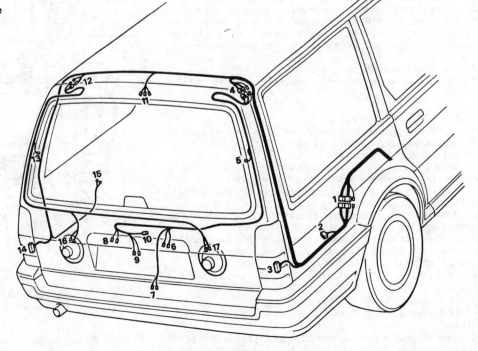

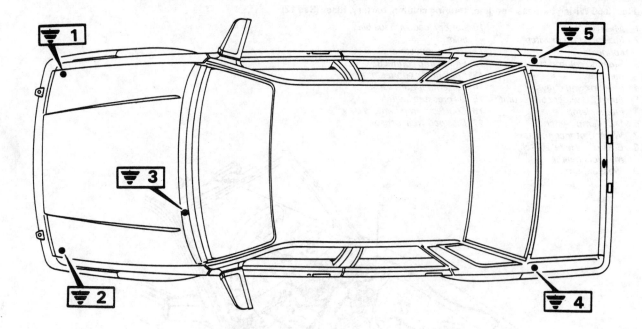

Fig. 13.58 Earthing points on 1986 and later models (Sec 12)

1 Front right-hand inner wing
2 Front left-hand inner wing

3 Bulkhead, top left behind
   facia panel

4 Rear left-hand wing behind
   trim panel

5 Rear right-hand wing behind
   trim panel

153   To remove this type of unit, remove the knobs, bezels and finisher, and then press the centre of the exposed clips inwards and withdraw the unit, with masking plate. Disconnect the wiring plugs and leads. Refitting is a reversal of removal.

Radio/cassette player (electronic tune type) – removal and refitting

Warning: *Before removing an electronic radio/cassette player, ensure the security code (if so equipped) is de-activated.*

154   Using a small screwdriver, prise the blanking plates from each side of the unit.
155   Bend a piece of welding rod of stiff wire to form two U-shaped tools and insert them into the two holes at each end of the unit panel.
156   Push the tools fully in and incline them outwards to compress the retaining clips, then pull the radio/cassette unit from the console using the tools as handles. Pull smoothly and gently to avoid disconnecting the tools.
157   Disconnect the wiring plugs and leads.
158   To refit, simply press the unit into position until the clips are heard to engage.

Cassette holder – removal

159   The cassette holder, which is located below the radio/cassette player, can be removed after the facia panel has first been withdrawn as described in Chapter 12, Section 35.
160   Should a cassette tray fail to open when pushed in, it may be possible to release it using one of the following methods.
161   Insert a thin blade along the upper edge of the tray and attempt to ease the tray downwards.
162   If the tray still fails to release, eject the tray which is operating correctly, pinch its sides and withdraw it.
163   Locate the holes at the back of the cassette holder which service the jammed tray, and press upwards. At the same time depress the release pad to free the catch. Withdraw the tray.
164   Failure to engage the tray guides on their rails will cause jamming of a tray.

Radio aerial (windscreen pillar mounted) – removal and refitting

165   Carefully prise up the cover trim at the base of the aerial and undo the baseplate retaining screws. Note the screening braid location on one of the screws.

166   From inside the car, disconnect the aerial lead at the rear of the radio and withdraw the lead from the console and facia.
167   Tie a length of string to the end of the aerial lead as an aid to refitting, and withdraw the aerial and lead upwards and out of the windscreen pillar.
168   Refitting is the reverse sequence to removal. Use the string to pull the aerial lead down the pillar and through behind the facia.

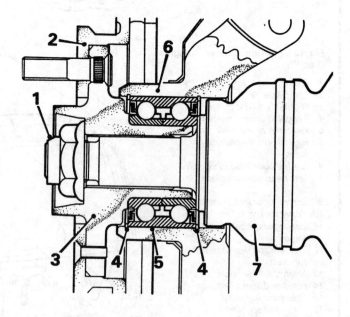

Fig. 13.59 Sectional view of front hub with one-piece bearing (Sec 13)

1 Hub nut
2 Brake disc
3 Drive flange
4 Circlip

5 Bearing
6 Hub
7 Driveshaft

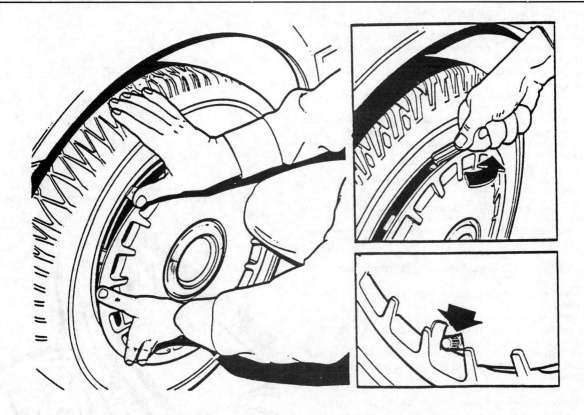

Fig. 13.60 Removal and refitting of later type wheel trims (Sec 13)

Wiring harness – 1986-on
169 Typical wiring harness layouts are shown in 13.52 to 13.57.

Earthing points – 1986-on
170 It is very important that the vehicle earthing points are kept rust and corrosion free, and the fixing which secures the various circuit earth wires at each of the five locations is kept tight.

---

13 Suspension and steering

Front suspension – knocking
1 If a loud knock from the front suspension results when a front wheel goes sharply to the full rebound position, the noise can be minimised by fitting modified replacement cup and washer assemblies to the upper end of the suspension strut (see Fig. 11.1, Chapter 11).
2 These components are available from Austin Rover parts dealers.

Camber angle (later models)
3 Later models have had the front suspension camber angle changed from negative to positive by moving the suspension strut lower fixing bolt hole inward by 0.04 in (1.0 mm). On earlier models with negative camber, the bolt holes were in line.
4 This modification does not alter the servicing procedures given in Chapter 11, but for the purposes of the Specifications in this Supplement, the camber angle is grouped under the heading 'strut, type B', to avoid confusion.

Front hub bearings (one-piece type) – description and renewal
5 One-piece front hub bearings were introduced on later models. Early versions of the one-piece bearings used ball-bearings, while later versions use taper-roller bearings, and are the preferred type, although both types are fully interchangeable.
6 Note that neither type of one-piece bearing is interchangeable with the early two-piece bearings. The hub bearing (driveshaft) nut torque wrench setting given in the Specifications section of Chapter 11 remains the same for all types.
7 Apply the handbrake and remove the front roadwheel.
8 Have an assistant apply the footbrake hard. Unscrew the driveshaft-to-hub nut.
9 Unbolt the brake caliper and tie it up out of the way. There is no need to disconnect the hydraulic hose.
10 Remove the two screws which secure the brake disc to the hub flange. Remove the disc.
11 Extract the screws and remove both halves of the disc shield.
12 Unscrew the nut from the tie-rod end balljoint taper pin, and using a suitable splitter tool disconnect the balljoint from the steering arm.
13 Unscrew and remove the two bolts which hold the base of the suspension strut to the hub carrier.
14 Unscrew the nut from the suspension track control arm balljoint, and using a lever prise the arm downwards to separate the arm from the hub carrier.
15 Remove the hub carrier assembly.
16 A press will now be required to press the drive flange from the hub. Alternatively, a long bolt and nut with distance pieces can be used to draw it out, but on no account attempt to hammer it out. The bearing inner track and ballrace (outer section) will come away with the drive flange, and can be removed with a puller.
17 Extract the two circlips and press the bearing from the hub. Never attempt to re-use the old bearing.
18 Press the bearing into the hub until it contacts the circlip, then fit the second circlip.
19 Support the bearing inner track and press the drive flange into the hub.
20 Refitting the hub carrier is a reversal of removal, but observe the following points.
21 Tighten all nuts and bolts to the specified torque.
22 Fit a new driveshaft nut, tighten to the specified torque and stake the nut into the driveshaft groove.
23 Once the brake caliper has been refitted, apply the footbrake hard two or three times.

Wheel trims (1989-on) – removal and refitting
24 The wheel trims fitted to 1989-on models can be difficult to

remove and refit, due to their more positive fit tolerance. The following procedure should be used.

25   To remove a wheel trim, insert the special tool supplied with the vehicle between the trim and the wheel rim, and using a twisting action, lever the trim outwards.

26   Repeat the operation in several positions around the rim until the trim is released.

27   To refit the trim, position the wheel with the inflation valve at its lowest point.

28   Offer the trim to the wheel, with the large cut-out over the valve stem, and apply equal pressure to both sides of the wheel trim.

29   Apply pressure to the top of the trim until the trim is fully engaged with the wheel.

Steering wheel (1989-on, all models) – removal and refitting

30   The procedure is as described in Chapter 11, Section 16, but the steering wheel is secured in place by a bolt, and not a nut.

31   Note also the bolt has a different torque wrench setting (see Specifications).

Steering column (1989-on, all models) – removal and refitting

32   The procedure is as described in Chapter 11, Section 17 with the following differences.

33   After lifting out the inner columns, the top and bottom bushes must be tapped out from the outer column using a suitable drift.

34   Note also that there is a spacer under the top bush.

35   When fitting new bushes, apply Loctite 424 to their outer surfaces, and note that the chamfered edge of the bushes faces towards the centre of the outer column.

Power-assisted steering – description and maintenance

36   As from 1985, power-assisted steering became optionally available on 1.6 litre models.

37   Check the fluid level only when the system is cold and the engine switched off.

38   Wipe the filler cap clean and then unscrew and remove it.

39   Clean the dipstick with non-fluffy cloth or tissue, then refit the cap. Remove the cap once more and check that the fluid mark is between the MIN and MAX marks on the dipstick.

40   Top up if necessary with the specified fluid to the MAX mark.

41   The tension of the power steering pump drivebelt must be checked regularly. The belt should deflect by 0.5 in (12.5 mm) when moderate thumb pressure is applied to the belt mid-way along its longest run.

42   To adjust the belt tension, first release the fuel pump inlet hose and the expansion tank hose from their clips.

43   Remove the timing belt upper cover.

44   Loosen the high pressure hydraulic pipeline nut and the pump adjustment and mounting bolts.

45   Pull the pump away from the engine by gripping the mounting bracket to apply tension to the belt. Tighten the bolts when the correct tension is achieved. Do not attempt to tension the belt by levering on the reservoir.

46   On completion, tighten the hydraulic pipeline and bleed the system as described later.

47   To renew the belt, follow the same procedure, but push the pump in towards the engine as far as it will go in order to slip the belt from the pulleys.

Power-assisted steering pump – removal and refitting

48   Disconnect the cooling system expansion tank hose and fuel pump inlet hose from their clips and remove the timing belt upper cover.

49   Remove the pump drivebelt.

50   Place a container under the pump and then disconnect the pressure hose union from the pump. Seal the hose and the pump.

51   Remove the pump mounting and adjustment bolts and nuts, and disconnect the return hose from the reservoir pipe. Seal the hose and pipe.

52   Remove the pump and mounting bracket.

53   Unbolt and remove the pulley.

54   Mark the relationship of the pump to its mounting bracket for reference purposes when refitting, and separate the pump and bracket.

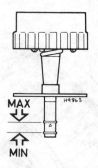

Fig. 13.61 Power steering fluid reservoir dipstick (Sec 13)

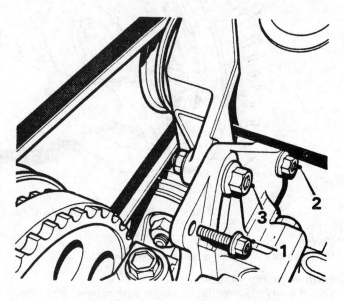

Fig. 13.62 Power steering pump fixings (Sec 13)

1   Camshaft belt upper cover bolt

2   Pump mounting bolt

3   Adjuster bolt (belt tension)

55   A faulty pump should be renewed, not overhauled.

56   Refitting is a reversal of removal, but note the following points.

57   The pump pressure hose should be set at 30° towards the right-hand side of the vehicle.

58   Apply thread locking fluid to the threads of the mounting bracket bolts.

59   Tension the drivebelt, and bleed the system, as described elsewhere in this Section.

Power-assisted steering gear – removal and refitting

60   Disconnect the battery.

61   Unscrew and remove the coupling pinch-bolt from the bottom of the steering column shaft.

62   Slacken the front roadwheel nuts, raise the front of the vehicle and remove the roadwheels.

63   Unscrew the tie-rod end balljoint nuts and, using a suitable balljoint splitter tool, disconnect the tie-rod end balljoints from the steering arms.

64   Remove the suspension crossmember support bracket from the driver's side.

65   Disconnect the fluid hoses from the rack pipe. Lockwashers may be fitted, in which case bend their tabs flat. Anticipate the loss of some fluid.

66   Unscrew the steering rack mounting bolts and manoeuvre the rack assembly out from under the front wheel arch on the driver's side.

67   Worn or faulty power steering gear should be renewed, not

overhauled, due to the need for special tools and the difficulty of obtaining individual spare parts.

68   Refitting is a reversal of removal, but observe the following points.

69   With the front roadwheels and steering wheel in the straight-ahead position, turn the rack pinion until the tie-rod end balljoints will just drop into the eyes of the steering arms when the rack housing is correctly located for bolting into position. When this situation is achieved, the pinion should be aligned with the steering shaft coupling so that the pinion groove and coupling holes are also in line to accept the pinch-bolt.

70   Tighten all nuts and bolts to the specified torque, except the crossmember support bracket, which should be tightened after the vehicle is lowered to the floor.

71   Bleed the system as described in the following paragraphs.

Power-assisted steering – bleeding

72   The system will require bleeding if the system hoses or pipe lines have been uncoupled and reconnected or new components fitted.

73   Make sure that the fluid reservoir is topped up to the MAX mark. Start the engine and turn the steering slowly several times from lock to lock.

74   Top up the fluid as necessary and repeat.

75   Make a final check of the fluid level when the system has stood inoperative for several hours and is cold.

Manual steering gear – centralising

76   On models built from 1986, the rack centralising hole (Chapter 11, Section 22, paragraph 13) is no longer incorporated.

77   When refitting this later rack and pinion steering gear, refer to paragraph 69 of this Section for details of rack centralising method.

Front wheel alignment – special precautions

78   It is very important that, when checking and adjusting the front wheel alignment (toe), as described in Chapter 11, Section 24, the following special conditions are observed.

79   If a measuring gauge is being used, then the vehicle must have been pushed in a forward direction for at least three metres with the front roadwheels in the straight-ahead position before carrying out any checks.

80   If scuff plates are being used, then the rear roadwheels must be standing on packing pieces of a thickness equivalent to that of the scuff plates.

81   Make sure that the front and rear trim heights are as specified on both sides of the vehicle. If necessary, load or jack one side of the vehicle to achieve similar trim heights on both sides.

Rear axle – modification

82   From VIN 168414, the pivot bushes in the rear axle trailing arms are fitted the other way round, with the flange shoulder towards the outside. When renewing pivot bushes, ensure that they are fitted accordingly (refer to item 2, Fig. 11.2).

Self-levelling rear suspension – description and testing

83   This is optionally available on most models in the Montego range. The system reacts to vehicle loading and automatically maintains the specified vehicle trim heights.

84   The self-levelling units are sealed struts fitted in place of the normal rear suspension units. A pump in the damper operates under the action of the suspension to raise the rear of the car until normal trim height is regained. On an undulating road, this process will be carried out within one mile. When the additional load is removed, the suspension remains at the correct level.

85   To test the operation of the self-levelling struts, measure and record the rear suspension trim height on each side with the vehicle at kerb weight. The trim height is the distance from the centre of the hub to the lower edge of the wheel arch. Compare the figures obtained with

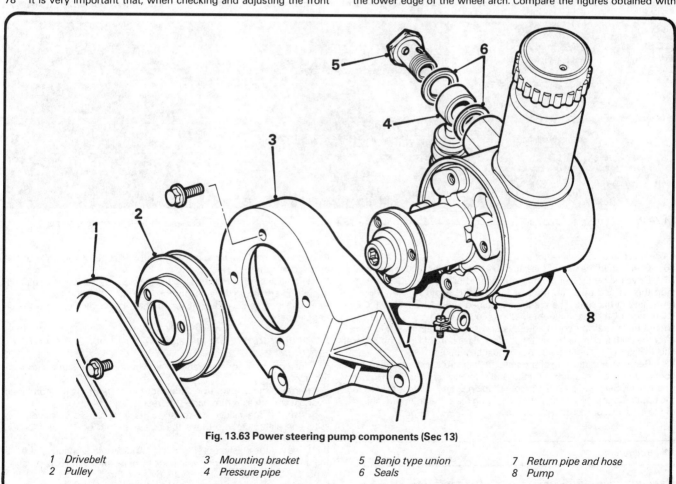

Fig. 13.63 Power steering pump components (Sec 13)

1 Drivebelt	3 Mounting bracket	5 Banjo type union	7 Return pipe and hose
2 Pulley	4 Pressure pipe	6 Seals	8 Pump

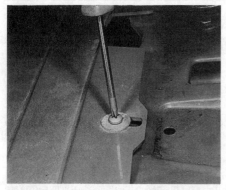

14.2 Remove the screws securing the upper edge of the grille to the cross panel

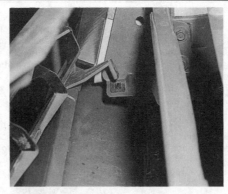

14.3 Lift the grille upward to disengage the clips

14.5A Prise out the speaker grille ...

14.5B ... and undo the four door trim screws

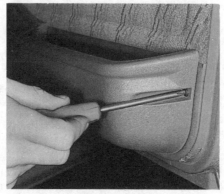

14.6A Remove the screws from the door bin ...

14.6B ... then lift off the door bin and panel

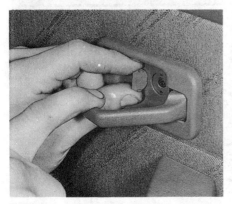

14.7A Prise out the plastic cover ...

14.7B ... remove the screw ...

14.7C ... and lift off the escutcheon plate

those given in the Specifications of Chapter 11.

86   Evenly load the rear of the car with 440 to 550 lbs (200 to 250 kg).

87   Drive the car over undulating roads for at least 3 miles (5 km), then stop the car and measure the rear trim heights again with the driver remaining in his seat. The readings obtained should be within 0.4 in (10.0 mm) of the readings obtained previously. Both sides of the car should have risen by equal amounts and should retain their position after the car has stopped. A unit that is low, does not hold its position, or shows signs of excessive oil leakage must be renewed.

88   Removal and refitting of the self-levelling struts is the same as for conventional rear suspension units, and reference should be made to Chapter 11, Sections 13 and 14. Note that, on Estate models, access to the upper mounting is gained after removing the panel from the middle of the luggage compartment side trim.

14 Bodywork

Radiator grille (1989-on) – removal and refitting

1   Open the bonnet.

2   Remove the screws securing the upper edge of the grille to the cross panel (photo).

3   Lift the grille upward to disengage the clips at each lower, outer end, and withdraw the grille panel (photo).

4   Refit in reverse order, ensuring that the lower clips are in full engagement.

Front door trim panel (all models, 1989-on) – removal and refitting

5   Prise out the door speaker grille and undo the four screws under it (photos).

6   Remove the screws from the door bin then lift off the combined door bin and speaker trim panel (photos).

7   Prise out the plastic cover from the interior release handle escutcheon plate screw, remove the screw and lift off the escutcheon plate (photos).

8   Prise out the plastic cover from the window winder handle, remove the nut and lift off the handle and backing plate (photos). This operation is not applicable where electric windows are fitted.

9   Remove the screws from, and lift off, the armrest (photo).

10   Carefully, using a flat-bladed tool and starting from the lower edge,

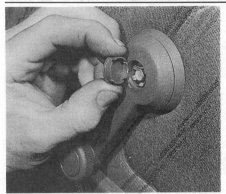

14.8A Prise out the plastic cover ...

14.8B ... remove the nut ...

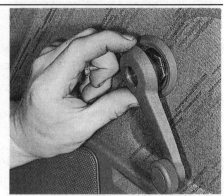

14.8C ... lift off the handle ...

14.8D ... and backing plate

14.9 Remove the screws from the armrest

14.10 Remove the trim panel

14.14A Remove the screws securing the capping to the door ...

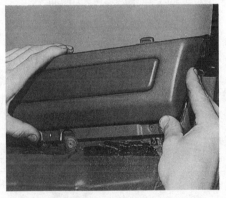

14.14B ... and lift off the capping upward

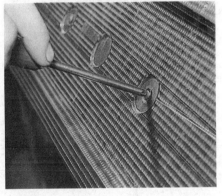

14.19 Remove the two self-tapping screws under the number plate

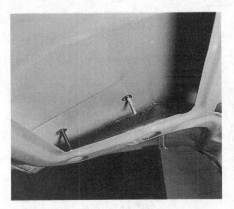

14.20 Remove the nuts from the threaded spigots

14.21 Withdraw the appliqué panel

prise out the plastic 'screws' securing the trim panel to the door, then lift off the panel (photo).
11   Refit in reverse order.

Front door capping strip (1989-on) – removal and refitting
12   Remove the door trim panel as described earlier.
13   Where fitted, prise out and disconnect the door 'tweeter' speaker.
14   Remove the screws securing the capping to the door, and lift off the capping upwards to disengage the seal (photos).
15   Refit in reverse order.

Rear door trim panel (1989-on) – removal and refitting
16   The procedure is as described for the front doors, but there is no speaker grille.

Rear door capping strip (1989-on) – removal and refitting
17   The procedure is as described for the front doors, but there is no 'tweeter' speaker.

Appliqué panel (Saloon models, 1989-on) – removal and refitting
18   Prise off the rear number plate which is secured by double-sided tape.
19   Remove the two self-tapping screws under the number plate (photo).
20   Open the boot lid and remove the nuts from the threaded spigots at each end of the appliqué panel, inside the double skin (photo).
21   Withdraw the appliqué panel (photo).
22   Refitting is a reversal of removal, but use new double-sided tape, which is available from your dealer. (The use of normal double-sided tape, such as would be used for carpets, is not recommended, and would result in the number plate dropping off).

Headlining (Saloon models, 1989-on) – removal and refitting
23   The procedure is similar to that described for Estate models, except that the front seats must be moved fully rearwards and reclined,

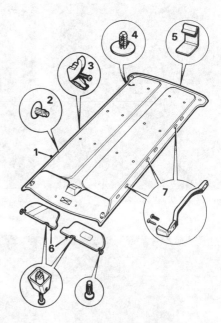

Fig. 13.64 Headlining and associated components (Sec 14)

1 Headlining	5 Edge clip
2 Fixing plug	6 Sunvisors and fixings
3 Coat hook	7 Grab handle
4 Support button	

and the headlining lowered from the front to release the rear end from the plastic trim strip.
24   Manoeuvre the headlining from the vehicle through one of the front doors.

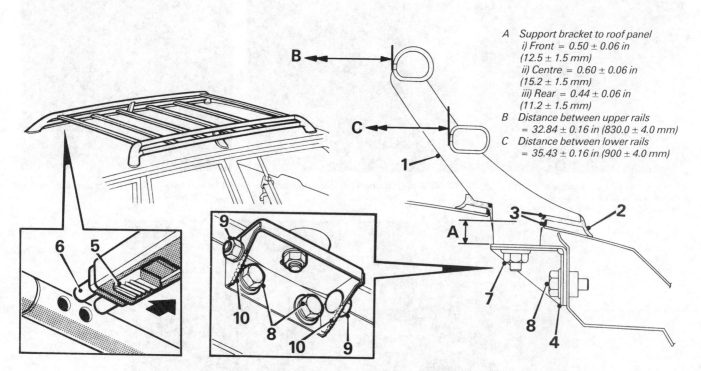

A   Support bracket to roof panel
    i) Front = 0.50 ± 0.06 in
    (12.5 ± 1.5 mm)
    ii) Centre = 0.60 ± 0.06 in
    (15.2 ± 1.5 mm)
    iii) Rear = 0.44 ± 0.06 in
    (11.2 ± 1.5 mm)
B   Distance between upper rails
    = 32.84 ± 0.16 in (830.0 ± 4.0 mm)
C   Distance between lower rails
    = 35.43 ± 0.16 in (900 ± 4.0 mm)

Fig. 13.65 Integral roof rack fittings and dimensions (Sec 14)

1 Post and side rails	5 Spring-loaded button	9 Support bracket nuts
2 Gasket	6 Locating pins	10 Tack welds
3 Sealant	7 Side rail retaining nut	
4 Mounting bracket	8 Height adjustment screw	

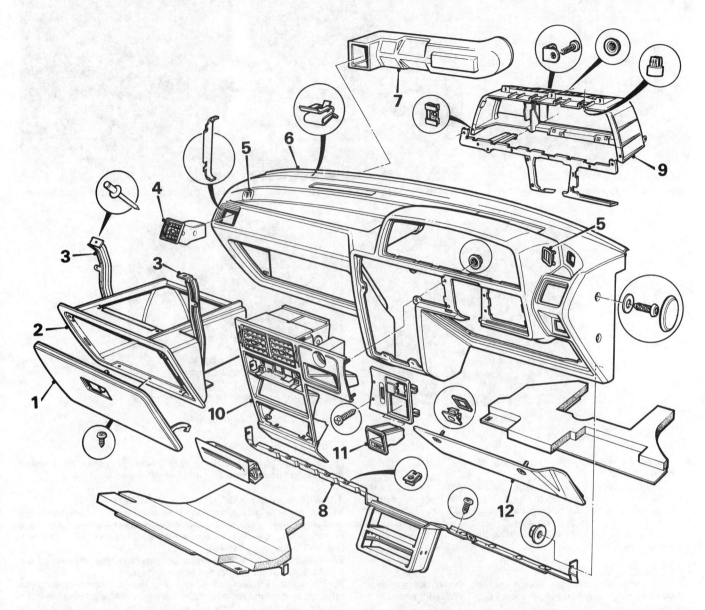

Fig. 13.66 Facia panel on 1989-on models (Sec 14)

1 Glovebox lid	4 Outer vent	7 Cross duct	10 Radio and centre vent
2 Glovebox	5 Side window demister	8 Lower stiffening rail	housing
3 Glovebox supports	escutcheon	9 Instrument carrier	11 Coin tray
	6 Facia panel		12 Fusebox cover

Headlining (Estate models) – removal and refitting

25   Disconnect the battery.

26   If fitted, fold down the rear-facing seats, then remove the head restraints from the remaining seats.

27   Release the door seal from the front 'A' pillars.

28   Remove the sunvisors and their brackets, the interior and load space lamps, and the front and rear grab handles.

29   Carefully release the plastic screws and buttons securing the headlining to the roof panel and the rear quarter.

30   Remove the rear edge clips then lower the lining and carefully manoeuvre it out of the open tailgate.

31   Refit in reverse order.

Integral roof rack (Estate models) – removal and refitting

32   Remove the headlining as described earlier.

33   Depress the spring-loaded clips at the ends of the roof rack cross-rails and remove the cross-rails.

34   Remove the side-rail securing nuts and carefully ease the side rails

and gaskets from the roof. Clean all sealant from the holes in the roof and the side-rail posts.

Caution: *It is essential that correct and accurate refitting of the roof rack is achieved to prevent the cross-rails becoming detached in use, to avoid water leaks and prevent damage to the roof.*

35   Apply sealant (Expandite SR 51 or equivalent) to the base of each rack post, then use an adhesive (Dunlop S 758 or equivalent) to stick new gaskets to the posts.

36   Apply sealant around each roof panel hole.

37   Check the distance between the roof panel and each support bracket measuring at the centre-line of each hole. Dimensions are given in Fig. 13.65.

38   To adjust the brackets, loosen the bracket securing nuts and move the bracket up or down as necessary, tightening the nuts on completion.

39   Fit the side-rail assemblies to the roof, then fit and tighten the securing nuts only finger-tight.

40   Measure the distance between the inside faces of the rails which must be as shown in Fig. 13.65.

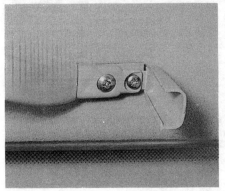

14.58 Grab handle retaining screw and cover

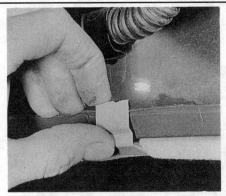

14.59 Removing a headlining retaining clip

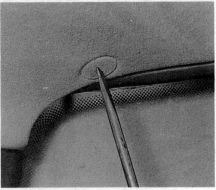

14.60 Removing a headlining retaining button

14.66A Tailgate strut upper balljoint

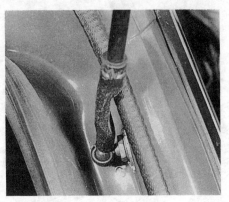

14.66B Tailgate strut lower balljoint

41   Adjust as necessary, then tighten the securing nuts to 16 lbf ft (22 Nm).

42   If the correct dimension cannot be obtained it may be necessary to grind down the track welds on the front edge of the front and centre brackets, loosen the nuts and reset the bracket positions. Tighten and tack-weld the nuts on completion.

43   Check for water leaks before refitting the headlining, which is a reverse of removal.

Front seats – loose backrest

44   On some high-mileage 1989-on models, the backrest-to-seat squab bolts may loosen and allow excess play between the backrest and the seat squab.

45   This problem can be overcome by fitting hexagon-headed bolts in place of the Torx type bolts (available from Austin Rover dealers) and tightening them to a torque of 29 to 40 lbf ft (40 to 50 Nm).

Door seals – inadequate retention

46   If the door seals tend to come adrift from the top of the door frame, they can be stuck more firmly in position by using double-sided tape, available from Austin Rover dealers.

Centre console (1989-on) – removal and refitting

47   Disconnect the battery.

48   Remove the gear lever knob on manual transmission models.

49   Prise out the panel at the rear of the console and remove the console retaining screws under it.

50   Prise out the handbrake cover panel and remove the remaining screws securing the console to the floorpan.

51   Lift the console slightly and disconnect the plugs to the window lift switches, and the gear selector illumination bulb on automatic transmission models.

52   Lift out the console.

53   Refit in reverse order.

Facia (1989-on) – removal and refitting

54   The procedure is basically as described in Chapter 12, Section 35.

55   Refer to the relevant Sections of this Supplement for removal of the radio and heater controls, which differ from earlier models.

Tailgate (Estate models) – removal and refitting

56   Open the tailgate and release the interior trim moulding by prising out the retaining buttons. Use two screwdrivers or a forked tool for this operation, but note that the buttons are easily broken and some new ones will probably be needed.

57   Release the trim from the upper retaining tags, disconnect the speaker leads (where fitted) and remove the trim.

58   Spring back the plastic covers over the grab handles, undo the retaining screws and remove the handles from the headlining (photo).

59   Peel back the rubber weatherstrip around the upper edge of the tailgate aperture, then release the clips securing the headlining (photo).

60   Unscrew the headlining retaining buttons as necessary (photo), and lower the headlining sufficiently to gain access to the tailgate hinge retaining nuts.

61   Disconnect the wiring multi-plugs and cable ties at all the tailgate electrical components. Tape the multi-plugs to the wiring harness, then tie drawstrings to each harness end. Similarly, disconnect the washer hose and tie a drawstring to its end.

62   Release the grommets in the tailgate and pull the wiring and washer hose out of the tailgate until the drawstrings emerge. Untie the strings, but leave them in place.

63   Have an assistant support the tailgate, then release the support strut upper balljoints, prising off the retaining clips with a screwdriver.

64   Mark the hinge outlines on the body, then reach through the lowered headlining and undo the hinge retaining nuts. Remove the tailgate from the car.

65   Refitting is the reverse sequence to removal. Use the drawstrings to pull the wiring and washer hose through into the tailgate. Adjust the position of the hinges to achieve correct alignment and ease of closure.

Tailgate support struts (Estate models) – removal and refitting

66   Support the tailgate in the open position and prise out the strut upper and lower balljoint retaining clips using a screwdriver (photos).

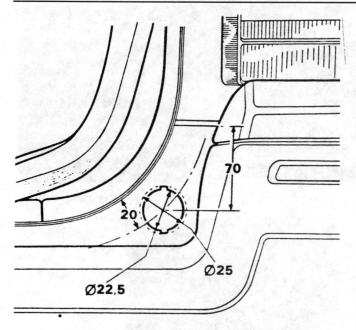

Fig. 13.67 Tailgate lift assister fitting diagram (Sec 14)
All dimensions in mm

67   Withdraw the support strut from the car.
68   Refitting is the reverse sequence to removal.

Tailgate lift assisters (Estate models) – fitting

69   Should the tailgate open only as far as the safety catch when the remote release control is operated, a pair of lift assisters (available from your dealer) may be fitted in the following way.
70   Drill a hole as shown in Fig. 13.67 on each side of the luggage area. The diameter of the holes is critical.
71   De-burr the edges of the holes, and paint.
72   Apply underbody protective wax to the assemblies and the edges of the holes.
73   Fit the assisters, screwing them to their minimum length before closing the tailgate for the first time. Unscrew each assister half a turn and check the lift of the tailgate. Unscrew the assisters in half-turn increments until the tailgate operation is satisfactory.

Tailgate lock (Estate models) – removal and refitting

74   Remove the tailgate interior trim moulding, as described in paragraphs 56 and 57 of this Section.
75   Release the retaining clip and disconnect the operating rod from the lock lever.
76   Undo the two retaining bolts and remove the lock from the tailgate (photo).
77   Refitting is the reverse sequence to removal. If necessary, adjust the striker plate position after refitting the lock, so that the tailgate shuts and locks without slamming.

Tailgate lock cylinder (Estate models) – removal and refitting

78   Remove the tailgate interior trim moulding, as described in paragraphs 56 and 57 of this Section.
79   Release the retaining clip and disconnect the operating rod from the cylinder lock lever (photo).
80   Undo the retaining nuts, remove the lock lever and withdraw the lock cylinder from the number plate lamp housing.
81   Refitting is the reverse sequence to removal.

Tailgate release cable and lever (Estate models) – removal and refitting

82   Fold back the rear seat cushion and the luggage compartment carpet.
83   Remove the front and rear sill carpet retainers on the driver's side.

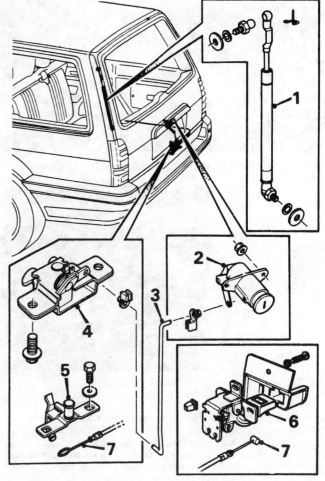

Fig. 13.68 Tailgate lock and support strut components (Sec 14)

1 Strut	5 Striker
2 Lock cylinder	6 Remote release lever
3 Operating rod	7 Remote release cable
4 Latch	

84   Slacken all the driver's seat retaining bolts and remove the bolts from the runner nearest the door.
85   Remove the oddments tray.
86   Undo the release lever retaining screws, lift up the trim moulding and disconnect the release cable. Remove the release lever from the car.
87   Fold back the side of the front and rear carpets, disconnect the cable from the retaining clips and pull the cable into the luggage compartment.
88   Undo the screws and withdraw the tailgate aperture lower trim moulding (photo).
89   Undo the striker plate retaining bolts (photo). Lift up the striker plate and disconnect the cable from the release lever. Withdraw the cable from the car.
90   Refitting is the reverse sequence to removal, but adjust the striker plate position so that the tailgate shuts and locks without slamming.

Rear bumper (Estate models) – removal and refitting

91   Undo the rear wheel arch liner retaining screws and remove the wheel arch liners.
92   Undo the nut securing each bumper support bracket to the rear valance.
93   From inside the car, fold the rear panel of the luggage compartment floor and remove the floor covering.
94   Undo the two bolts securing each bumper rear support bracket to the body. Lift away the rear bumper assembly.
95   Refitting is the reverse sequence to removal.

14.76 Tailgate lock and retaining bolts

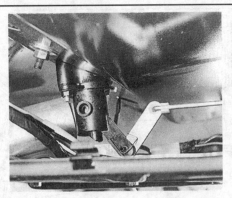

14.79 Tailgate lock, lock lever and fixing nuts

14.88 Removing tailgate aperture lower trim moulding

14.89 Tailgate lock striker

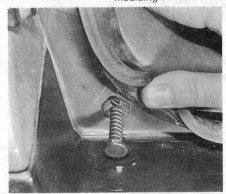

14.102 Tailgate lower trim mouldings and weatherseal

Rear seat (Estate models) – removal and refitting

96   With the rear doors open, fold the cushion or cushions forward, undo the retaining screws and withdraw the seat cushion(s).
97   Remove the outer hinge bracket covers, release the seat squab retaining latches and lower the squab. Undo the outer bracket retaining screws.
98   If a one-piece squab is fitted, remove the squab. If a split squab is fitted, undo the retaining screw and extract the clip securing the centre pivots. Withdraw the relevant squab.
99   Refitting is the reverse sequence to removal.

Luggage area side trim moulding (Estate models) – removal and refitting

100   If fitted, fold the rear facing seat, fold the rear seat cushion, lower the squab and remove the squab release handle.
101   Detach the rear door weatherstrip seal from along the side trim moulding.
102   Open the tailgate, and remove the guide plate and the three screws securing the tailgate aperture lower trim moulding on the side being worked on (photo).
103   Undo the upper and lower rear seat belt mounting bolts, and detach the guide escutcheon from the side moulding.
104   Undo the upper and lower rear facing seat belt mounting bolts, detach the guide escutcheon from the side moulding and release the seat belt support strut at its bottom end.
105   Release the fastener securing the moulding to the rear of the wheel arch. Using a forked tool or suitable alternative, carefully release the four clips securing each end of the panel.
106   Pull the sill edge retaining clips from the body brackets and lift the moulding to clear the four bottom edge retainers from the body slots.
107   Disengage the support strut, pass the seat belt webbing through the side trim moulding and remove the moulding.
108   Refitting is the reverse sequence to removal.

Rear seat belts (Saloon models) – removal and refitting
Outer belt
109   Remove the rear seat cushion.
110   Undo the seat belt lower mounting and rear squab extension retaining bolt.

111   Push the squab extension upwards and remove it from the car.
112   Extract the seat belt upper guide bracket plastic cover, then undo the bolt and remove the guide and spacers.
113   Release the door weatherstrip seal from the rear body pillar and remove the finisher trim.
114   Release the parcel shelf trim extension plastic retaining clip and withdraw the extension.
115   Detach the shelf trim finisher from the seat belt, and remove the shelf trim extension by sliding it along the belt.
116   From within the boot, fold back the front of the boot carpet, undo the seat belt locking buckle anchorage bolts, and remove the locking buckle.
117   Refitting is the reverse sequence to removal.
Centre belt
118   Remove the rear seat cushion, then, from inside the boot, fold back the front of the boot carpet.
119   Undo the seat belt centre belt anchorage bolts and remove the belts from the car.
120   Refitting is the reverse sequence to removal.

Rear seat belts (Estate models) – removal and refitting

121   Remove the luggage area side trim moulding, as described earlier in this Section.
122   Remove the seat belt guide bracket plastic cover, then undo the retaining bolt and withdraw the belt guide and spacers.
123   Release the belt guide escutcheon from the side moulding, open the escutcheon and slide out the webbing. Pass the belt through the side moulding.
124   Undo the reel retaining bolt and remove the belt assembly.
125   Undo the seat belt locking buckle retaining bolt and remove the locking buckle. Note the position of the rear facing seat belt where these are fitted.
126   Refitting is the reverse sequence to removal.
127   To remove the centre belt, fold the rear facing seat (if fitted) and the rear seat, and remove the backrest(s) as necessary.
128   Undo the locking buckle and anchor bracket retaining bolts, and remove the complete centre seat belt.
129   Refitting is the reverse sequence to removal.

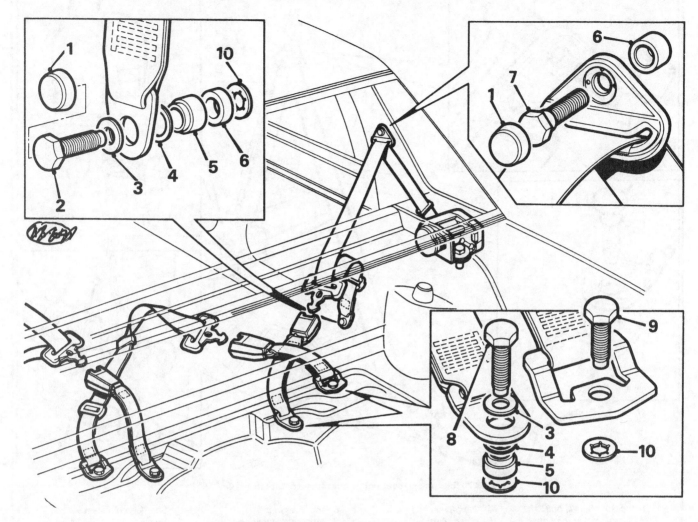

Fig. 13.69 Rear seat belt anchorage points – Saloon (Sec 14)

1 Plastic cap	4 Anti-rattle washer	7 Upper mounting bolt	9 Centre belt bolt
2 Lower mounting bolt	5 Shouldered distance piece	8 Locking buckle bolt	10 Retaining washer (transit)
3 Plain washer	6 Spacer		

Rear facing seat belts (Estate models) – removal and refitting

130 Fold the rear seat and the rear facing seat, then remove the rear seat backrest.

131 Remove the upper guide bracket plastic cover, then undo the retaining bolt.

132 Remove the plastic cover, then undo the lower anchorage retaining bolt.

133 Release the belt guide escutcheon from the side moulding, open the escutcheon and slide it off the belt.

134 Remove the relevant grab handle and the rearmost headlining retaining buttons.

135 Release the tailgate weatherstrip seal from the upper edge of the aperture, then extract the headlining retaining clips.

136 Undo the bolt at the top of the support strut, then support the headlining by refitting the retaining clips.

137 Remove the luggage area side trim moulding, as described earlier in this Section.

138 Undo the seat belt support strut lower retaining bolt and the belt reel retaining bolt from inside the side panel.

139 Pass the seat belt through the side moulding and remove the belt assembly.

140 Undo the locking buckle anchorage bolt and remove the locking buckle.

141 Refitting is a reversal of removal.

Remote control door mirror – removal, refitting and glass renewal

142 Remove the door trim panel, as described in Chapter 12.

143 Slide the plastic cover from the mirror control lever.

144 Remove the internal escutcheon panel and extract the exposed mirror fixing screws.

145 Remove the retaining plate and take off the mirror assembly.

146 Refitting is a reversal of removal.

147 If the mirror glass is to be renewed, this can be done without the need to remove the mirror from the door.

148 Wear a protective glove and push the outboard end of the mirror towards the front of the car, until the mirror casing begins to move away from the glass. Insert the fingers between the glass and the casing and prise the glass gently. When the gap is wide enough, insert a thin screwdriver to release the socket from the pivot ball.

149 Disconnect the tension spring and remove the glass.

150 To fit the glass, first apply silicone grease to the socket, then fit the glass to the retaining arm and connect the spring.

151 Gently apply even pressure to push the glass into position, engaging the socket with the ball.

Fuel filler flap remote control lever and cable – removal and refitting

152 On Saloon models, remove the rear seat cushion and peel back the carpet from the right-hand side of the luggage compartment.

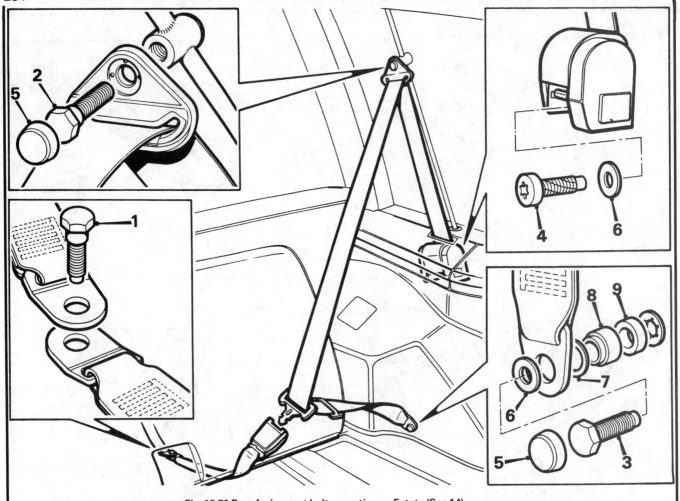

Fig. 13.70 Rear-facing seat belt mountings – Estate (Sec 14)

1	Locking buckle bolt	4	Inertia reel bolt	6	Plain washer	8	Shouldered distance piece
2	Upper mounting bolt	5	Plastic cap	7	Anti-rattle washer	9	Spacer
3	Lower mounting bolt						

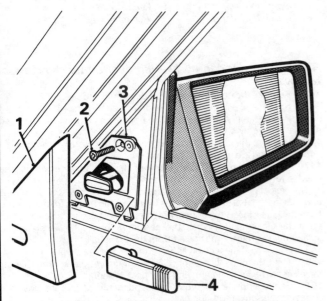

Fig. 13.71 Remote control door mirror (Sec 14)

1	Escutcheon panel	3	Retaining plate
2	Screw	4	Control lever plastic cover

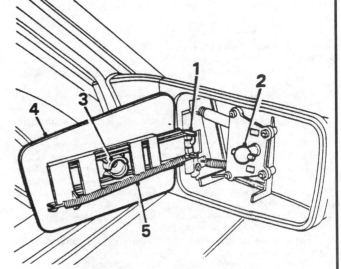

Fig. 13.72 Door mirror attachment (Sec 14)

1	Mirror arm	4	Glass
2	Ball	5	Tension spring
3	Socket		

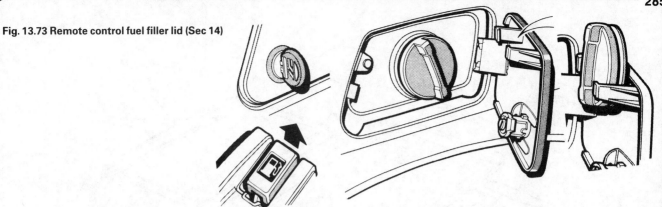

Fig. 13.73 Remote control fuel filler lid (Sec 14)

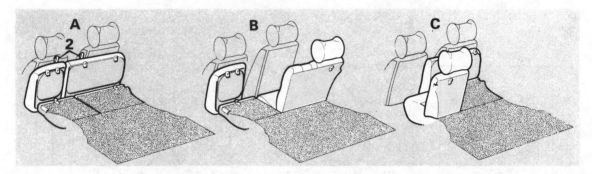

Fig. 13.74 Asymmetrical split rear seat – Estate (Sec 14)

A   Maximum load area
B   Luggage plus two
    passengers
C   Luggage plus one passenger
2   Seat loops

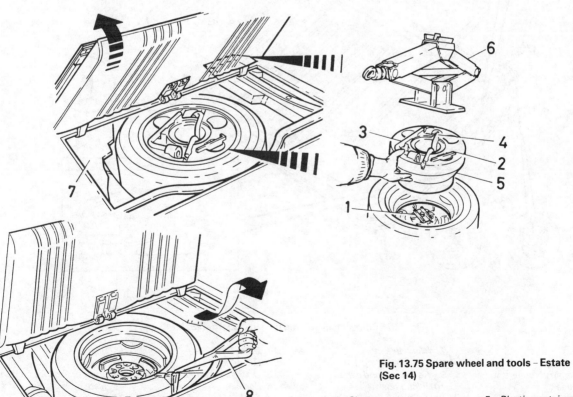

Fig. 13.75 Spare wheel and tools – Estate
(Sec 14)

1   Clamp
2   Wheel trim tool
3   Wheel brace
4   Jack handle
5   Plastic container
6   Jack
7   Carpet
8   Strap (wheel withdrawal)

286

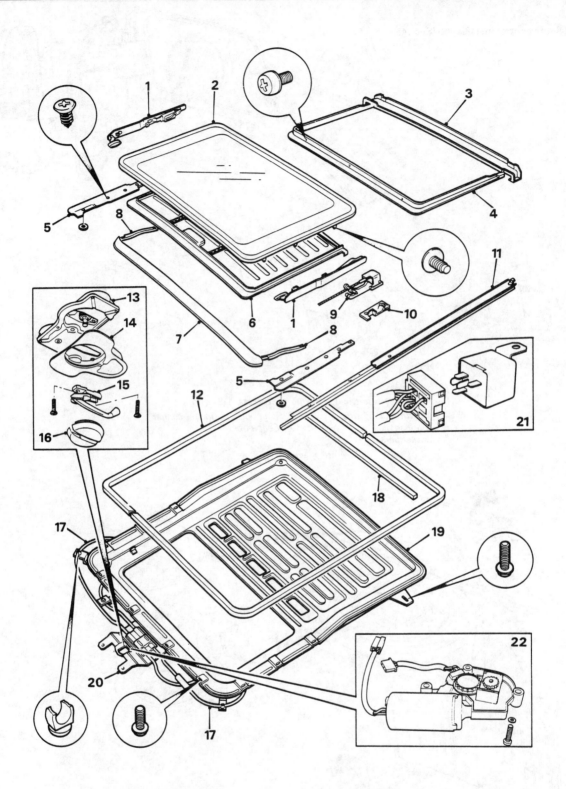

Fig. 13.76 Components of the tilt/slide sunroof (Sec 14)

1 Side arm
2 Glass panel
3 Rear drain channel
4 Side shield
5 Slide rail
6 Visor
7 Wind deflector
8 Deflector lifting arm
9 Rear guide and cable
10 Pilot guide
11 Guide rail
12 Tray seal
13 Drivegear
14 Escutcheon
15 Operating handle
16 Bezel
17 Cable guide and run-out tubes
18 Rear tray seal
19 Tray
20 Front bracket and mounting plate
21 Relay
22 Electric motor/gearbox drive assembly

153 On Estate models, fold the rear seat and peel back the luggage area floor covering. Where fitted, erect the rear facing seats and remove the right-hand side trim from the luggage area.
154 Remove the sill carpet retainers from the driver's side.
155 Slacken the driver's seat belt bolts and then remove the bolts from the outboard seat runner.
156 Remove the oddments tray.
157 Extract the fuel flap release lever fixing screws, raise the moulding and disconnect the release lever from the cable.
158 Peel back the carpets to free the cable by releasing the retaining clips. Withdraw the cable into the boot (Saloon) or luggage area (Estate).
159 Open the fuel filler flap, release the retaining clip and then pull the cable from the rear of the wheel arch into the boot or luggage area.
160 Refitting is a reversal of removal.

Asymmetrical split rear seat
161 On later Estates, the rear seat is split 60/40 in order that, by folding the sections independently, load carrying area is provided while still retaining rear passenger accommodation. Removal operations are similar to those described in paragraphs 96 to 99 of this Section.

Rear facing child seats
162 Optionally available on Estate models, the seats should be regarded as being for occasional use only. They must only be used with the forward facing rear seat erected and locked. Seat belts must always be used. The seats are not suitable for adults.

Spare wheel and tools (Estate models)
163 The spare wheel and tools are located below the luggage compartment floor. Access is obtained by raising the floor or seat backrest if rear facing seats are fitted.
164 The tools are located in a plastic moulding within the spare wheel.

Grab handles – removal and refitting
165 To remove a grab handle from the headlining above the door opening, flip back the end covers with a small screwdriver to expose the fixing screws. Remove the screws and the handle.
166 Refitting is a reversal of removal.

Plastic components – repair
167 With the use of more and more plastic body components by the vehicle manufacturers (eg bumpers, spoilers, and in some cases major body panels), rectification of more serious damage to such items has become a matter of either entrusting repair work to a specialist in this field, or renewing complete components. Repair of such damage by the DIY owner is not really feasible owing to the cost of the equipment and materials required for effecting such repairs. The basic technique involves making a groove along the line of the crack in the plastic using a rotary burr in a power drill. The damaged part is then welded back together by using a hot air gun to heat up and fuse a plastic filler rod into the groove. Any excess plastic is then removed and the area rubbed down to a smooth finish. It is important that a filler rod of the correct plastic is used, as body components can be made of a variety of different types (eg polycarbonate, ABS, polypropylene).
168 Damage of a less serious nature (abrasions, minor cracks etc) can be repaired by the DIY owner using a two-part epoxy filler repair material, like Holts Body + Plus or Holts No Mix which can be used directly from the tube. Once mixed in equal proportions (or applied direct from the tube in the case of Holts No Mix), this is used in similar fashion to the bodywork filler used on metal panels. The filler is usually cured in twenty to thirty minutes, ready for sanding and painting.
169 If the owner is renewing a complete component himself, or if he has repaired it with epoxy filler, he will be left with the problem of finding a suitable paint for finishing which is compatible with the type of plastic used. At one time the use of a universal paint was not possible owing to the complex range of plastics encountered in body component applications. Standard paints, generally speaking, will not bond to plastic or rubber satisfactorily, but Holts Professional Spraymatch paints to match any plastic or rubber finish can be obtained from dealers. However, it is now possible to obtain a plastic body parts finishing kit which consists of a pre-primer treatment, a primer and coloured top coat. Full instructions are normally supplied with a kit, but basically the method of use is to first apply the pre-primer to the component concerned and allow it to dry for up to 30 minutes. Then the

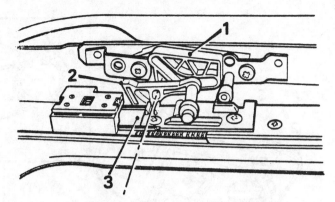

Fig. 13.77 Alignment of side arm and lifting arm (Sec 14)

1 Side arm
2 Lifting arm
3 Pilot plate

primer is applied and left to dry for about an hour before finally applying the special coloured top coat. The result is a correctly coloured component where the paint will flex with the plastic or rubber, a property that standard paint does not normally possess.

Tilt/slide sunroof components (1987 and later models) – removal and refitting
Glass panel and seal
170 Open the panel to the halfway position, then release the screws which secure the side shield and rear drain channel to the glass panel.
171 Slide the cover, drain channel, and the visor into the recess above the headlining.
172 Close the glass panel then tilt the rear edge.
173 Extract the screws which hold the glass panel to the side arms.
174 Lift off the glass panel and remove the seal.
175 When refitting, first push the seal onto the glass panel with the rubber surface facing inwards. Cut the seal to form a butt joint.
176 Set the operating mechanism in the closed position and locate the glass panel on the side arms. Fit the screws, align the panel with the roof and then tighten the screws.
177 Open and close the panel and then check that in the closed position, the front edge is flush with, or up to 0.04 in (1.0 mm) below, the level of the roof. The rear edge should be within the same tolerance above the level of the roof.
178 Open the panel to the halfway position, locate the side shield and rear drain channel and fit the fixing screws.
Sunroof visor
179 Remove the glass panel as previously described.
180 Remove the pilot plates and side arms.
181 Remove the slide and collect the flat washer from under the front of each side rail.
182 Disengage the rear guides from their channels and allow them to hang in the roof aperture.
183 Press the wind deflector downwards, then slide the cover and rear chain channel forwards and out through the roof.
184 Slide the visor forwards and out through the roof aperture, taking care not to damage the guides.
185 Refitting is a reversal of removal; remember to fit the flat washer under each side rail.
Sunroof drivegear (manually operated type)
186 Remove the glass panel as previously described, then set the operating mechanism in the closed position.
187 Check that the handle is in the freeplay position, and then remove the handle bezel and escutcheon.
188 Remove the finisher from around the sunroof aperture, and remove both windscreen sunvisors and clips.
189 Remove the grab handle from the passenger side headlining, and the stud from the driver's side.
190 Release and pull down the front edge of the headlining and extract the screws which hold the operating handle.
191 Extract the fixing screw and withdraw the drivegear.
192 To refit, position the side arm assembly so that the hole in the lifting lever and the triangular cut-out in the side arm are aligned with the slot in the pilot plate, as shown in Fig. 13.77.

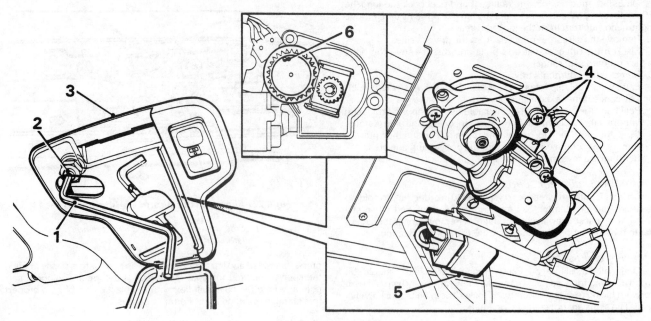

Fig. 13.78 Sunroof drive motor (Sec 14)

1	Emergency hand crank	3	Switch panel	5	Relay
2	Hand crank drive	4	Motor mounting screws	6	Gearwheel timing hole

193   Carry out the alignment on the opposite side.
194   Smear the drivegear with silicone grease and engage the cables. Fit and tighten the mounting screw.
195   Fit the operating handle, secure the headlining and fit the grab handle and visors.
196   Fit the finisher, handle escutcheon and bezel, and the glass panel.

Sunroof drivegear (electrically operated type)
197   Remove the glass panel as previously described.
198   Check that the sunroof mechanism is switched off in the closed position. If there is any doubt about closure, use the emergency hand crank, which is stored in the glovebox.
199   Open the cover, release the switch panel, disconnect the switch leads and remove the switch panel.
200   Release and pull away the headlining.
201   Extract the screw and release the relay from the motor bracket (Fig. 13.78).
202   Remove the mounting screws and withdraw the motor/gearbox drive assembly, noting that the gear timing hole is adjacent to the terminals when the mechanism is in the roof-closed mode.
203   Commence refitting by aligning the guides to the pilot plates, using 5.0 mm diameter pins to retain the alignment.
204   Lubricate the motor drivegears.
205   Support the motor, connect the multi-plug and switch on the

ignition. Actuate the roof switch to bring the gear timing hole to the position shown in Fig. 13.78. Switch off the ignition.
206   Align the gearbox slides and engage the drive cables, then fit the motor/gearbox drive assembly. Withdraw the temporary pins.
207   Refit the headlining, switch panel and glass panel.

Rear guide and cable assemblies
208   Remove the glass panel as previously described.
209   Set the operating mechanism in the closed position.
210   Remove the drivegear (manual) or motor/gearbox (electrically operated).
211   Extract the screws and remove the pilot plates and front slide rails.
212   Remove both side arm assemblies and withdraw the cable and rear guides.
213   Before refitting, inject grease into the cable guide tubes and insert the cables.
214   Locate the rear guides, fit the side arm assemblies, front slide rails and pilot plates.
215   Align the side arms as described in paragraph 192.
216   Fit the drivegear or motor/gearbox according to type. Refit the glass panel.

Sunroof – complete
217   Remove the glass panel as previously described.

14.244 Pull off the control knobs

14.245 Prise the cover panel forward ...

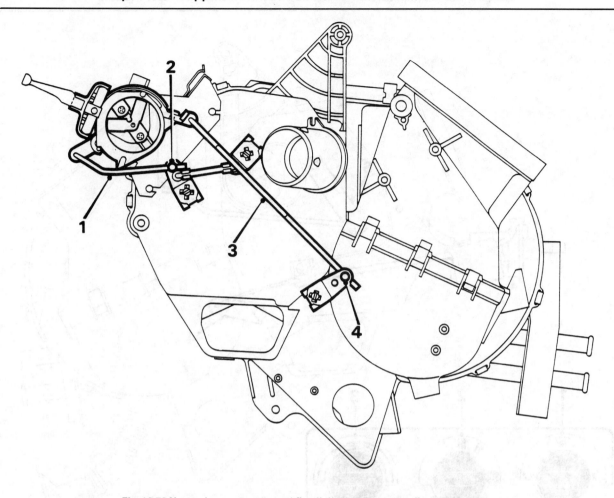

Fig. 13.79 Heater (non-vacuum type) flap linkage adjustment diagram (Sec 14)

1   *Link rod (temperature*   2   *Pinch-bolt and trunnion*   3   *Link rod (air distribution)*   4   *Pinch-bolt and trunnion*
    control)

Manually operated type
218   Remove the handle bezel, escutcheon and the handle itself.
Electrically operated type
219   Open the cover and release the switch panel. Disconnect the switch and relay and take off the switch panel.
All models
220   Disconnect the drain tubes from the sunroof tray.
221   Locate a block of wood above the front bracket and drill out the two pop-rivets.
222   Fully recline the two front seats and then extract all the screws which hold the sunroof tray to the roof.
223   Lower the tray and withdraw it through the front passenger door.

Discard the tray seal.
224   Refitting is a reversal of removal but before starting, fit a strip of 20.0 x 20.0 mm compriband seal (available from your dealer) around the edge of the sunroof tray.

Non-vacuum type heater – general
225   Later models are equipped with a modified type of heater unit, whereby the air intake and temperature flaps are operated by cables instead of the vacuum actuators used previously.
226   Apart from the cable locations shown in Fig. 13.79 and the cable adjustment procedure contained in the following paragraphs, the heater is the same in construction and layout as the earlier type. All operations

14.246A Remove the screws securing the backpanel ...

14.246B ... and drop the back panel forward

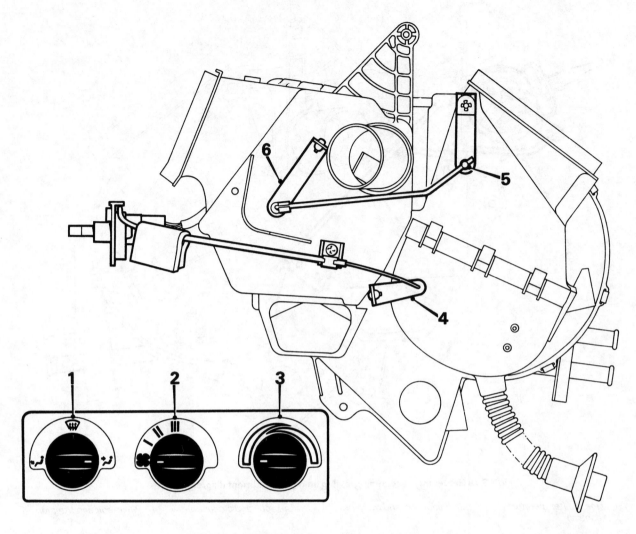

Fig. 13.80 Heater controls on 1989-on models (Sec 14)

1   Air distribution control   3   Air temperature control   5   Air inlet flap lever and   6   Matrix air outlet flap lever
2   Blower motor switch   4   Air distribution flap lever   pinch-bolt

contained in Chapter 12 are still applicable, but note the cable locations and attachments when carrying out the work.

Non-vacuum type heater control linkage – adjustment
Air distribution lever
227   Raise the control lever fully and slacken the pinch-bolt which holds the control link rod to the flap lever trunnion.
228   Hold the control lever fully up and the flap lever fully down and tighten the pinch-bolt.
Temperature control lever
229   Disconnect the battery and remove the facia panel, as described in Chapter 12.
230   Raise the control lever fully and slacken the pinch-bolt which holds the link rod to the trunnion on the lower flap lever which is adjacent to the facia.
231   Hold the control lever fully up and the flap lever towards the vehicle interior and tighten the pinch-bolt.
232   Refit the facia and connect the battery.

Heater (1989-on) – general description
233   The heater on 1989-on models is basically the same as the non-vacuum type fitted to earlier models, but has rotary controls as opposed to levers.

Heater control cables (1989-on) – removal and refitting
234   Disconnect the battery.

235   Remove the centre console and fascia as described.
236   Remove the heater control cover as described in paragraphs 243 to 246.
237   Remove the outer cable clips from the control panel.
238   Release the cable operating racks from the control panel, then disconnect the cables from the racks.
239   Remove the cable clips from the heater casing, disconnect the cables from the flap levers and withdraw the cables.
240   Disconnect the blower motor switch multi-plug and withdraw the control panel.
241   Refitting is a reversal of removal, fully closing the heater flaps before locating the control cables in the racks, which must also be in the fully closed position before fitting the cables to them.
242   Adjust the heater flap by loosening the lever pinch-bolt and pushing the air intake flap lever fully rearwards. Push the matrix air outlet flap lever rearwards and tighten the pinch-bolt.

Heater control panel (1989-on) – removal and refitting
243   Disconnect the battery.
244   Pull off the control knobs (photo).
245   Prise the cover panel forward (photo).
246   Remove the screws securing the back panel to the facia panel, and drop the back panel forwards (photos).
247   Disconnect the lead and withdraw the panel.
248   Refit in reverse order.

Key to main wiring diagram (all models) up to 1986

Symbols used in the wiring diagrams

1. Component earthed by its lead
2. Component earthed by its fixing
3. If fitted
4. Line connector
5. Sealed joint
6. Printed circuit connector
7. Printed circuit

Connections to other circuits

A Headlamp wash
B Radio/cassette
C Electric windows
D Central locking

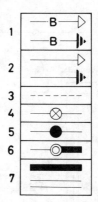

Colour code

P	Purple	B	Black
R	Red	G	Green
S	Slate	K	Pink
U	Blue	LG	Light green
W	White	N	Brown
Y	Yellow	O	Orange

When two colour code letters are shown together, the first denotes the main wire colour and the second denotes the tracer colour

No	Component	Grid ref.	No	Component	Grid ref.
1	Alternator	A1	115	Heated rear screen switch	C2
3	Battery	A1	116	Heated rear screen	C2
4	Starter motor solenoid	A1	118	Windscreen washer/wiper switch	B3
5	Starter motor	A1	150	Heated rear screen warning lamp	A2
6	Lighting switch – main	B1	152	Hazard warning lamp	B1
7	Headlamp dip switch	B1	153	Hazard warning switch	B1
8	Headlamp dip beam	B3	165	Handbrake warning lamp switch	B4
9	Headlamp main beam	B3	166	Handbrake warning lamp	A4
10	Main beam warning lamp	A4	174	Starter solenoid relay	A1
11	RH sidelamp	B3	178	Radiator cooling fan thermostat	C2
12	LH sidelamp	B3	179	Radiator cooling fan motor	C2
14	Panel illumination lamps	A4	182	Brake fluid level switch	B4
15	Number plate illumination lamps	B4	208	Cigar lighter illumination lamp	B3
16	Stop-lamps	B4	210	Panel illumination lamp rheostat/resistor	A3
17	RH tail lamp	B4	211	Heater control illumination	B2
18	Stop-lamp switch	B4	231	Headlamp relay	B1
19	Fusebox	A2,B2	232	Sidelamp warning lamp	A4
20	Interior lamps	B3	240	Heated rear screen relay	A2
21	Interior lamp door switches	C3	242	Two way interior lamp switch	B3
22	LH tail lamp	B4	246	Glovebox illumination lamp	B3
23	Horns	B1	247	Glovebox illumination switch	B3
24	Horn push	B1	256	Brake warning blocking diode	A4
26	Direction indicator switch	B1	265	Ambient temperature sensor	C1
27	Direction indicator warning lamp(s)	A4	267	Headlamp washer motor	B1
28	RH front direction indicator lamp	B3	286	Rear foglamp switch	A2
29	LH front direction indicator lamp	B3	287	Rear foglamp warning lamp	A2
30	RH rear direction indicator lamp	B4	288	Rear foglamp(s)	B4
31	LH rear direction indicator lamp	B4	294	Fuel cut-off solenoid	C2
32	Heater/fresh air motor switch	B3	297	Brake failure warning lamp	A4
33	Heater/fresh air motor	B3	308	Direction indicator/hazard flasher unit	C2
34	Fuel level gauge	A4	325	Brake pad wear warning lamp	A4
35	Fuel level gauge tank unit	B1	326	Brake pad wear sensor	A4
37	Windscreen wiper motor	B3	347	Electronic control unit – fuel management	C2
38	Ignition/starter switch	B1	355	Throttle switch	C1
39	Ignition coil	A1	359	Stepper motor	C2
40	Distributor	B1	367	Trailer indicator warning light	A3
42	Oil pressure switch	A4	368	Spare warning lamp	A4
43	Oil pressure warning lamp	A4	381	Knock sensor	B1
44	Ignition/no charge warning lamp	A3	382	Crankshaft sensor	C1
45	Headlamp flash switch	B1	386	One-touch control – window lift	C2
46	Coolant temperature gauge	A4	389	Column switch illumination	B3
47	Water temperature transducer	C1	393	Programmed ignition electronic control unit	C1
49	Reverse lamp switch	C4	398	Manifold heater relay	B1
50	Reverse lamps	B4	399	Manifold heater	B1
56	Clock	A2	400	Temperature switch	B1
57	Cigar lighter	B2	401	Interior lamp delay unit	C3
65	Luggage area lamp switch	B3	402	Windscreen wipe/wash programmed control	B3
66	Luggage area lamp	B3	403	Auxiliary ignition relay	A1
75	Automatic gearbox starter inhibitor switch	A1	410	Centre console illumination	B2
76	Automatic gearbox selector indicator lamp	B4	413	Fusible link	A1
77	Windscreen washer motor	B1	415	Fuel level LED	A4
82	Switch illumination lamp(s)	A2	416	Coolant temperature LED	A4
95	Tachometer	A4	426	Heater vacuum solenoid	C3
			427	Heater vacuum solenoid switch	C3

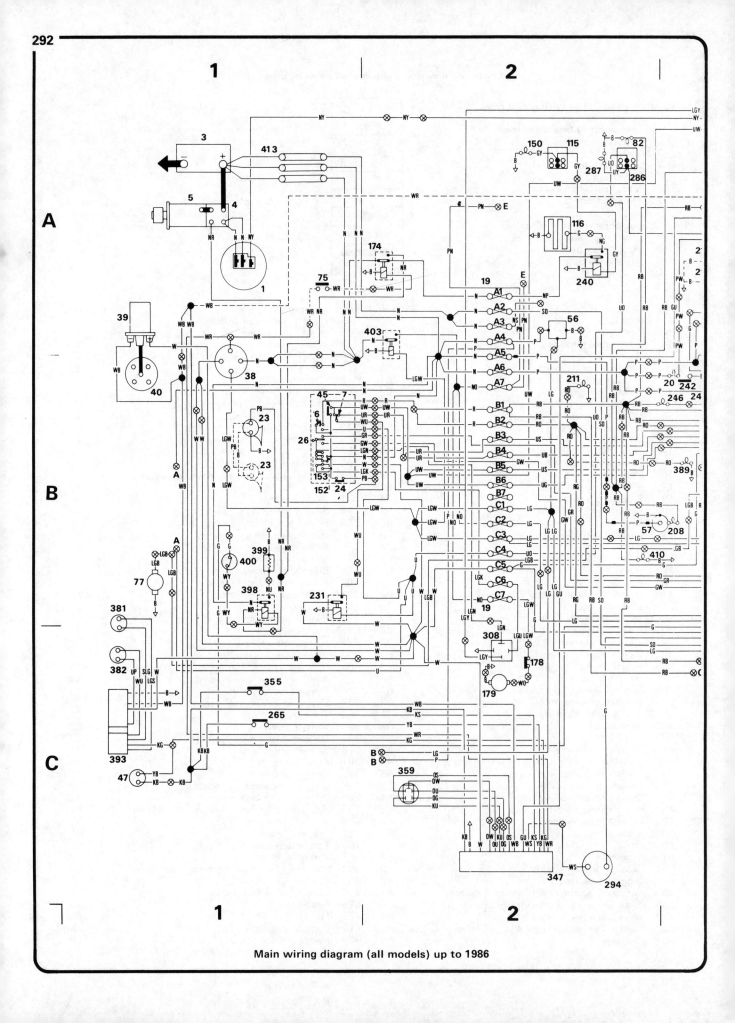

Main wiring diagram (all models) up to 1986

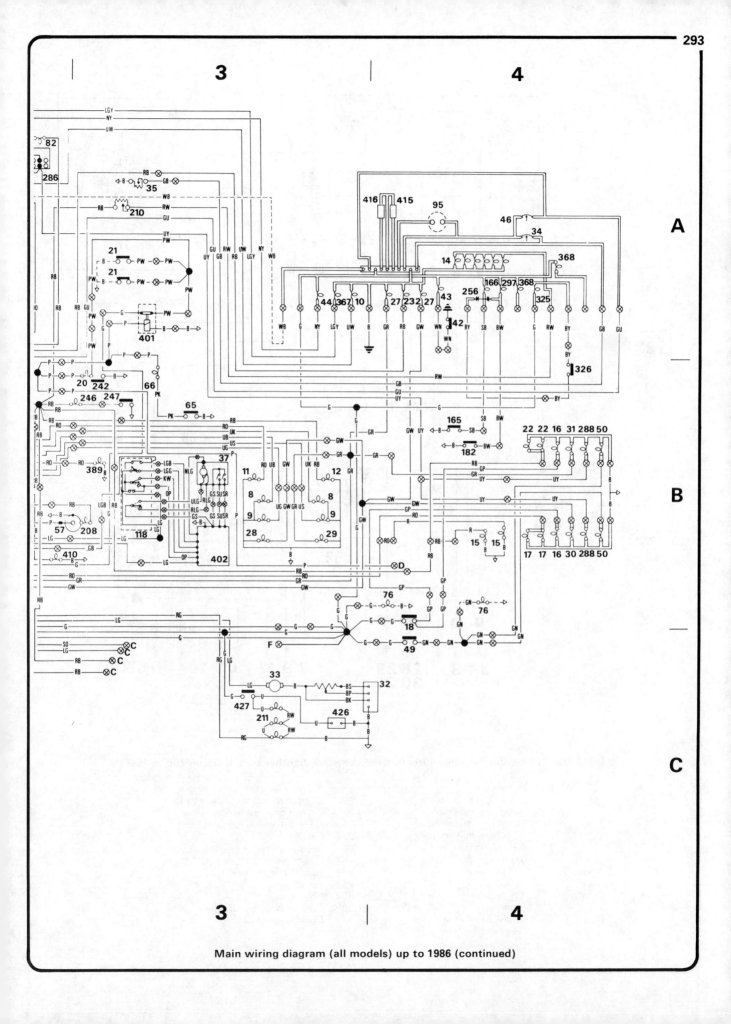

Main wiring diagram (all models) up to 1986 (continued)

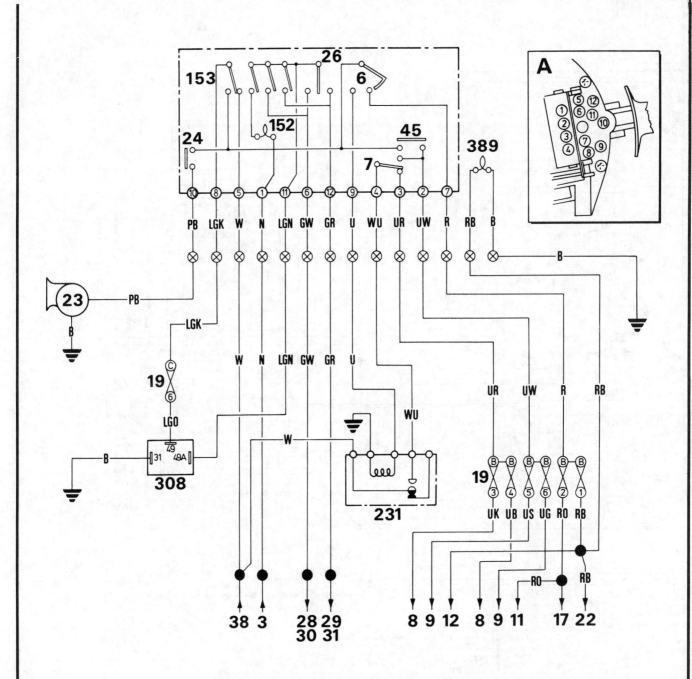

Supplementary wiring diagram – direction indicator and lighting switch circuit (up to 1986)

3	Battery	28	RH front direction indicator lamp
6	Lighting switch – main	29	LH front direction indicator lamp
7	Headlamp dip switch	30	RH rear direction indicator lamp
8	Headlamp dip beam	31	LH rear direction indicator lamp
9	Headlamp main beam	38	Ignition switch
11	RH sidelamp	45	Headlamp flasher switch
12	LH sidelamp	152	Hazard warning light
17	RH tail lamp	153	Hazard warning switch
19	Fusebox	231	Headlamps relay
22	LH tail lamp	308	Flasher unit
23	Horn	389	Column switch lamp – fibre optic
24	Horn push switch	A	Switch connectors
26	Direction indicator switch		

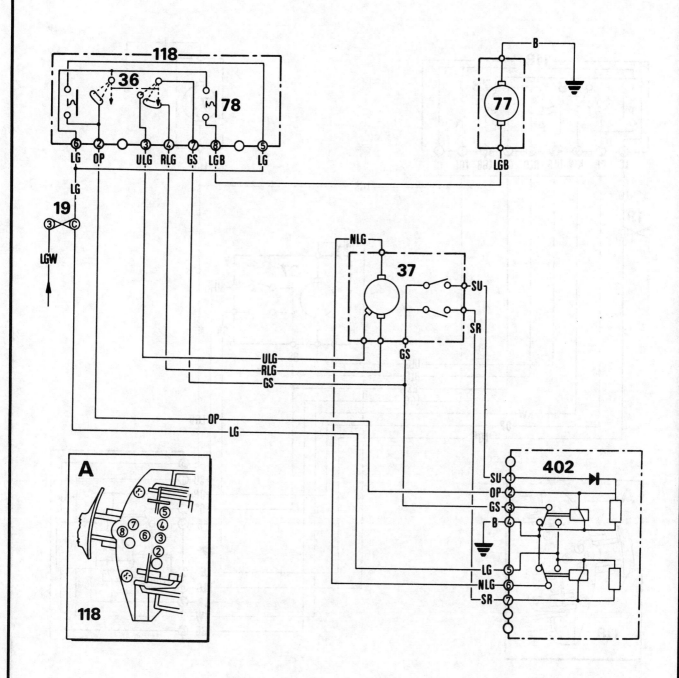

Supplementary wiring diagram – wash/wipe with flick wipe circuit (up to 1986)

19	Fusebox		78	Screen washer switch
36	Windscreen wiper switch and flick wipe		118	Combined windscreen wipe/wash switch
37	Windscreen wiper motor		402	Windscreen wiper programmed control
77	Screen washer pump		A	Switch connector – wipe/wash and flick wipe only

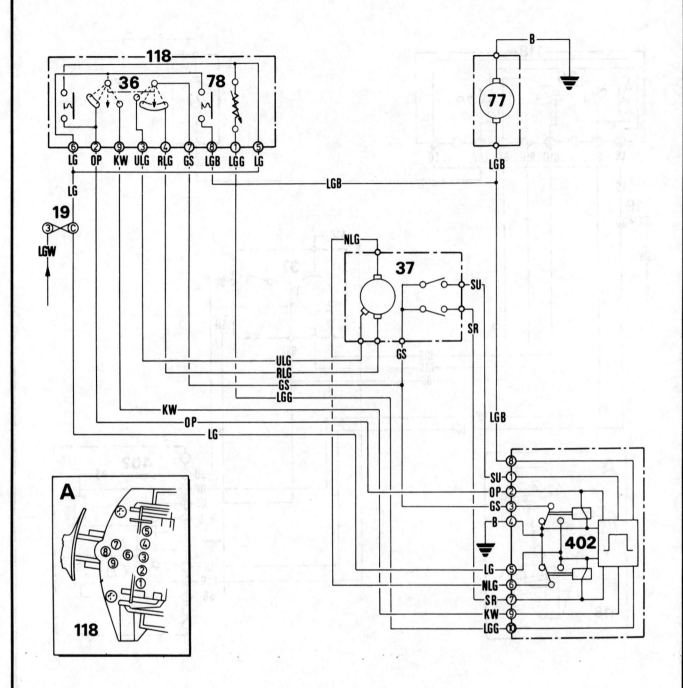

Supplementary wiring diagram – wash/wipe with variable intermittent delay circuit (up to 1986)

19	Fusebox	118	Combined windscreen wipe/wash and intermittent wiper switch
36	Windscreen wiper switch and flick wipe		
37	Windscreen wiper motor	402	Windscreen wipe/wash programmed control
77	Screen washer pump	A	Switch connectors – wipe/wash and intermittent wipe only
78	Screen washer switch with intermittent wiper control		

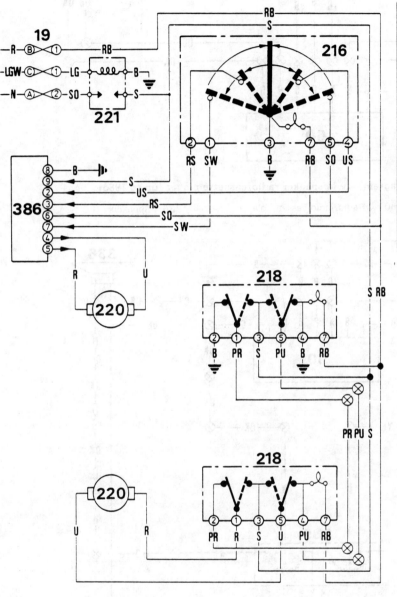
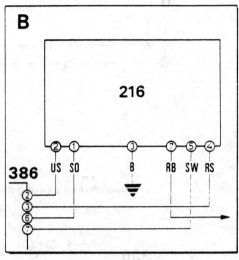
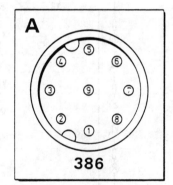

Supplementary wiring diagram – electric front window circuit (up to 1986)

19	Fusebox	221	Window lift relay
216	Window lift switch – driver's	386	One-touch window lift unit
218	Window lift switch – passenger's	A	Multi-plug connections – one-touch unit
220	Window lift motor	B	Switch connections – LH steering

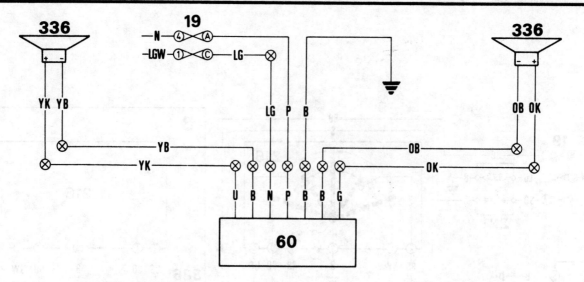

Supplementary wiring diagram – twin speaker radio/cassette player (up to 1986)

19   Fusebox   60   Radio/radio cassette player   336   Speakers

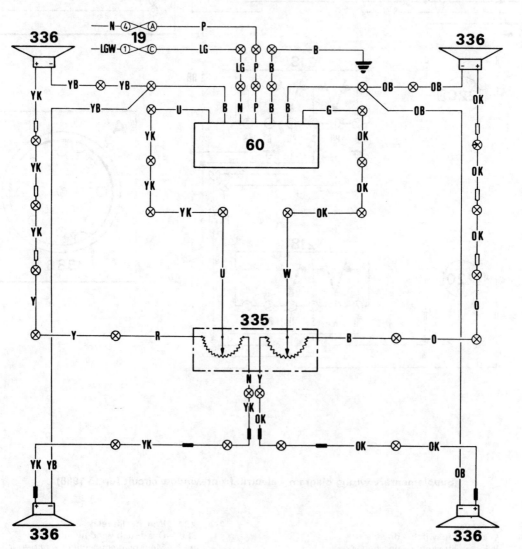

Supplementary wiring diagram – four speaker radio/radio cassette player (up to 1986)

335   Balance control   336   Speakers

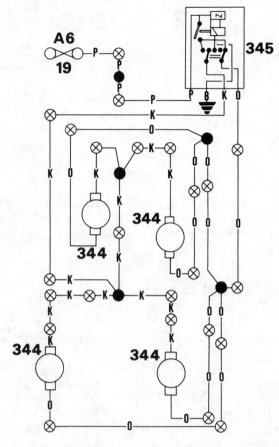

Supplementary wiring diagram – central door locking circuit (up to 1986)

19	Fusebox	345	Driver's door lock control unit
344	Door lock solenoid		

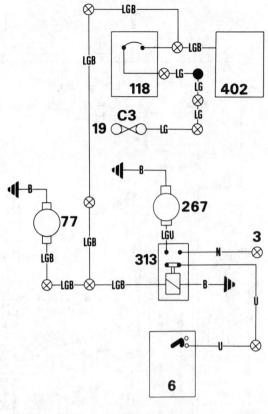

Supplementary wiring diagram – headlamp washer circuit (up to 1986)

3	Battery	267	Headlamp washer pump
6	Lighting switch	313	Headlamp washer relay
19	Fusebox	402	Programme control unit – wash/wipe
77	Windscreen washer pump		
118	Windscreen wiper/washer switch		

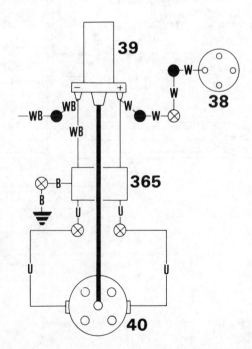

Supplementary wiring diagram – breakerless ignition circuit (1.3 models)

38	Ignition switch	39	Ignition coil	40	Distributor	365	Ignition amplifier

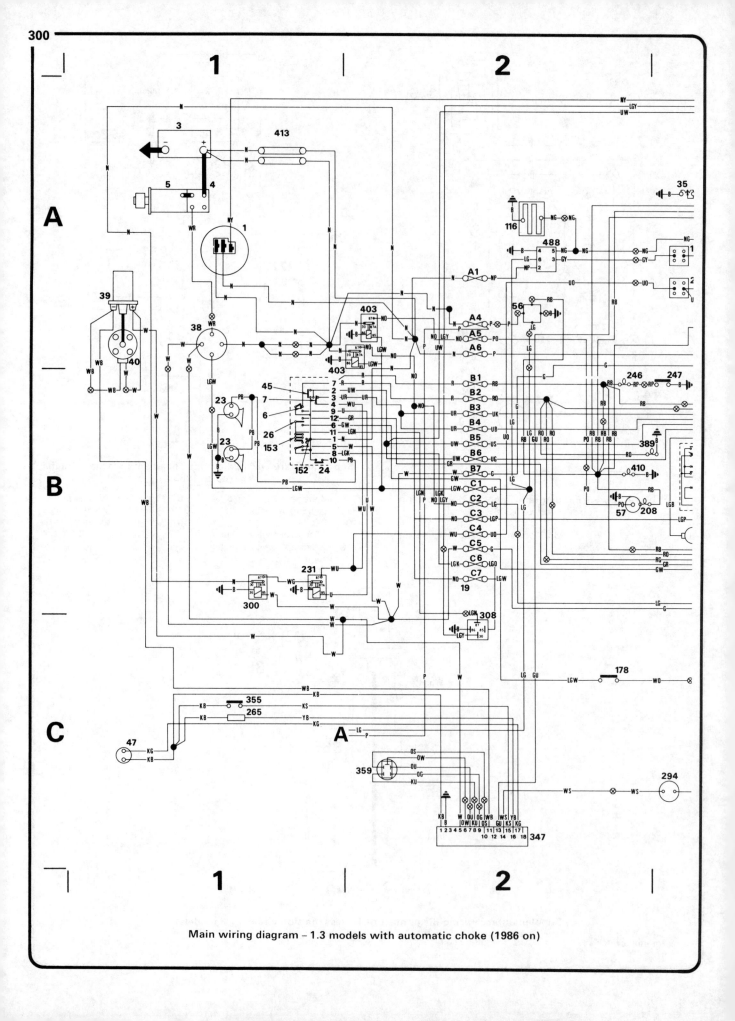

Main wiring diagram – 1.3 models with automatic choke (1986 on)

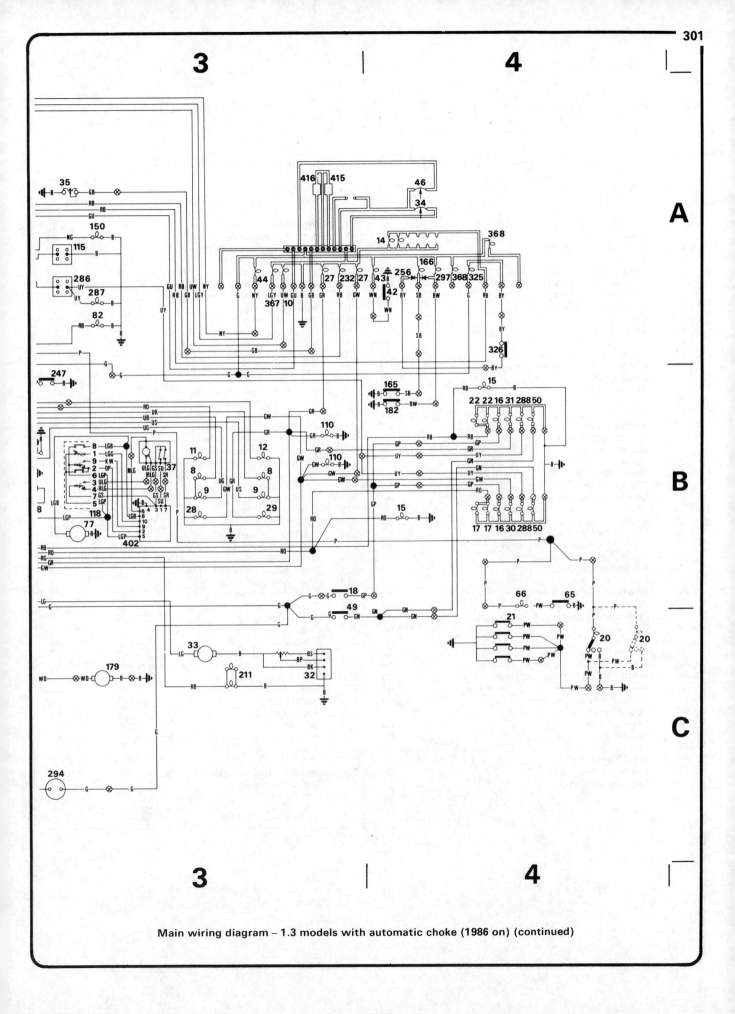

Main wiring diagram – 1.3 models with automatic choke (1986 on) (continued)

Key to main wiring diagram – 1.3 models with automatic choke (1986 on)

No	Description	Grid reference	No	Description	Grid reference
1	Alternator	A1	77	Windscreen washer motor	B3
3	Battery	A1	82	Switch illumination lamp(s)	A3
4	Starter motor solenoid	A1	110	Side repeater flashers	B3
5	Starter motor	A1	115	Heated rear screen switch	A3
6	Lighting switch – main	B1	116	Heated rear screen	A2
7	Headlamp dip switch	B1	118	Windscreen washer/wiper switch	B3
8	Headlamp dip beam	B3	150	Heated rear screen warning lamp	A3
9	Headlamp main beam	B3	152	Hazard warning lamp	B1
10	Main beam warning lamp	A3	153	Hazard warning switch	B1
11	RH parking lamp	B3	165	Handbrake warning lamp switch	B4
12	LH parking lamp	B3	166	Handbrake warning lamp	A4
14	Panel illumination lamps	A4	178	Radiator cooling fan thermostat	C2
15	Number plate illumination lamps	B4	179	Radiator cooling fan motor	C3
16	Stop-lamps	B4	182	Brake fluid level switch	B4
17	RH tail lamp	B4	208	Cigar lighter illumination lamp	B2
18	Stop-lamp switch	B3	211	Heater control illumination	C3
19	Fusebox	A2, B2	231	Headlamp relay	B1
20	Interior lamps	C4	232	Parking lamp warning lamp	A3
21	Interior lamp door switches	C4	246	Glovebox illumination lamp	B2
22	LH tail lamp	B4	247	Glovebox illumination switch	B3
23	Horns	B1	256	Brake warning blocking diode	A4
24	Horn push	B1	265	Ambient air temperature sensor	C1
26	Direction indicator switch	B1	286	Fog rearguard lamp switch	A3
27	Direction indicator warning lamp(s)	A3	287	Fog rearguard warning lamp	A3
28	RH front direction indicator lamp	B3	288	Fog rearguard lamp(s)	B4
29	LH front direction indicator lamp	B3	294	Fuel cut-off solenoid	C3
30	RH rear direction indicator lamp	B4	297	Brake failure warning lamp	A4
31	LH rear direction indicator lamp	B4	300	Ignition switch relay	B1
32	Heater/fresh air motor switch	C3	308	Direction indicator/hazard flasher unit	C2
33	Heater/fresh air motor	C3	325	Brake pad wear warning lamp	A4
34	Fuel level gauge	A4	326	Brake pad wear sensor	A4
35	Fuel level gauge tank unit	A3	347	ECU – fuel management	C2
37	Windscreen wiper motor	B3	355	Accelerator pedal switch	C1
38	Ignition/starter switch	A1	359	Stepper motor	C2
39	Ignition coil	A1	367	Trailer indicator warning light	A3
40	Distributor	A1	368	Spare warning lamp	A4
42	Oil pressure switch	A4	389	Column switch illumination	B2
43	Oil pressure warning lamp	A4	402	Windscreen wipe/wash programmed control	B3
44	Ignition/no charge warning lamp	A3	403	Auxiliary ignition relay	A2
45	Headlamp flash switch	B1	410	Centre console illumination	B2
46	Coolant temperature gauge	A4	413	Fusible link	A1
47	Coolant temperature thermistor	C1	415	Fuel level LED	A3
49	Reverse lamp switch	C3	416	Coolant temperature LED	A3
50	Reverse lamps	B4	488	Heated rear screen timer and relay	A2
56	Clock	A2			
57	Cigar lighter	B2	**Supplementary diagram connections**		
65	Luggage area lamp switch	B4	A	Radio cassette player	C1
66	Luggage area lamp	B4			

Key to main wiring diagram – 1.3 models with manual choke (1986 on)

No	Description	Grid reference	No	Description	Grid reference
1	Alternator	A1	65	Luggage area lamp switch	C4
3	Battery	A1	66	Luggage area lamp	C4
4	Starter motor solenoid	A1	77	Windscreen washer motor	B3
5	Starter motor	A1	82	Switch illumination lamp(s)	A3
6	Lighting switch – main	B1	110	Side repeater flashers	B3
7	Headlamp dip switch	B1	115	Heated rear screen switch	A3
8	Headlamp dip beam	B3	116	Heated rear screen	A2
9	Headlamp main beam	B3	118	Windscreen washer/wiper switch	B3
10	Main beam warning lamp	A3	150	Heated rear screen warning lamp	A3
11	RH parking lamp	B3	152	Hazard warning lamp	B1
12	LH parking lamp	B3	153	Hazard warning switch	B1
14	Panel illumination lamps	A4	165	Handbrake warning lamp switch	B4
15	Number plate illumination lamps	B4	166	Handbrake warning lamp	A4
16	Stop-lamps	B4	178	Radiator cooling fan thermostat	C2
17	RH tail lamp	B4	179	Radiator cooling fan motor	C3
18	Stop-lamp switch	C3	182	Brake fluid level switch	B4
19	Fusebox	A2, B2	211	Heater control illumination	C3
20	Interior lamps	C4	212	Choke control warning light switch	A4
21	Interior lamp door switches	C4	213	Choke control warning light	A4
22	LH tail lamp	B4	231	Headlamp relay	B1
23	Horns	B1	232	Parking lamp warning lamp	A3
24	Horn push	B1	246	Glovebox illumination lamp	B2
26	Direction indicator switch	B1	247	Glovebox illumination switch	B3
27	Direction indicator warning lamp(s)	A3	256	Brake warning blocking diode	A4
28	RH front direction indicator lamp	B3	286	Fog rearguard lamp switch	A3
29	LH front direction indicator lamp	B3	287	Fog rearguard warning lamp	A3
30	RH rear direction indicator lamp	B4	288	Fog rearguard lamp(s)	B4
31	LH rear direction indicator lamp	B4	297	Brake failure warning lamp	A4
32	Heater/fresh air motor switch	C3	300	Ignition switch relay	B1
33	Heater/fresh air motor	C3	308	Direction indicator/hazard flasher unit	C2
34	Fuel level gauge	A4	325	Brake pad wear warning lamp	A4
35	Fuel level gauge tank unit	A3	326	Brake pad wear sensor	A4
37	Windscreen wiper motor	B3	367	Trailer indicator warning light	A3
38	Ignition/starter switch	A1	368	Spare warning light	A4
39	Ignition coil	A1	389	Column switch illumination	B3
40	Distributor	A1	402	Windscreen wipe/wash programmed control	B3
42	Oil pressure switch	A4	403	Auxiliary ignition relay	A1
43	Oil pressure warning lamp	A4	410	Centre control illumination	B2
44	Ignition/no charge warning lamp	A3	413	Fusible link	A1
45	Headlamp flash switch	B1	415	Fuel level LED	A3
46	Coolant temperature gauge	A4	416	Coolant temperature LED	A3
47	Coolant temperature thermistor	A3	488	Heated rear screen timer and relay	A2
49	Reverse lamp switch	C3			
50	Reverse lamps	B4	**Supplementary diagram connections**		
56	Clock	A2	A	Radio cassette player	C1

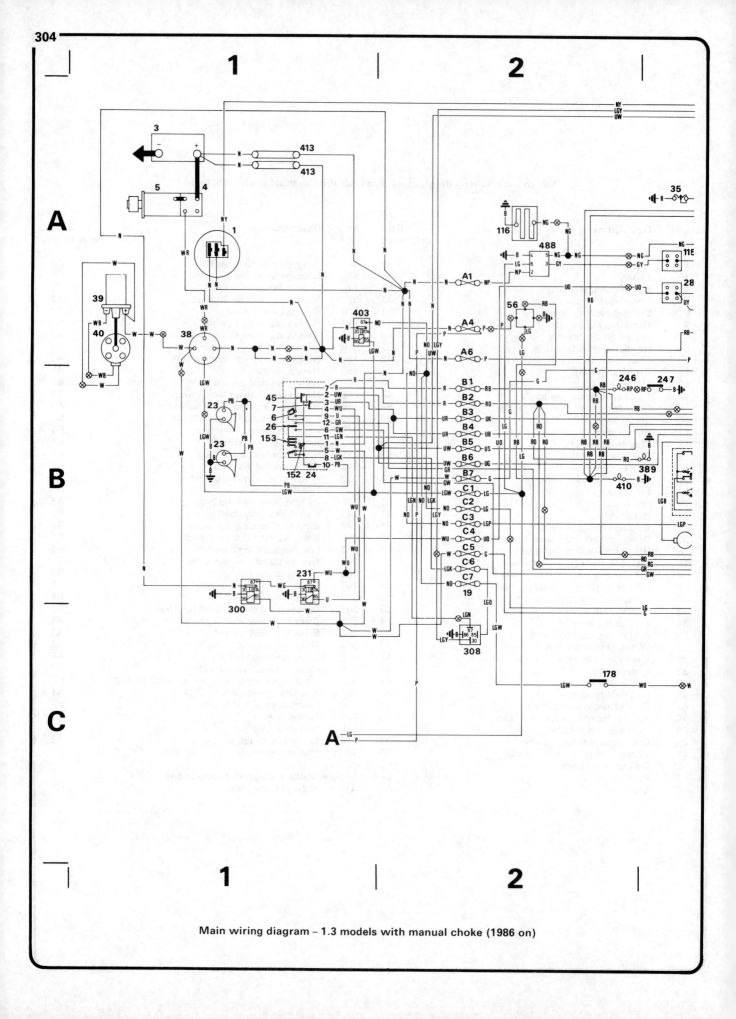

Main wiring diagram – 1.3 models with manual choke (1986 on)

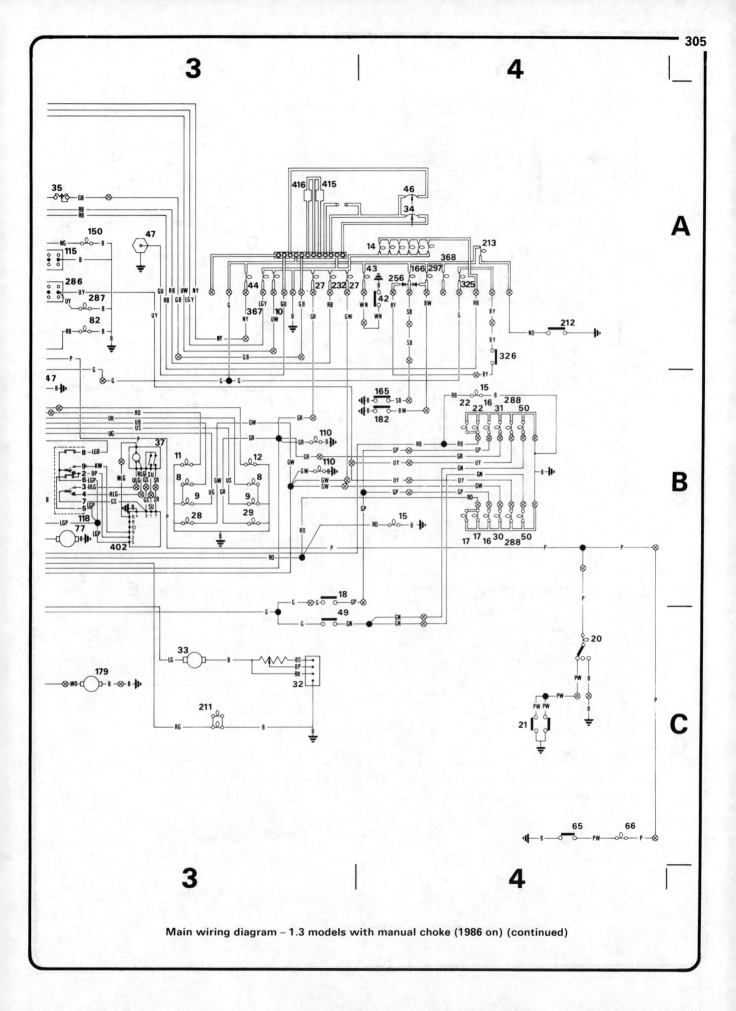

Main wiring diagram – 1.3 models with manual choke (1986 on) (continued)

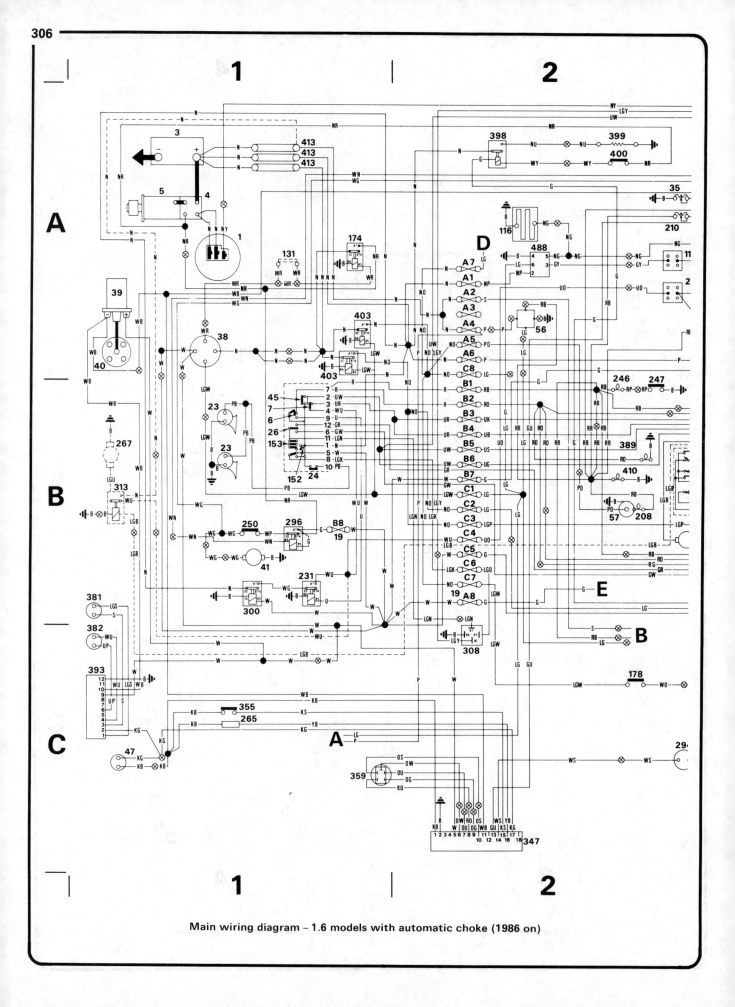

Main wiring diagram – 1.6 models with automatic choke (1986 on)

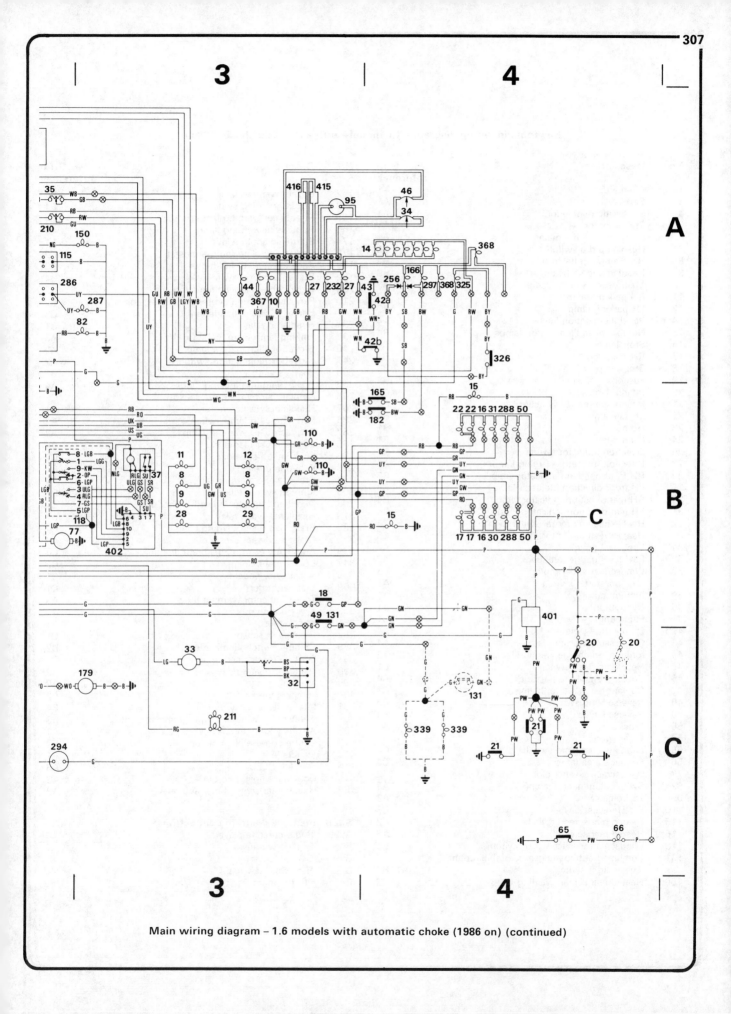

Main wiring diagram – 1.6 models with automatic choke (1986 on) (continued)

Key to main wiring diagram – 1.6 models with automatic choke (1986 on)

No	Description	Grid reference
1	Alternator	A1
3	Battery	A1
4	Starter motor solenoid	A1
5	Starter motor	A1
6	Lighting switch – main	B1
7	Headlamp dip switch	B1
8	Headlamp dip beam	B3
9	Headlamp main beam	B3
10	Main beam warning lamp	A3
11	RH parking lamp	B3
12	LH parking lamp	B3
14	Panel illumination lamps	A4
15	Number plate illumination lamps	B4
16	Stop-lamps	B4
17	RH tail lamp	B4
18	Stop-lamp switch	B3
19	Fusebox	A2, B1, B2
20	Interior lamps	C4
21	Interior lamp door switches	C4
22	LH tail lamp	B4
23	Horns	B1
24	Horn push	B1
26	Direction indicator switch	B1
27	Direction indicator warning lamp(s)	A3
28	RH front direction indicator lamp	B3
29	LH front direction indicator lamp	B3
30	RH rear direction indicator lamp	B4
31	LH rear direction indicator lamp	B4
32	Heater/fresh air motor switch	C3
33	Heater/fresh air motor	C3
34	Fuel level gauge	A4
35	Fuel level gauge tank unit	A2
37	Windscreen wiper motor	B3
38	Ignition/starter switch	A1
39	Ignition coil	A1
40	Distributor	A1
41	Fuel pump	B1
42b	Oil pressure switch – 1.6	A4
43	Oil pressure warning lamp	A4
44	Ignition/no charge warning lamp	A3
45	Headlamp flash switch	B1
46	Coolant temperature gauge	A4
47	Coolant temperature thermistor	C1
49	Reverse lamp switch	B3
50	Reverse lamps	B4
56	Clock	A2
57	Cigar lighter	B2
65	Luggage area lamp switch	C4
66	Luggage area lamp	C4
77	Windscreen washer motor	B2
82	Switch illumination lamp(s)	A3
95	Tachometer	A3
110	Side repeater flashers	B3
115	Heated rear screen switch	A2
116	Heated rear screen	A2
118	Windscreen washer/wiper switch	B3
131	Combined automatic gearbox inhibitor and reverse light switch	A1, B3
150	Heated rear screen warning lamp	A3
152	Hazard warning lamp	B1
153	Hazard warning switch	B1
165	Handbrake warning lamp switch	B4
166	Handbrake warning lamp	A4
174	Starter solenoid relay	A1
178	Radiator cooling fan thermostat	C2
179	Radiator cooling fan motor	C3
182	Brake fluid level switch	B4
208	Cigar lighter illumination lamp	B2
210	Panel illumination lamp rheostat/resistor	A2
211	Heater control illumination	C3
231	Headlamp relay	B1
232	Parking lamp warning lamp	A3
246	Glovebox illumination lamp	A2
247	Glovebox illumination switch	A2
250	Inertia switch	B1
256	Brake warning blocking diode	A4
265	Ambient air temperature sensor	C1
267	Headlamp washer motor	B1
286	Fog rearguard lamp switch	A3
287	Fog rearguard warning lamp	A2
288	Fog rearguard lamp(s)	B4
294	Fuel cut-off solenoid	C2
296	Fuel pump relay	B1
297	Brake failure warning lamp	A4
300	Ignition switch relay	B1
308	Direction indicator/hazard flasher unit	C2
313	Headlamp wash relay	B1
325	Brake pad wear warning lamp	A4
326	Brake pad wear sensor	A4
339	Automatic gearbox quadrant illumination	C4
347	ECU – fuel management	C2
355	Accelerator pedal switch	C1
359	Stepper motor	C1
367	Trailer indicator warning light	A3
368	Spare warning lamp	A4
381	Knock sensor	B1
382	Crankshaft sensor	C1
389	Column switch illumination	B2
393	ECU – Programmed ignition	C1
398	Manifold heater relay	A2
399	Manifold heater	A2
400	Temperature switch	A2
401	Interior lamp delay unit	B4
402	Windscreen wipe/wash programmed control	B3
403	Auxiliary ignition relay	A1
410	Centre console illumination	B2
413	Fusible link	A1
415	Fuel level LED	A3
416	Coolant temperature LED	A3
488	Heated rear screen timer and relay	A2

Supplementary diagram connections

A	Radio cassette player	C1
B	Electric windows	C2
C	Central locking	B4
D	Rear washer/wipe	A2
E	Electric mirror	B2

Key to main wiring diagram – 1.6 models with manual choke (1986 on)

No	Description	Grid reference	No	Description	Grid reference
1	Alternator	A1	131	Combined automatic gearbox inhibitor and reverse light switch	C4
3	Battery	A1	150	Heated rear screen warning lamp	A3
4	Starter motor solenoid	A1	152	Hazard warning lamp	B1
5	Starter motor	A1	153	Hazard warning switch	B1
6	Lighting switch – main	B1	165	Handbrake warning lamp switch	B4
7	Headlamp dip switch	B1	166	Handbrake warning lamp	A4
8	Headlamp dip beam	B3	174	Starter solenoid relay	A1
9	Headlamp main beam	B3	178	Radiator cooling fan thermostat	C2
10	Main beam warning lamp	A3	179	Radiator cooling fan motor	C3
11	RH parking lamp	B3	182	Brake fluid level switch	B4
12	LH parking lamp	B3	208	Cigar lighter illumination lamp	B3
14	Panel illumination lamps	A4	211	Heater control illumination	C3
15	Number plate illumination lamps	B4	212	Choke control warning light switch	A4
16	Stop-lamps	B4	213	Choke control warning light	A4
17	RH tail lamp	B4	231	Headlamp relay	B1
18	Stop-lamp switch	B3	232	Side lamp warning lamp	A3
19	Fusebox	A2, B1, B2	246	Glovebox illumination lamp	B2
20	Interior lamps	C4	247	Glovebox illumination switch	B3
21	Interior lamp door switches	C4	250	Inertia switch	B1
22	LH tail lamp	B4	256	Brake warning blocking diode	A4
23	Horns	B1	267	Headlamp washer motor	B1
24	Horn push	B1	286	Fog rearguard lamp switch	A3
26	Direction indicator switch	B1	287	Fog rearguard warning lamp	A3
27	Direction indicator warning lamp(s)	A3, A4	288	Fog rearguard lamp(s)	B4
28	RH front direction indicator lamp	B3	296	Fuel pump relay	B1
29	LH front direction indicator lamp	B3	297	Brake failure warning lamp	A4
30	RH rear direction indicator lamp	B4	300	Ignition switch relay	B1
31	LH rear direction indicator lamp	B4	308	Direction indicator/hazard flasher unit	C2
32	Heater/fresh air motor switch	C3	313	Headlamp wash relay	B1
33	Heater/fresh air motor	C3	325	Brake pad wear warning lamp	A4
34	Fuel level gauge	A4	326	Brake pad wear sensor	A4
35	Fuel level gauge tank unit	A3	339	Automatic gearbox quadrant illumination	C4
37	Windscreen wiper motor	B3	367	Trailer indicator warning light	A3
38	Ignition/starter switch	A1	368	Spare warning lamp	A4
39	Ignition coil	A1	381	Knock sensor	B1
40	Distributor	A1	382	Crankshaft sensor	C1
41	Fuel pump	B1	389	Column switch illumination	B3
42	Oil pressure switch	A4	393	ECU – programmed ignition	C1
43	Oil pressure warning lamp	A4	398	Manifold heater relay	A2
44	Ignition/no charge warning lamp	A3	399	Manifold heater	A3
45	Headlamp flash switch	B1	400	Temperature switch	A3
46	Coolant temperature gauge	A4	402	Windscreen wipe/wash programmed control	B3
47	Coolant temperature thermistor	C1	403	Auxiliary ignition relay	A1
49	Reverse lamp switch	C3	410	Centre console illumination	B2
50	Reverse lamps	B4	413	Fusible link	A1
56	Clock	A2	415	Fuel level LED	A3
57	Cigar lighter	B2	416	Coolant temperature LED	A3
65	Luggage area lamp switch	C4	488	Heated rear screen relay and timer	A2
66	Luggage area lamp	C4			
77	Windscreen washer motor	B3		**Supplementary diagram connections**	
82	Switch illumination lamp(s)	A3	A	Radio cassette player	C1
110	Side repeater flashers	B3	B	Electric windows	C3
115	Heated rear screen switch	A3	C	Central locking	B4
116	Heated rear screen	A2	D	Rear washer/wiper	A2
118	Windscreen washer/wiper switch	B3			

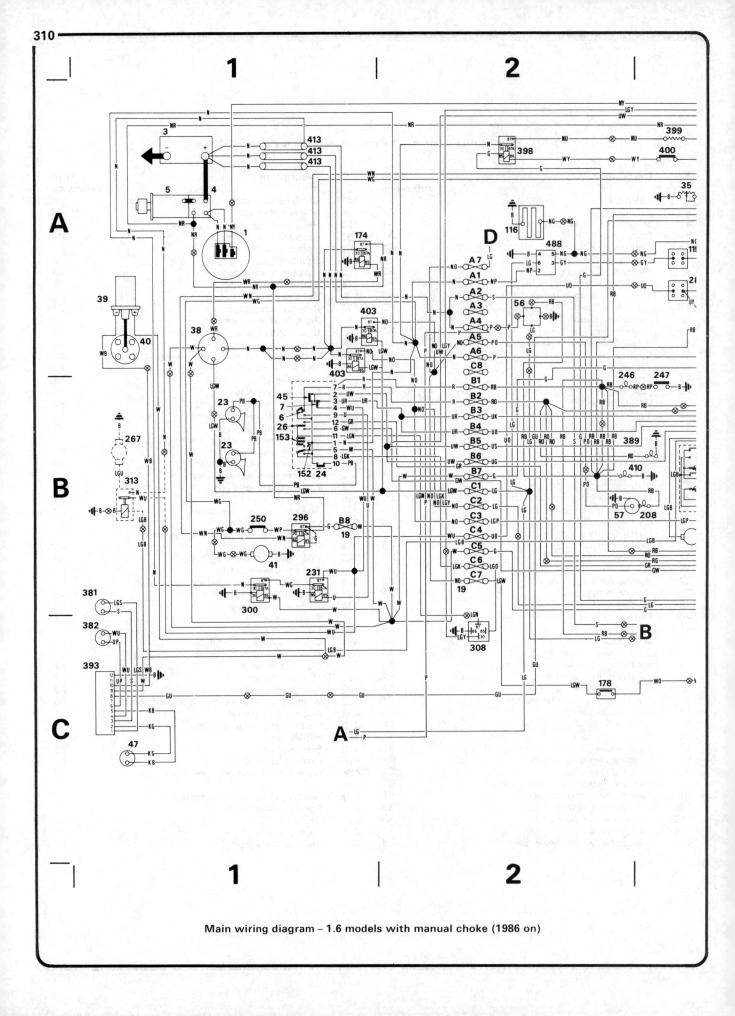

Main wiring diagram – 1.6 models with manual choke (1986 on)

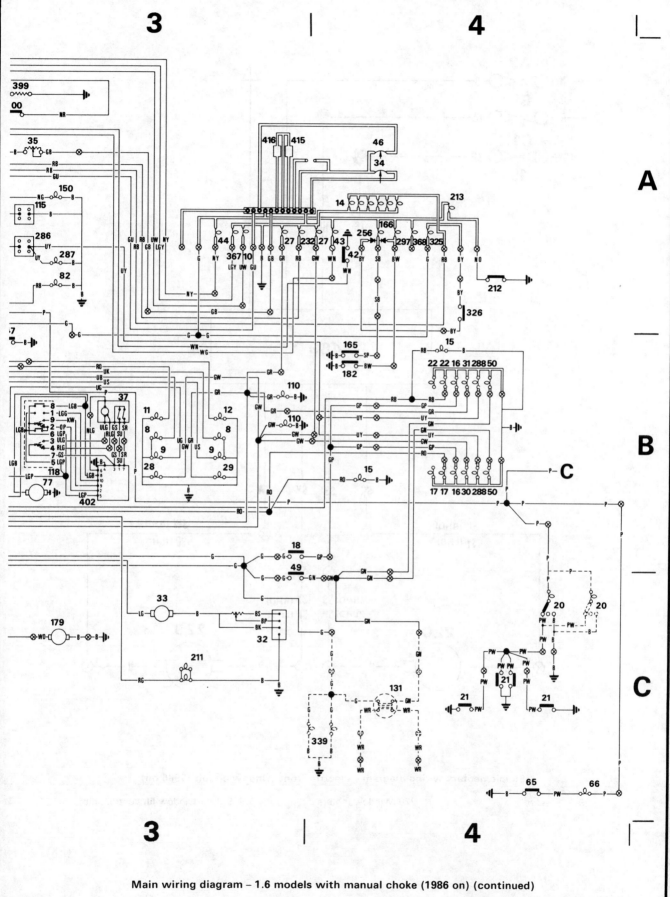

Main wiring diagram – 1.6 models with manual choke (1986 on) (continued)

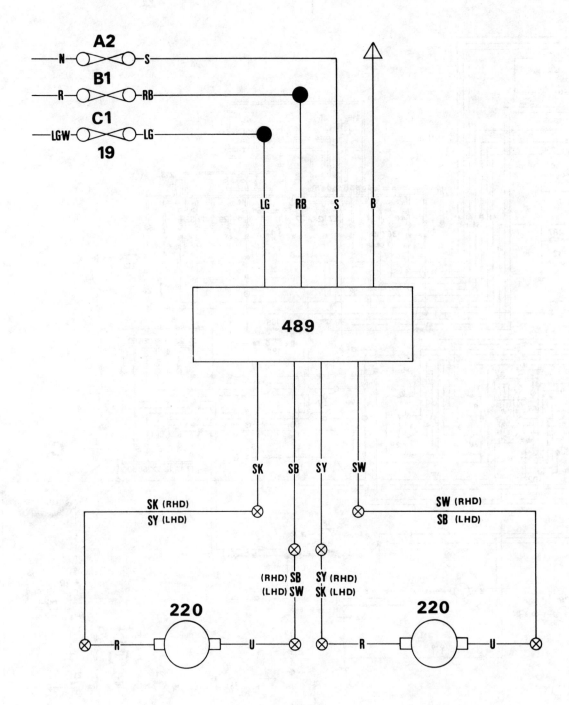

Supplementary wiring diagram – electric front window circuit (1986 on)

9   Fusebox   220 Window motors   489 Two-window lift control unit

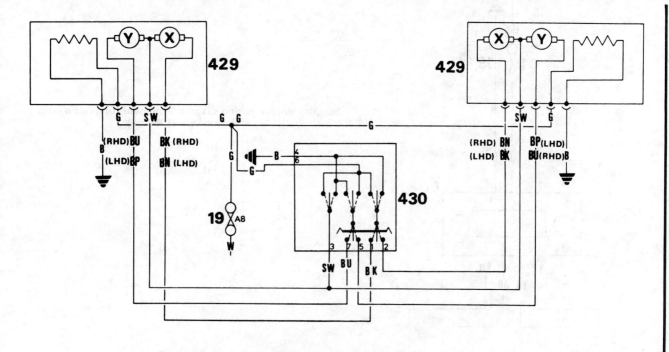

Supplementary wiring diagram – electric door mirrors (1986 on)

19   Fusebox
429 Mirrors

430 Mirror switch
X   Vertical movement motor

Y   Horizontal movement motor

**Supplementary wiring diagram –
manual tune radio/cassette player
with four speakers (1986 on)**

19   Fusebox
60   Radio/cassette player
335 Balance control
336 Speakers
A   Connectors on Estate
     version only

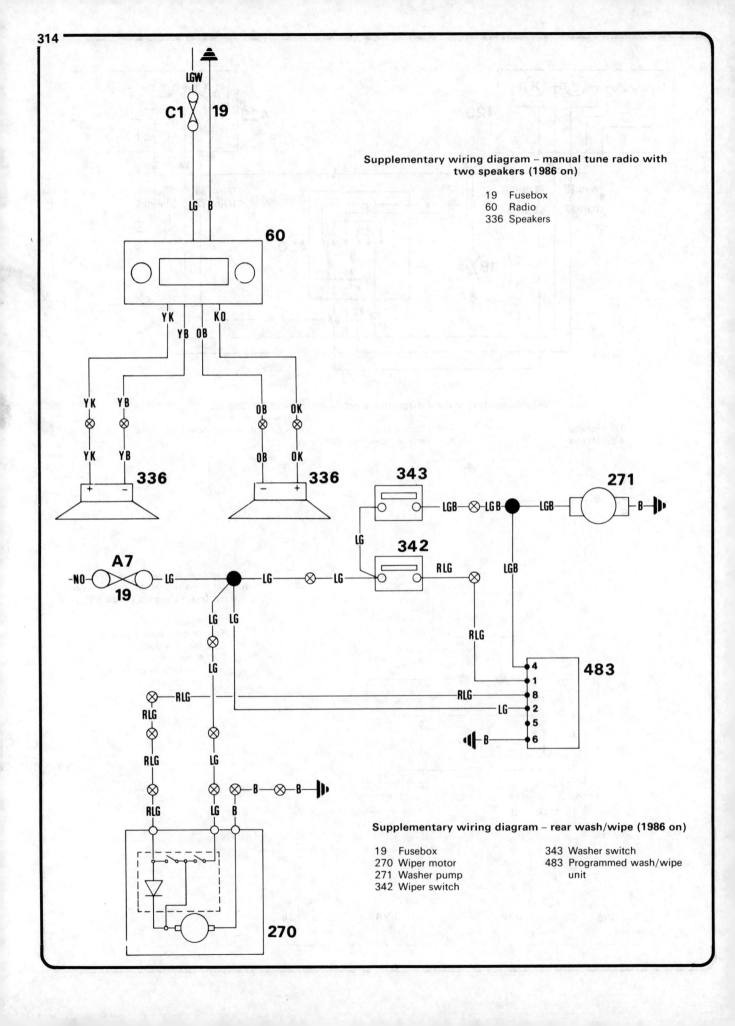

314

Supplementary wiring diagram – manual tune radio with
two speakers (1986 on)

19 Fusebox
60 Radio
336 Speakers

Supplementary wiring diagram – rear wash/wipe (1986 on)

19 Fusebox
270 Wiper motor
271 Washer pump
342 Wiper switch

343 Washer switch
483 Programmed wash/wipe
unit

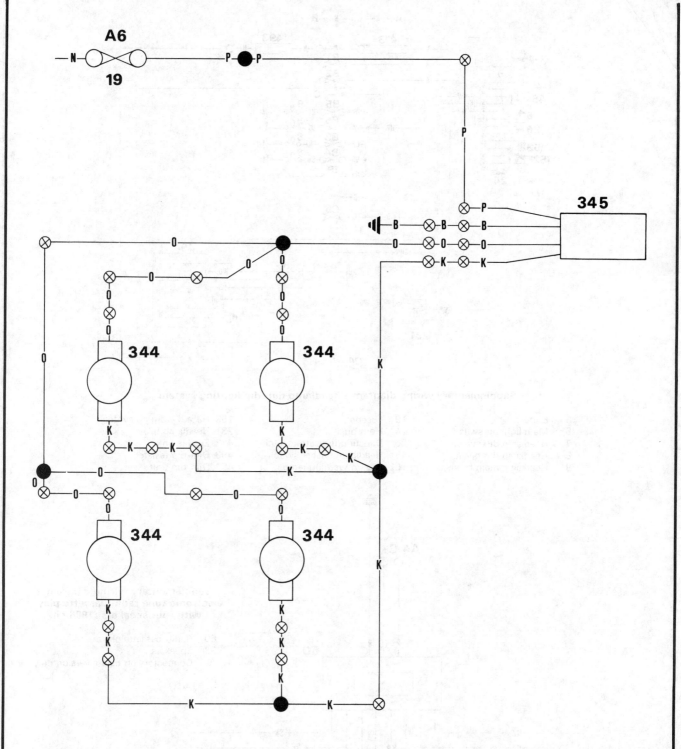

Supplementary wiring diagram – central door locking (1986 on)

19   Fusebox   344 Door lock motor   345 Driver's door lock control unit

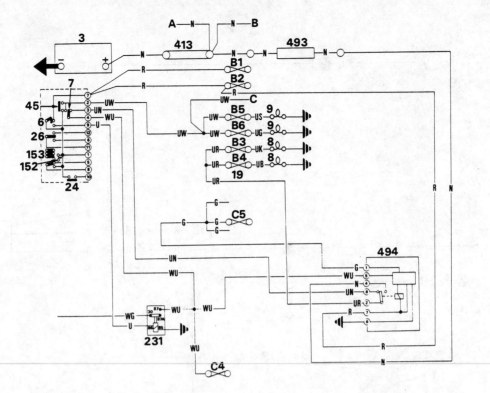

Supplementary wiring diagram – headlamp dim-dip lighting system

3	Battery	19	Fusebox	153	Hazard warning switch
6	Main lighting switch	24	Horn push	231	Headlamp relay
7	Headlamp dip switch	26	Direction indicator switch	413	Fusible link
8	Headlamp dip beam	45	Headlamp flasher switch	493	Dim-dip resistor
9	Headlamp main beam	152	Hazard warning lamp	494	Dim-dip unit

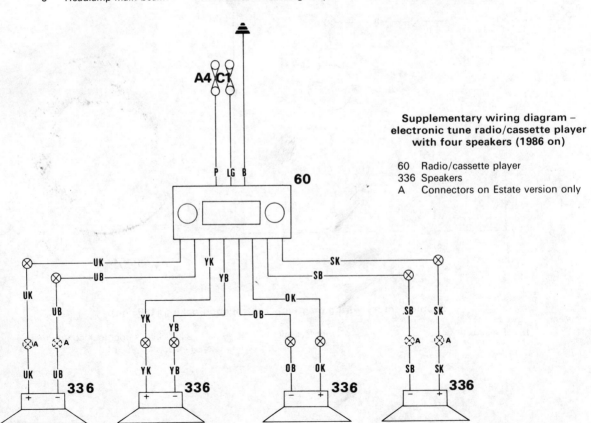

**Supplementary wiring diagram –
electronic tune radio/cassette player
with four speakers (1986 on)**

60	Radio/cassette player
336	Speakers
A	Connectors on Estate version only

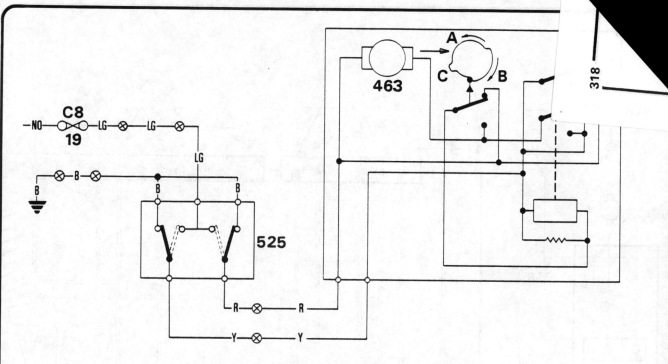

Supplementary wiring diagram – electric sunroof circuit

19	Fusebox	525	Switch	A	Open
463	Motor	574	Changeover relay	B	Close
				C	Tilt

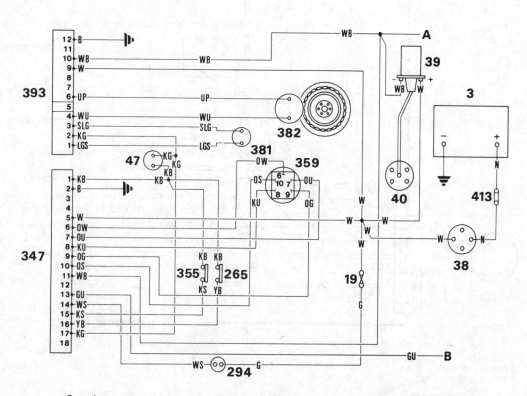

Supplementary wiring diagram – programmed ignition circuit (1.6 models)

3	Battery	47	Coolant temperature thermistor	355	Throttle switch	393	Programmed ignition ECU
19	Fusebox			359	Stepper motor	413	Fusible link
38	Ignition switch	265	Ambient temperature sensor	381	Knock sensor	A	To tachometer
39	Ignition coil			382	Crankshaft sensor and reluctor disc	B	To coolant temperature gauge
40	Distributor cap	294	Fuel cut-off solenoid				
		347	Fuel ECU				

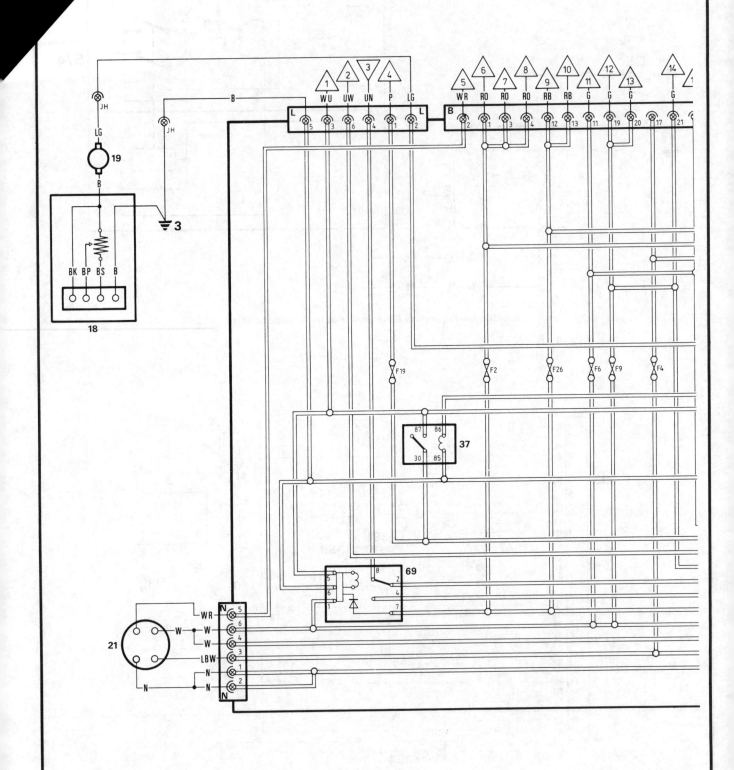

Main wiring diagram – 1.6 litre models, 1989-on

Main wiring diagram – 1.6 litre models, 1989-on (continued)

Main wiring diagram – 1.6 litre models, 1989-on (continued)

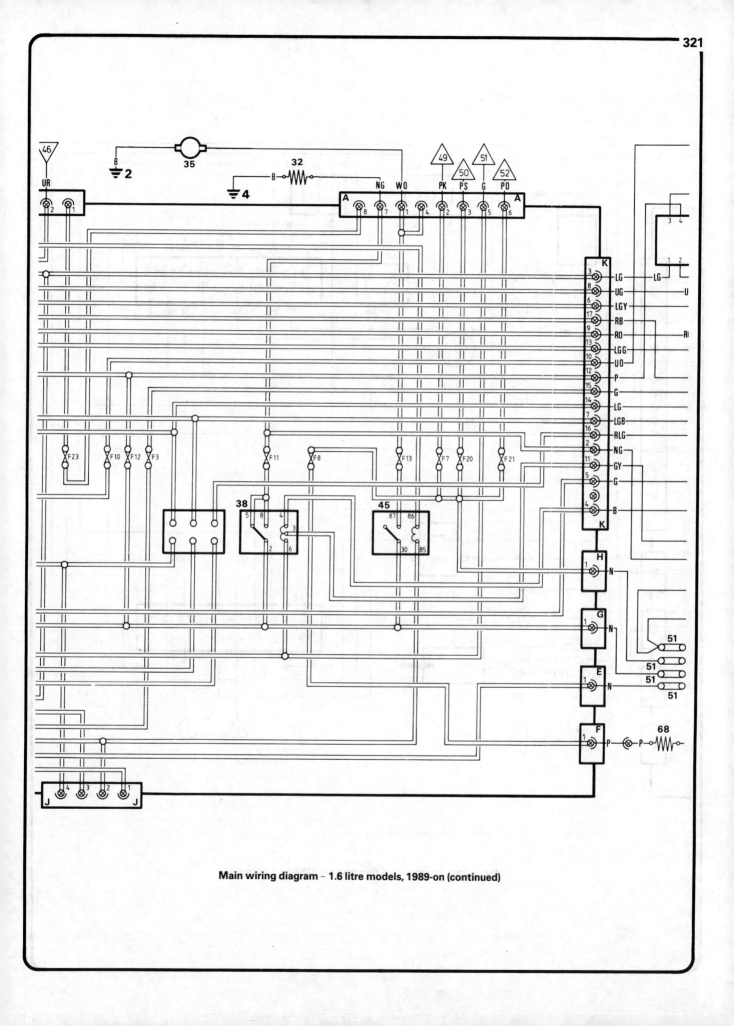

Main wiring diagram – 1.6 litre models, 1989-on (continued)

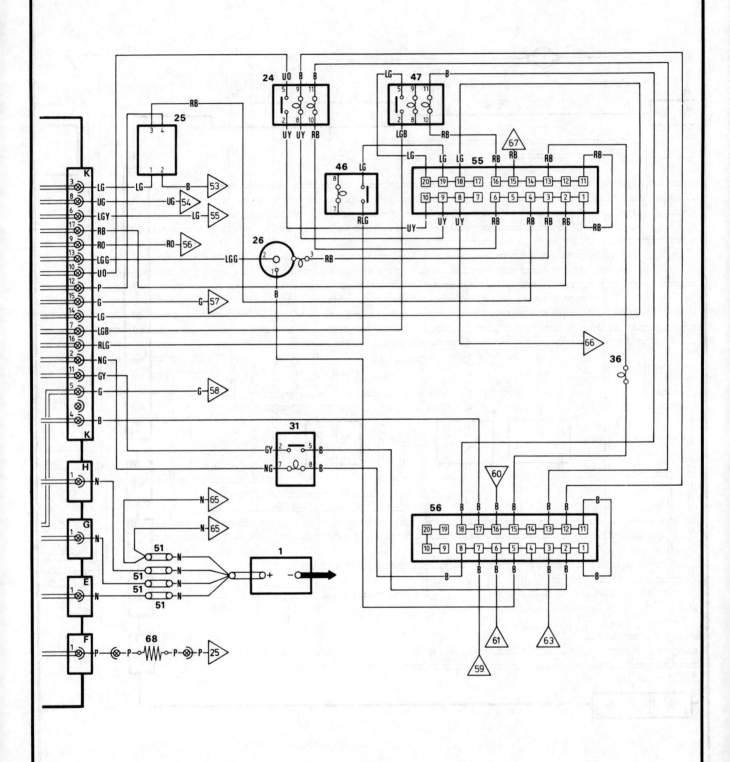

Main wiring diagram – 1.6 litre models, 1989-on (continued)

Key to main wiring diagram – 1.6 litre models 1989-on

No	Description	No	Description
1	Battery	37	Head lamp relay
11	Fuse box	38	Heated rear screen timer
18	Heater or fresh air motor switch(es) or rheostat(s)	44	Ignition switch relay
19	Heater motor	45	Auxiliary circuits relay
21	Ignition/start switch	46	Rear screen wiper switch (Estate only)
24	Foglamp switch	47	Rear screen wash switch (Estate only)
25	Clock	49	Interior lamp delay unit
26	Cigar lighter	51	Fusible links
27	Flasher relay	55	Facia lamp header
31	Heated rear screen switch	56	Facia earth header
32	Heated rear screen element	68	Dim-dip resistor
35	Radiator cooling fan motor	69	Dim-dip unit
36	Heater control illumination		

Connections to other circuits

No	Description	No	Description
1	Exterior lamps/wipers	34	Exterior lamps/wipers
2	Exterior lamps/wipers	35	Radio/cassette
3	Exterior lamps/wipers	36	Exterior lamps/wipers – Estate only
4	Exterior lamps/wipers	37	Exterior lamps/wipers – Estate only
5	Engine	38	Engine
6	Exterior lamps/wipers	39	Radio/cassette
7	Radio/cassette	40	Interior lamps
8	Exterior lamps/wipers	41	Interior lamps
9	Exterior lamps/wipers	42	Interior lamps
10	Exterior lamps/wipers	43	Interior lamps
11	Mirrors	44	All window lift circuits
12	Exterior lamps/wipers	45	Exterior lamps/wipers
13	Exterior lamps/wipers	49	All window lift circuits
14	Exterior lamps/wipers	50	All window lift circuits
15	Mirrors	53	Exterior lamps/wipers
16	Exterior lamps/wipers	54	Instruments
17	Central door locking	55	Instruments
18	Exterior lamps/wipers	56	instruments
19	All window lift circuits	57	Instruments
20	Exterior lamps/wipers	58	Mirrors
21	Exterior lamps/wipers	59	Instruments
22	Exterior lamps/wipers	60	Mirrors
23	Exterior lamps/wipers	61	4 door window lift
24	Exterior lamps/wipers	63	Exterior lamps/wipers
25	Power distribution – Dim-dip resistor	65	Engine
26	Exterior lamps/wipers	66	Exterior lamps/wipers
27	Exterior lamps/wipers	67	Instruments
28	Exterior lamps/wipers	68	All window lift circuits
29	Exterior lamps/wipers	69	Interior lamp circuit
32	All window lift circuits	70	Interior lamp circuit
33	Exterior lamps/wipers	71	Interior lamp circuit

Colour code

No	Description	No	Description
B	Black	P	Purple
G	Green	R	Red
K	Pink	S	Slate
LG	Light green	U	Blue
N	Brown	W	White
O	Orange	Y	Yellow

Fusible link	⊸☐⊸
Fuse	⊸✕⊸
Diode	▸◂
Sealed joint	⊤
Input connection	◁
Output connection	▷
Earth point	⏚

Symbols used in circuit diagrams

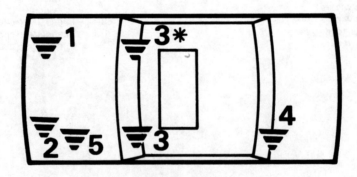

Earth point locations – 1989-on

No	Description	No	Description
1	Front RH inner wing	4	Rear LH wing behind trim panel
2	Front LH inner wing	5	Battery negative terminal
3	Top LH bulkhead behind facia panel		

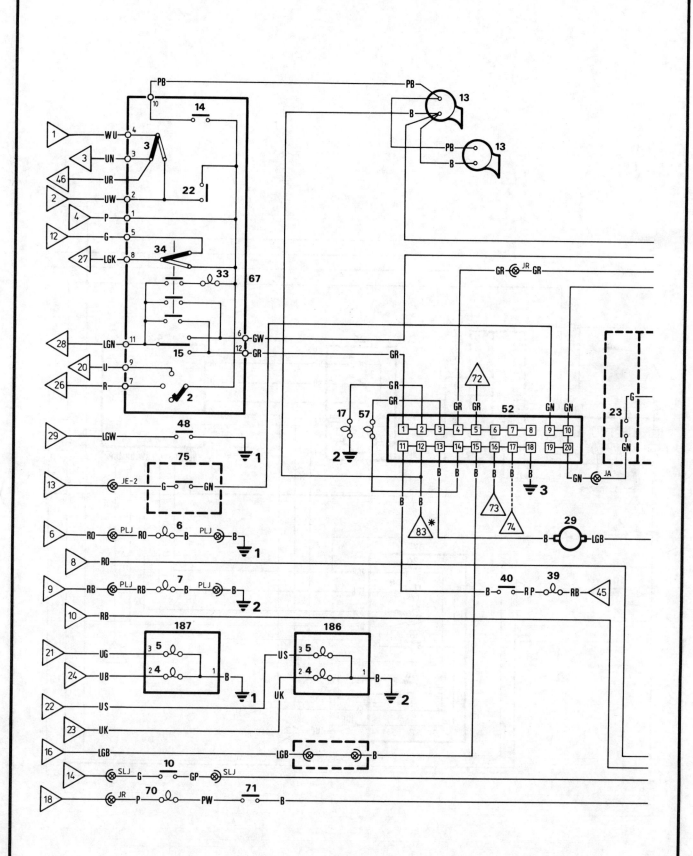

Wiring diagram for exterior lamps/wipers – Saloon, 1989-on

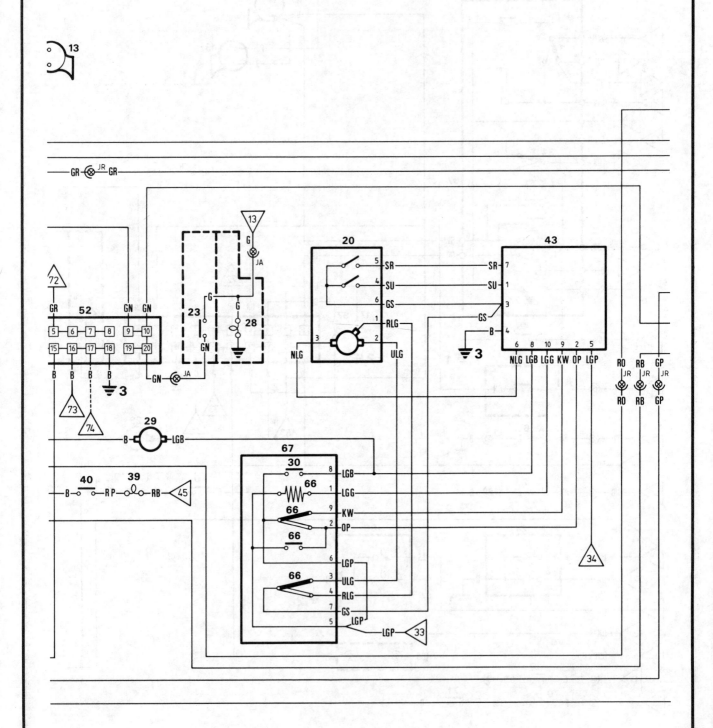

Wiring diagram for exterior lamps/wipers – Saloon, 1989-on (continued)

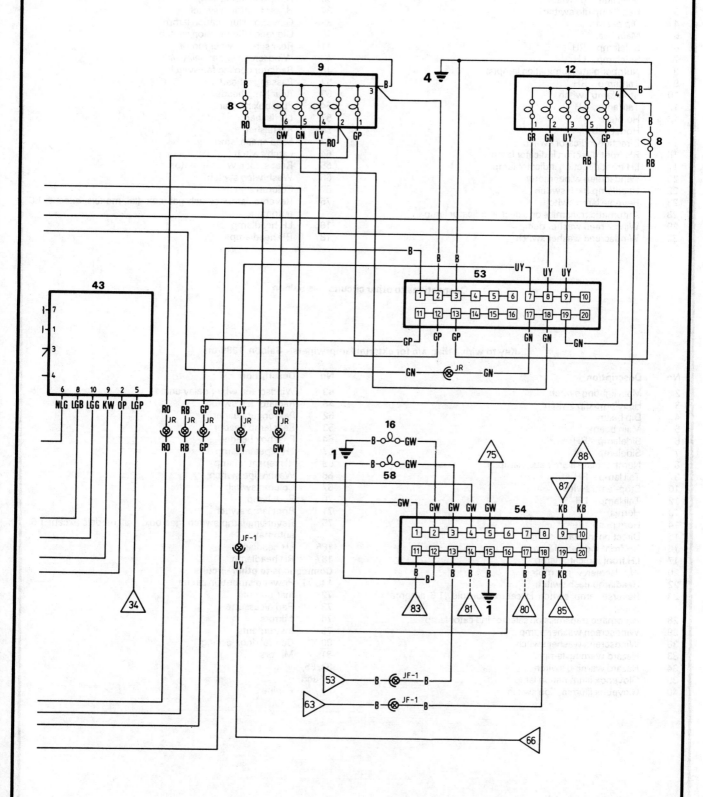

Wiring diagram for exterior lamps/wipers – Saloon, 1989-on (continued)

Key to wiring diagram for exterior lamps/wipers – Estate, 1989-on

No	Description	No	Description
2	Main lighting switch	33	Hazard warning lamp
3	Headlamp dip switch	34	Hazard warning switch
4	Dip beam	39	Glovebox illumination lamp
5	Main beam	40	Glovebox illumination switch
6	Sidelamp – RH	41	Rear screen wiper motor
7	Sidelamp – LH	43	Windscreen wiper delay unit
8	Number plate illumination lamp(s)	48	Radiator cooling fan switch
9	Tail lamp – RH	52	Passenger header
10	Stop-lamp switch	53	Rear lamp header
12	Tail lamp – LH	54	Fusebox header
13	Horn(s)	57	LH flasher
14	Horn push	58	RH flasher
15	Direction indicator switch	63	Load space lamp
16	RH front direction indicator lamp	64	Latch switch
17	LH front direction indicator lamp	65	Rear screen washer motor
20	Windscreen wiper motor	66	Wash/wipe switch
22	Headlamp flash switch	67	Column switch
23	Reverse lamp switch	75	Reverse lamp switch-gearbox (all models except 1.6 automatic)
28	Automatic transmission selector indicator lamp	186	LH headlamp
29	Windscreen washer pump	187	RH headlamp
30	Windscreen washer switch		

Connections to other circuits – as Saloon

Key to wiring diagram for exterior lamp/wipers – Saloon, 1989-on

No	Description	No	Description
2	Main lighting switch	43	Windscreen wiper delay unit
3	Headlamp dip switch	48	Radiator fan switch
4	Dip beam	52	Passenger header
5	Main beam	53	Rear lamp header
6	Sidelamp – RH	54	Fusebox header
7	Sidelamp – LH	57	LH repeater lamp
8	Number plate illumination lamp(s)	58	RH repeater lamp
9	Tail lamp – RH	66	Wash/wipe switch
10	Stop-lamp switch	67	Column switch
12	Tail lamp – LH	70	Boot lamp
13	Horn(s)	71	Boot lamp switch
14	Horn push	75	Reversing lamp switch (gearbox) – all models except 1.6 automatic
15	Direction indicator switch	186	LH headlamp
16	RH front indicator lamp	187	RH headlamp
17	LH front indicator lamp		**Connections to other circuits**
20	Windscreen wiper motor	1 to 71	Power distribution circuit
22	Headlamp flash switch	72	Instruments
23	Reverse lamp switch in centre console (1.6 automatic only)	73	Radio/cassette
28	Automatic transmission selector indicator lamp	74	Mirrors
29	Windscreen washer pump	75	Instruments
30	Windscreen washer switch	80	Central door locking
33	Hazard warning lamp	81	Mirrors
34	Hazard warning switch	83, 85, 87 and	
39	Glovebox illumination lamp	88	Engine
40	Glovebox illumination switch		

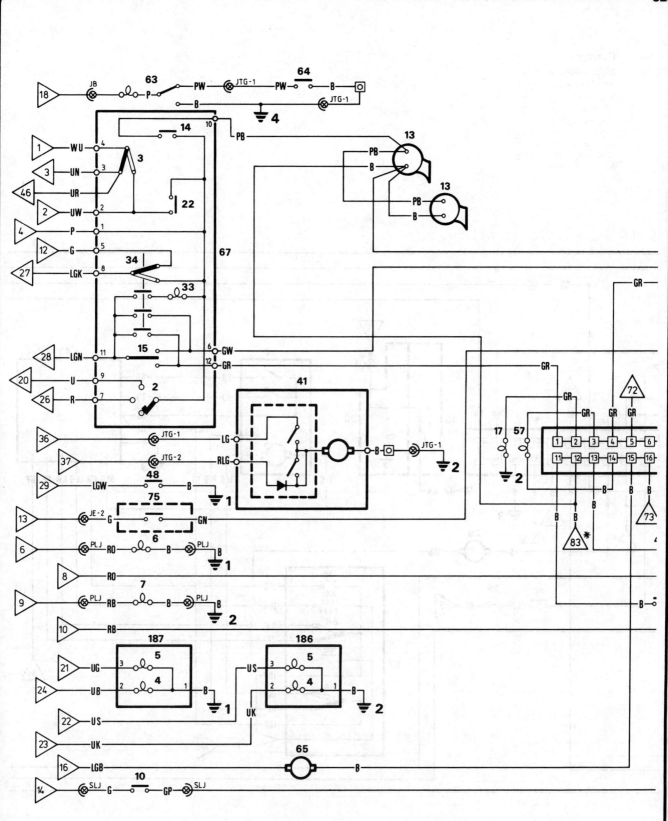

Wiring diagram for exterior lamps/wipers – Estate, 1989-on

Wiring diagram for exterior lamps/wipers – Estate, 1989-on (continued)

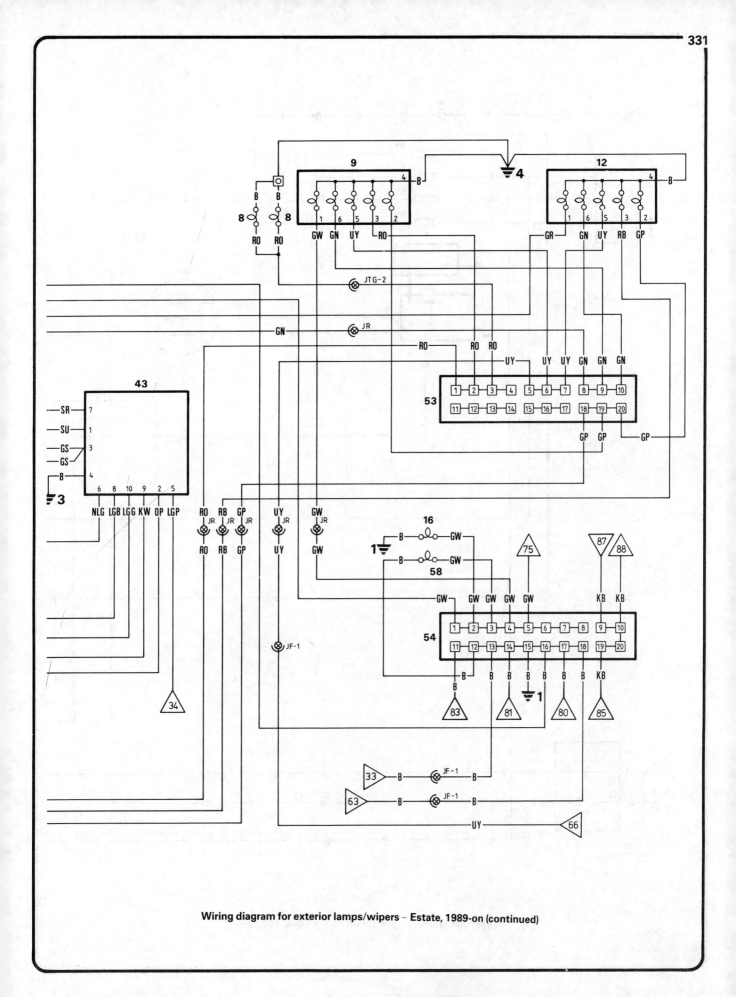

Wiring diagram for exterior lamps/wipers – Estate, 1989-on (continued)

Wiring diagram for engine circuit – 1.6 models, 1989-on

Wiring diagram for engine circuit – 1.6 models, 1989-on (continued)

Wiring diagram for engine circuit – 1.6 models, 1989-on (continued)

Key to wiring diagram for engine circuit – 1.6 models, 1989-on

No	Description	No	Description
1	Battery	99	P.T.C. relay
11	Fusebox	104	Multi-function unit
21	Ignition switch	105	Inhibitor switch
51	Fusible links	106	E.R.I.C ECU
54	Fusebox header	107	Coolant thermistor
80	Alternator	108	Ambient thermistor
81	Starter motor/solenoid	109	Throttle pedal switch
82	Starter relay	110	Diagnostic socket
83	Ignition coil	111	Stepper motor
86	Knock sensor	112	Fuel cut-off solenoid
87	Crankshaft sensor	113	Manifold heater
97	Fuel pump relay	194	Knock sensor screen
98	Main relay	195	Crank sensor screen

Connections to other circuits

No	Description
89	Instruments
91	Instruments
92	Instruments

Key to wiring diagram for instrument pack – 1.6 models, 1989-on

No	Description	No	Description
11	Fusebox	124	Temperature gauge
102	Fuel sensor	125	Fuel gauge
114	Oil pressure switch	126	Low fuel warning lamp
115	Brake fluid level switch	127	High temperature warning L.E.D.
116	Handbrake switch	128	Panel illumination
117	Pad wear sensor	129	Side lamp warning light
118	Choke switch	130	Main beam warning lamp
119	Ignition warning lamp	131	LH indicator warning lamp
120	Oil pressure warning lamp	132	RH indicator warning lamp
121	Brake system warning lamp	133	Trailer warning lamp
122	Choke warning lamp	134	Instrument illumination rheostat
123	Tachometer	135	Instrument pack

Connections to other circuits

No	Description	No	Description
1 to 71	Base circuit	89	Engine
72	Exterior lamps/wipers	91	Engine
75	Exterior lamps/wipers	96	Engine
86	Exterior lamps/wipers		

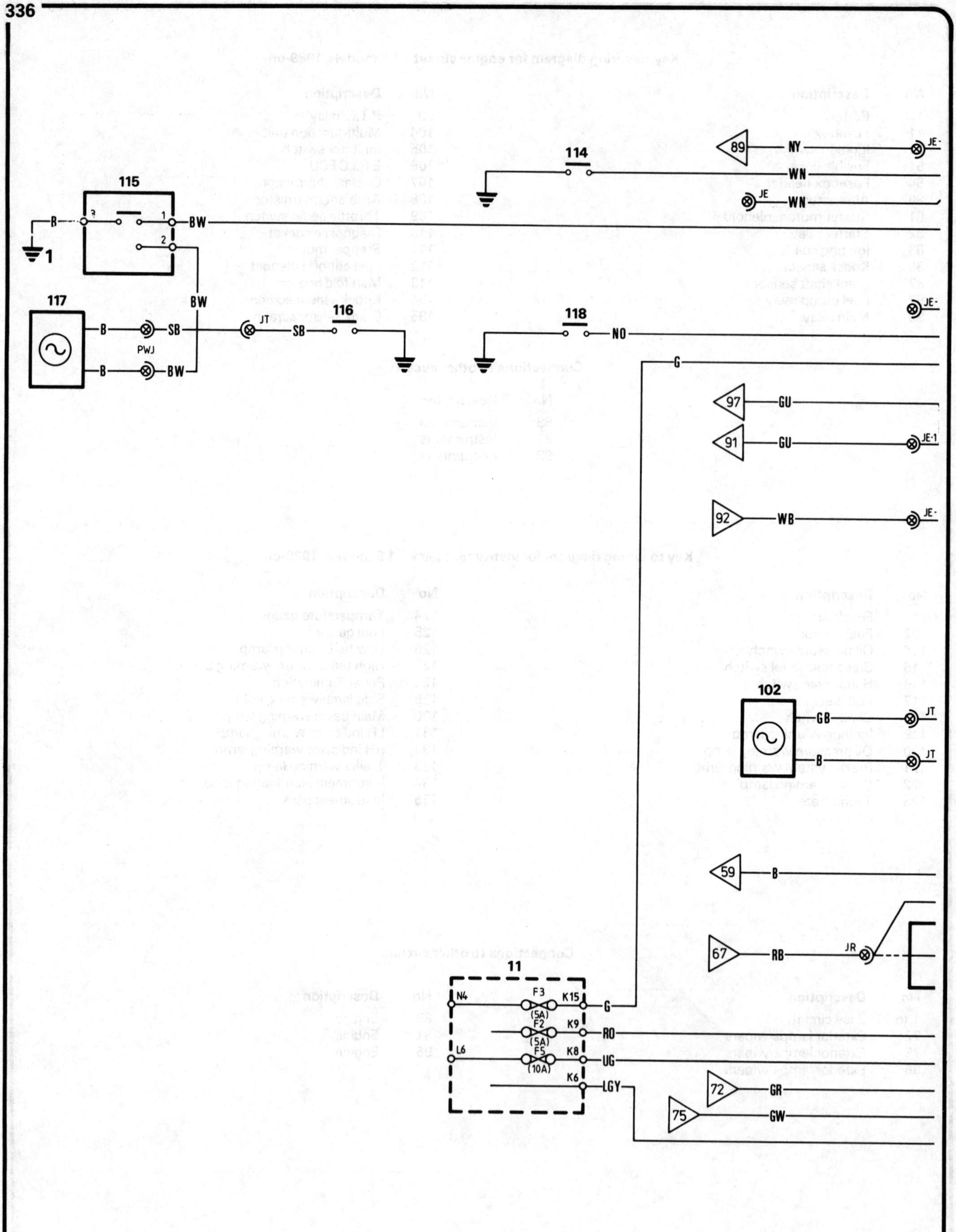

114

115

117

116

118

89 — NY

WN

JE — WN

JE-

97 — GU

91 — GU — JE-1

92 — WB — JE-

102 — GB — JT

B — JT

59 — B

67 — RB — JR

11

N4 — F3 K15 — G
(5A)
F2 K9 — RO
(5A)
L6 — F5 K8 — UG
(10A)
K6 — LGY

72 — GR

75 — GW

Wiring diagram for instrument pack – 1.6 models, 1989-on

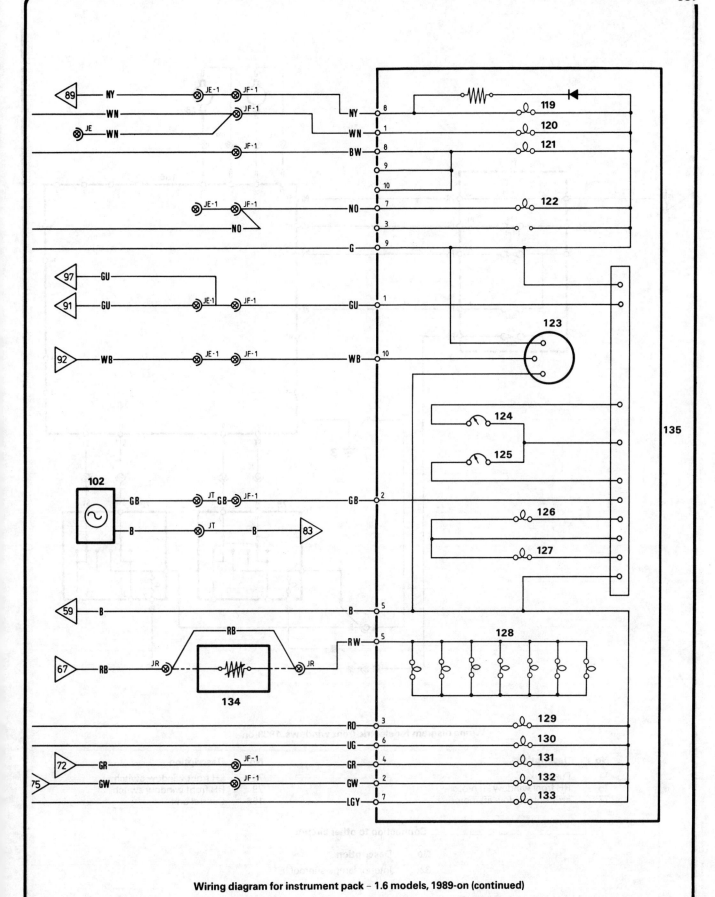

Wiring diagram for instrument pack – 1.6 models, 1989-on (continued)

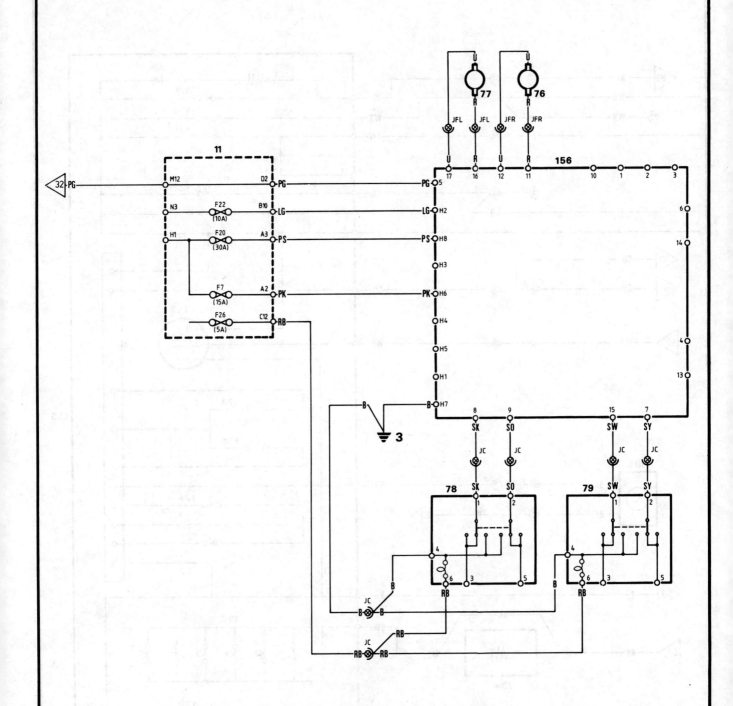

Wiring diagram for electric front windows, 1989-on

No	Description	No	Description
11	Fuse box	78	LH front window switch
76	RH front window lift motor	79	RH front window switch
77	LH front window lift motor	156	Control unit

Connection to other circuits

No	Description
32	Interior lamps/sunroof

A

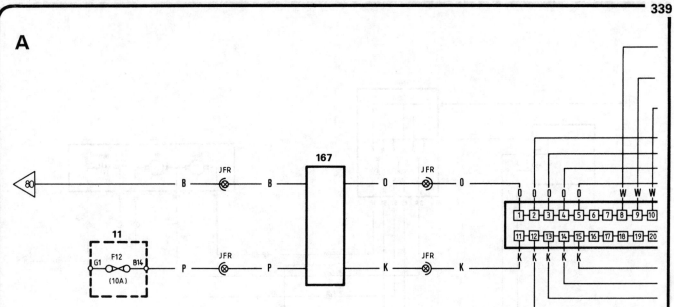

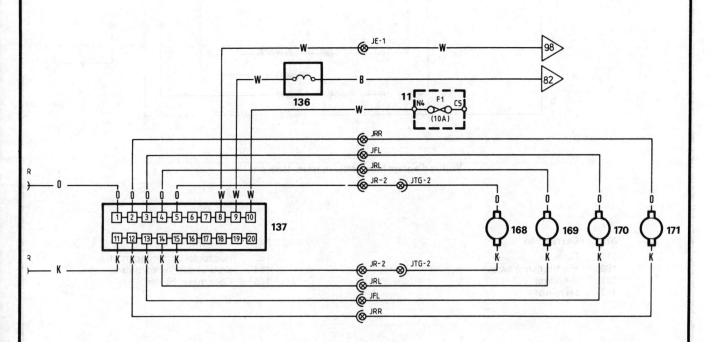

Wiring diagram for central door locking, 1989-on

No	Description	No	Description
11	Fusebox	168	Tailgate lock motor (Estate)
64	Tailgate lock switch (Estate)	169	LH rear motor
137	Central door locking header	170	Passenger door motor
167	Driver's door control unit	171	RH rear motor

Connections to other circuits

No	Description
80	Exterior lamps/wipers

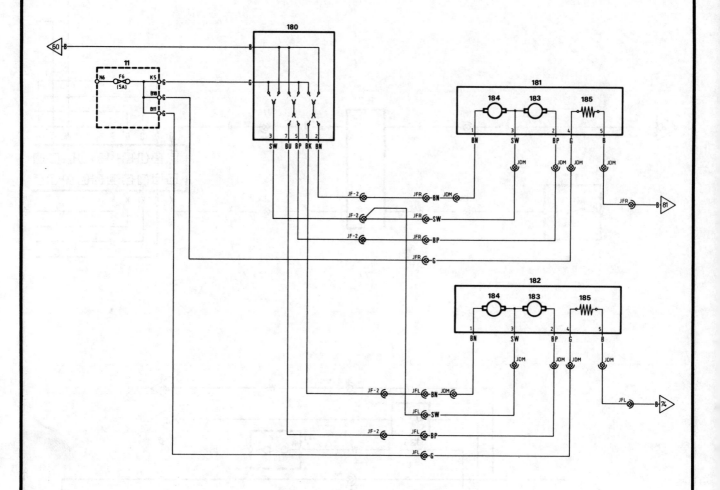

Wiring diagram for electric mirrors, 1989-on

No	Description	No	Description
11	Fusebox	183	Horizontal door mirror motor
180	Electric mirror switch	184	Vertical door mirror motor
181	RH mirror	185	Door mirror heater
182	LH mirror		

Connections to other circuits

No	Description	No	Description
60	Power distribution circuit	74	
		and 81	Exterior lamps/wipers

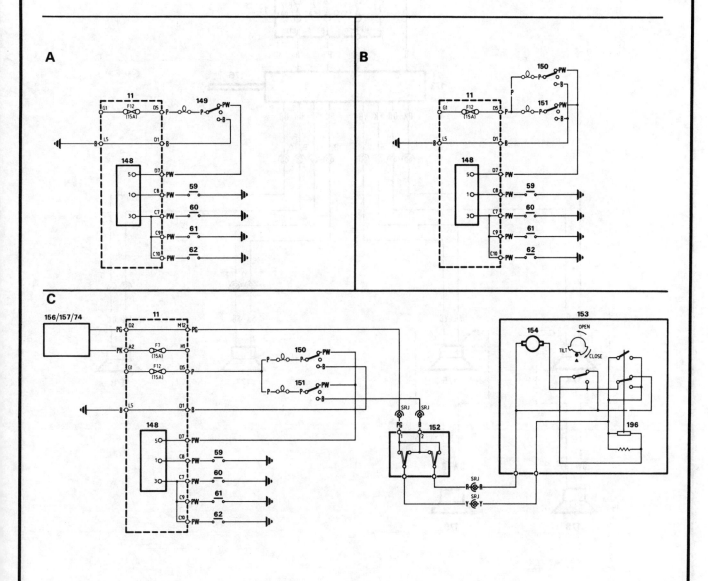

Wiring diagram – interior lamps/sunroof, 1989-on

No	Description	No	Description
A	Without sunroof	149	Front interior lamp
B	With mechanical sunroof	150	RH interior lamp
C	With electric sunroof	151	LH interior lamp
11	Fusebox	152	Sunroof switch
59	RH door switch	153	Sunroof unit
60	LH door switch	154	Sunroof motor
61	RH rear door switch	156	Two door window lift control unit
62	LH rear door switch	157	Four door window lift control unit
148	Interior lamp unit	196	Changeover relay

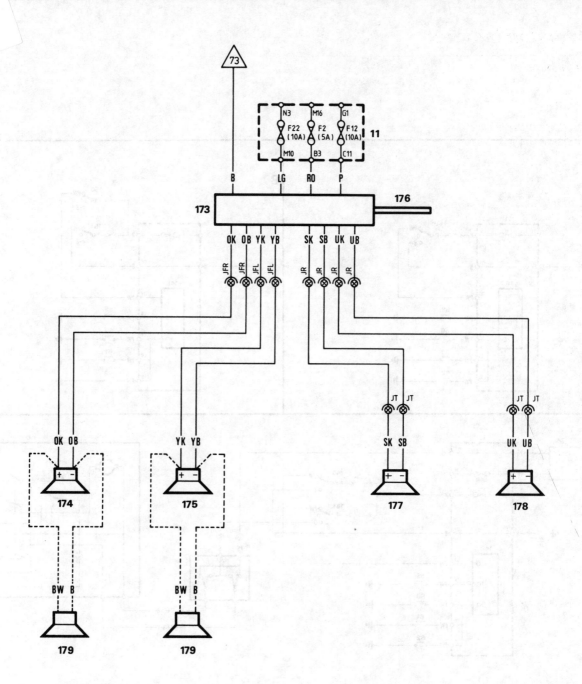

Wiring diagram for radio/cassette player, 1989-on

No	Description		No	Description
11	Fusebox		176	Aerial
173	Radio/cassette		177	Rear/right speaker
174	Front/right speaker		178	Rear/left speaker
175	Front/left speaker		179	Tweeter

Connections to other circuits

No	Description
73	Power distribution circuit

Conversion factors

Length (distance)

Inches (in)	X	25.4	= Millimetres (mm)	X	0.0394	= Inches (in)
Feet (ft)	X	0.305	= Metres (m)	X	3.281	= Feet (ft)
Miles	X	1.609	= Kilometres (km)	X	0.621	= Miles

Volume (capacity)

Cubic inches (cu in; in^3)	X	16.387	= Cubic centimetres (cc; cm^3)	X	0.061	= Cubic inches (cu in; in^3)
Imperial pints (Imp pt)	X	0.568	= Litres (l)	X	1.76	= Imperial pints (Imp pt)
Imperial quarts (Imp qt)	X	1.137	= Litres (l)	X	0.88	= Imperial quarts (Imp qt)
Imperial quarts (Imp qt)	X	1.201	= US quarts (US qt)	X	0.833	= Imperial quarts (Imp qt)
US quarts (US qt)	X	0.946	= Litres (l)	X	1.057	= US quarts (US qt)
Imperial gallons (Imp gal)	X	4.546	= Litres (l)	X	0.22	= Imperial gallons (Imp gal)
Imperial gallons (Imp gal)	X	1.201	= US gallons (US gal)	X	0.833	= Imperial gallons (Imp gal)
US gallons (US gal)	X	3.785	= Litres (l)	X	0.264	= US gallons (US gal)

Mass (weight)

Ounces (oz)	X	28.35	= Grams (g)	X	0.035	= Ounces (oz)
Pounds (lb)	X	0.454	= Kilograms (kg)	X	2.205	= Pounds (lb)

Force

Ounces-force (ozf; oz)	X	0.278	= Newtons (N)	X	3.6	= Ounces-force (ozf; oz)
Pounds-force (lbf; lb)	X	4.448	= Newtons (N)	X	0.225	= Pounds-force (lbf; lb)
Newtons (N)	X	0.1	= Kilograms-force (kgf; kg)	X	9.81	= Newtons (N)

Pressure

Pounds-force per square inch (psi; lbf/in^2; lb/in^2)	X	0.070	= Kilograms-force per square centimetre (kgf/cm^2; kg/cm^2)	X	14.223	= Pounds-force per square inch (psi; lbf/in^2; lb/in^2)
Pounds-force per square inch (psi; lbf/in^2; lb/in^2)	X	0.068	= Atmospheres (atm)	X	14.696	= Pounds-force per square inch (psi; lbf/in^2; lb/in^2)
Pounds-force per square inch (psi; lbf/in^2; lb/in^2)	X	0.069	= Bars	X	14.5	= Pounds-force per square inch (psi; lbf/in^2; lb/in^2)
Pounds-force per square inch (psi; lbf/in^2; lb/in^2)	X	6.895	= Kilopascals (kPa)	X	0.145	= Pounds-force per square inch (psi; lbf/in^2; lb/in^2)
Kilopascals (kPa)	X	0.01	= Kilograms-force per square centimetre (kgf/cm^2; kg/cm^2)	X	98.1	= Kilopascals (kPa)
Millibar (mbar)	X	100	= Pascals (Pa)	X	0.01	= Millibar (mbar)
Millibar (mbar)	X	0.0145	= Pounds-force per square inch (psi; lbf/in^2; lb/in^2)	X	68.947	= Millibar (mbar)
Millibar (mbar)	X	0.75	= Millimetres of mercury (mmHg)	X	1.333	= Millibar (mbar)
Millibar (mbar)	X	0.401	= Inches of water (inH$_2$O)	X	2.491	= Millibar (mbar)
Millimetres of mercury (mmHg)	X	0.535	= Inches of water (inH$_2$O)	X	1.868	= Millimetres of mercury (mmHg)
Inches of water (inH$_2$O)	X	0.036	= Pounds-force per square inch (psi; lbf/in^2; lb/in^2)	X	27.68	= Inches of water (inH$_2$O)

Torque (moment of force)

Pounds-force inches (lbf in; lb in)	X	1.152	= Kilograms-force centimetre (kgf cm; kg cm)	X	0.868	= Pounds-force inches (lbf in; lb in)
Pounds-force inches (lbf in; lb in)	X	0.113	= Newton metres (Nm)	X	8.85	= Pounds-force inches (lbf in; lb in)
Pounds-force inches (lbf in; lb in)	X	0.083	= Pounds-force feet (lbf ft; lb ft)	X	12	= Pounds-force inches (lbf in; lb in)
Pounds-force feet (lbf ft; lb ft)	X	0.138	= Kilograms-force metres (kgf m; kg m)	X	7.233	= Pounds-force feet (lbf ft; lb ft)
Pounds-force feet (lbf ft; lb ft)	X	1.356	= Newton metres (Nm)	X	0.738	= Pounds-force feet (lbf ft; lb ft)
Newton metres (Nm)	X	0.102	= Kilograms-force metres (kgf m; kg m)	X	9.804	= Newton metres (Nm)

Power

Horsepower (hp)	X	745.7	= Watts (W)	X	0.0013	= Horsepower (hp)

Velocity (speed)

Miles per hour (miles/hr; mph)	X	1.609	= Kilometres per hour (km/hr; kph)	X	0.621	= Miles per hour (miles/hr; mph)

Fuel consumption

Miles per gallon, Imperial (mpg)	X	0.354	= Kilometres per litre (km/l)	X	2.825	= Miles per gallon, Imperial (mpg)
Miles per gallon, US (mpg)	X	0.425	= Kilometres per litre (km/l)	X	2.352	= Miles per gallon, US (mpg)

Temperature

Degrees Fahrenheit = (°C x 1.8) + 32

Degrees Celsius (Degrees Centigrade; °C) = (°F – 32) x 0.56

* It is common practice to convert from miles per gallon (mpg) to litres/100 kilometres (l/100km), where mpg (Imperial) x l/100 km = 282 and mpg (US) x l/100 km = 235

Index